BIOELECTRONIC MEASUREMENTS

BIOELECTRONIC MEASUREMENTS

Dean A. DeMarre

David Michaels

Prentice-Hall, Inc., Englewood Cliffs, New Jersey 07632

Library of Congress Cataloging in Publication Data
DeMarre, Dean A.,
Bioelectronic measurements.
Includes bibliographies and index.
1. Medical electronics. 2. Human physiology—
Measurement. I. Michaels, David. II. Title.
R856.D44 616.07′54′0287 81-23460
ISBN 0-13-076398-5 AACR2

Editorial/production supervision: *Ellen Denning*
Manufacturing buyer: *Gordon Osbourne*

"Cover photographs courtesy of General Electric Co."

Printed in the United States of America

10 9 8 7 6 5 4 3 2 1

ISBN 0-13-076398-5

PRENTICE-HALL INTERNATIONAL, INC., *London*
PRENTICE-HALL OF AUSTRALIA PTY. LIMITED, *Sydney*
PRENTICE-HALL CANADA INC., *Toronto*
PRENTICE-HALL OF INDIA PRIVATE LIMITED, *New Delhi*
PRENTICE-HALL OF JAPAN, INC., *Tokyo*
PRENTICE-HALL OF SOUTHEAST ASIA PTE. LTD., *Singapore*
WHITEHALL BOOKS LIMITED, *Wellington, New Zealand*

CONTENTS

PREFACE xi

1. THE INSTRUMENTATION PROCESS 1

1.1 Introduction *1*
1.2 Instrumentation Process Objectives *1*
1.3 Transducer Effects *4*
1.4 Signal Processors *10*
1.5 Displays *16*
1.6 Conclusion *21*
1.7 Review Questions *22*
1.8 References *22*

2. MEDICAL INSTRUMENTATION SENSORS AND TRANSDUCERS 23

2.1 Introduction *23*
2.2 Pressure Transducers *25*
2.3 Electrodes *31*
2.4 Semipermeable Membrane Electrodes *34*
2.5 The Thermocouple and Thermister Transducer *35*
2.6 Piezoelectric Transducers *36*
2.7 Optical Transducers *37*
2.8 Infrared Detectors *38*
2.9 The Laser *39*

2.10 Fiber-Optic Transducers *39*
2.11 Review Questions *40*
2.12 References *40*

3. AMPLIFIERS USED IN INSTRUMENTATION **41**

3.1 Introduction *41*
3.2 Input Isolation *42*
3.3 DC Amplifiers *44*
3.4 Power Amplifiers *46*
3.5 Phase Splitters *47*
3.6 Differential Amplifiers *47*
3.7 Feedback *50*
3.8 Operational Amplifiers *52*
3.9 Electrometer Amplifiers *53*
3.10 Carrier Amplifiers *53*
3.11 Instrumentation Power Supplies *55*
3.12 Review Questions *57*
3.13 References *57*

4. RECORDERS AND DISPLAY DEVICES **58**

4.1 Introduction *58*
4.2 Oscillographic Recorders *58*
4.3 Galvanometric Recorders *59*
4.4 Potentiometric Recorders *60*
4.5 The X-Y Recorder *61*
4.6 Magnetic Recorders *63*
4.7 Dedicated Oscilloscopes *64*
4.8 The Storage Oscilloscope *65*
4.9 Review Questions *67*
4.10 References *67*

5. ELECTROCARDIOGRAPHY AND HEART SOUND MEASUREMENTS **68**

5.1 Introduction to Cardiac Muscle Physiology *68*
5.2 Introduction to the Electrocardiogram *73*
5.3 ECG Lead and Waveform Configurations *75*
5.4 Introduction to Continuous Monitoring of the ECG in the Coronary Care Unit *82*
5.5 Heart Sounds *88*
5.6 The Phonocardiogram *90*
5.7 Review Questions *93*
5.8 References *93*

6. BLOOD PRESSURE AND FLOW MEASUREMENT SYSTEMS 94

6.1 Introduction to Blood Pressure Mesurements *94*
6.2 Classification of Blood Pressure Measurements *95*
6.3 Sphygmomanometry *95*
6.4 Semiautomated and Automated Blood Pressure Measurement Systems *100*
6.5 Blood Flow Measurements: Hemodynamics and Hemodynamic Systems *103*
6.6 Measurement of Cardiac Output *105*
6.7 Cardiac Catheterization *112*
6.8 Review Questions *118*
6.9 References *118*

7. ELECTROENCEPHALOGRAMIC MEASUREMENTS 120

7.1 Introduction *120*
7.2 The Clinical EEG Examination *121*
7.3 Electronic Problems and Hazards Associated with the EEG *130*
7.4 Sleep Recordings and Patterns in EEG Measurements *132*
7.5 Review Questions *134*
7.6 References *134*

8. ELECTROMYOGRAMIC MEASUREMENTS 136

8.1 Introduction *136*
8.2 Physiological Response to Moderate Currents *139*
8.3 Electromyographic Bioelectronic Devices *143*
8.4 Electromyographic Examination Using Bioelectronic Measurements *146*
8.5 Biofeedback Electromyography *148*
8.6 Bioelectronic Device to Relieve Pain *149*
8.7 Electrotherapy *149*
8.8 A Wideband EMG Telemetry System *150*
8.9 Computers and the Electromyogram *151*
8.10 Review Questions *151*
8.11 References *151*

9. RESPIRATORY MEASUREMENTS 153

9.1 Introduction *153*
9.2 Mechanics of Breathing *155*
9.3 Lung Volume and Flow Rates *155*
9.4 Present Trends in Bioelectronic Spirometric Measuring Devices *159*

9.5 Automated Pulmonary Function Measurements *161*
9.6 The Cardiorespirogram *168*
9.7 Hybrid Respiration Signal Conditioner *169*
9.8 Review Questions *171*
9.9 References *171*

10. CLINICAL LABORATORY MEASUREMENTS 173

10.1 Introduction *173*
10.2 Mechanism of Blood *174*
10.3 Chemical Blood Tests *176*
10.4 Review Questions *194*
10.5 References *194*

11. ULTRASONIC MEASURING SYSTEMS 195

11.1 Introduction *195*
11.2 Ultrasonographic Techniques *198*
11.3 The Multielement Transducer System *205*
11.4 The Sector-Scan System *206*
11.5 Ultrasonics in Brain Disorders *208*
11.6 Ultrasonics in Cardiovascular Disease *210*
11.7 Ultrasonic Measurement of Cardiac Output *214*
11.8 Ultrasonics in Ophthalmology *215*
11.9 Obstetric Applications of Diagnostic Ultrasound *216*
11.10 Review Questions *223*
11.11 References *224*

12. RADIOLOGICAL AND NUCLEAR MEASUREMENTS 225

12.1 Introduction *225*
12.2 X-Rays *225*
12.3 Computer Axial Tomography *228*
12.4 Nuclear Cardiology *232*
12.5 Review Questions *234*
12.6 References *234*

13. THE INTENSIVE CARE UNIT 235

13.1 Introduction *235*
13.2 ICU/CCU Monitoring *238*
13.3 Biotelemetry *243*
13.4 Dysrhythmia Monitors *246*
13.5 Holter Recordings *250*
13.6 Review Questions *252*
13.7 References *252*

14. PERFORMANCE TESTING AND SAFETY 253

14.1 Introduction to Simulation Techniques *253*
14.2 Testing the Performance of Blood Pressure Transducers *256*
14.3 Cardiac Monitor, Heart Rate Meters, and Alarm Standards *259*
14.4 Ultrasonic Performance Testing *263*
14.5 Safety Hazards Associated with Cardiac Monitors and Biotelemetry *264*
14.6 Review Questions *268*
14.7 References *269*

GLOSSARY OF MICROPROCESSOR-BASED CONTROL SYSTEM TERMS 270

APPENDIX TABLES 283

INDEX 287

PREFACE

The authors have developed a text in biomedical electronic measurement as a one- or two-term course to be used in an undergraduate engineering technology program. This information may also serve paramedic personnel and physicians, providing them with a quick reference for biophysical measurement techniques. A knowledge of the fundamentals of human physiology is desirable for complete comprehension of the text material. The majority of the hospitals in the United States have established clinical engineering programs, using technicians and engineers in the repair and maintenance of medical equipment. This activity was the major motivating force for developing this text.

Each chapter is a discussion of the principles of operation, basic theory, and in most instances a description of a typical instrument or system currently in use. Additional illustrations, over one hundred review questions, other learning aids, and a glossary are included. When discussing the more sophisticated instruments, relevant human physiology is included.

This text differs from most books on medical electronic technology in that it stresses measurement technique rather than instrument theory. It is divided into common groups of measurement technique so that common physiological or biochemical parameters need not be reintroduced at each section.

The authors wish to acknowledge the assistance and good wishes of James D. Benner, Jr., Senior Design Engineer of Abbott Medical Electronics Co.; Eric Perron of Bio-Tek Instruments, Inc.; Milton Aronson of the publication *Medical Electronics*; Howard Louis, M. D., Ph.D.; Jim Utzerath of Midwest Analog and Digital, Inc.; Hal Fogg of Fogg Systems Co., Inc., Aurora, Colorado; Carl Hennige of General Electric Medical Products Systems, Folsom, California; Benedict Kingsley, Sc.D., consultant on ultrasound; Arthur S. Trappier, Sc.D., consultant on radiology; Jesse Crump, M. D.; Henry J. Wagner, Jr., M. D.; Robert Wightman; Michael K. Meinertz of Organon Teknika Corp., Oklahoma City, Oklahoma; Robert B. Pearson, M. D.; W. G. Kubicek, Ph.D.; William Gruen; Irene Oegielski of Narco Bio-Systems; Richard H. Swartz of the Medical Products Division, Sybron Corporation, Rochester, New York; Al Hancox, Senior Application Specialist of Beckman Instrument, Inc., Schiller Park, Illinois; Maurice R. Blais of LSE Corp., Waltham, Massachusetts; John F. Sweeney, President of LSE Corp., Waltham, Massachusetts.

The cover illustration was provided courtesy of General Electric, Medical Products Division, Milwaukee, Wisconsin, and the authors' design was completed by the Prentice-Hall Art Department. The authors are very grateful to one and all.

Last, but not least, the authors also wish to thank Hank Kennedy, David Boelio, and Barbara Mearns of Prentice-Hall, Inc., without whose efforts this project would not have been completed.

We dedicate this text to our wives, Marilyn and Alicia, and to all the technicians and engineers working in this field.

DEAN A. DEMARRE
DAVID MICHAELS

1

THE INSTRUMENTATION PROCESS

1.1 INTRODUCTION

The tasks in biomedical electronics may be simply stated as: the *development, implementation, and maintenance of tools of technology* to be used by medical and nursing staffs in solving problems of detection and measurement by living human systems. These tools are instruments or instrument systems designed to perform the process of measurement. As such, they are both integral and separable from the process.

An instrumentation process, therefore, has two elements: the *instrument* and the *object* upon which the instrument works. A typical instrumentation process is shown in Fig. 1.1. It includes the patient, who is the object of measurement, and the entire instrument system, from transducer through the signal conditioner to the controls and finally to the output recorders and display units.

In this chapter we discuss the elements of the instrumentation process individually and in combination and attempt to match the art of measurement to the science of instrumentation.

1.2 INSTRUMENTATION PROCESS OBJECTIVES

If at this time you are upset at the thought of calling a living human being an object, you are among the majority. From an instrumentation point of view, the entire human structure, a tissue sample from the structure, or an elimination from the structure may in fact be objectives of the measurement (see Fig. 1.2). From both a medical and a humanistic viewpoint, the living

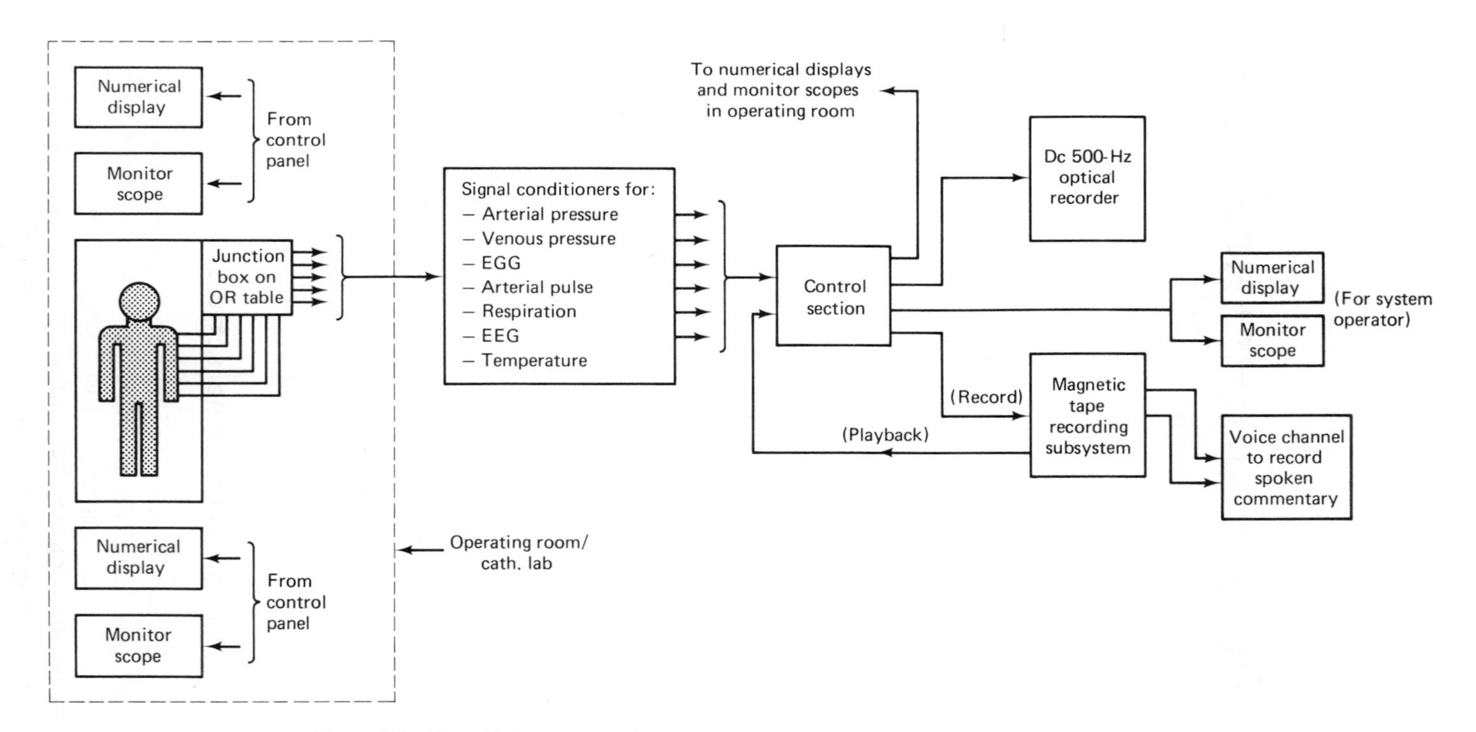

Figure 1.1 Overall instrumentation measuring process. (Courtesy of Hewlett-Packard Co., Palo Alto, Calif., *1969 Catalogue*, p. 57.)

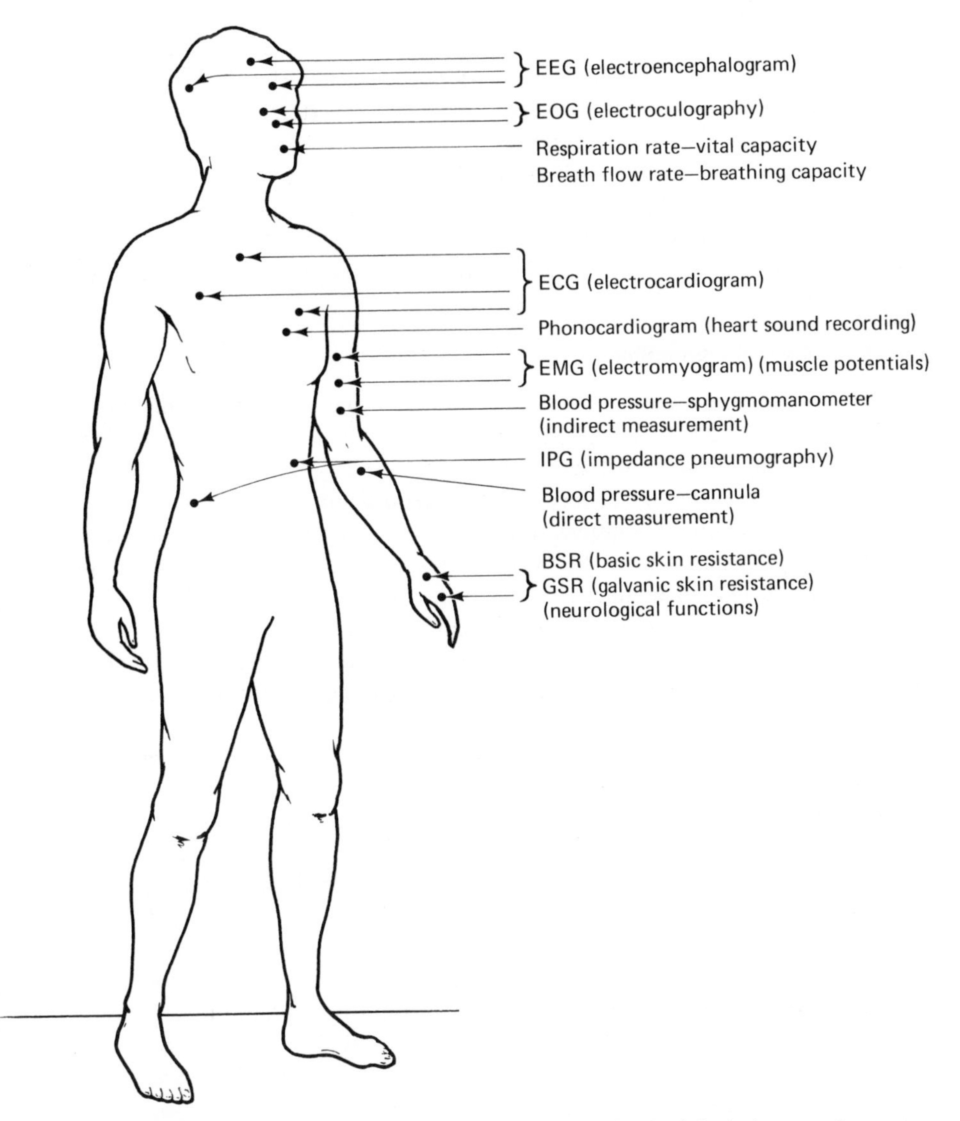

Figure 1.2 The human body and the major dynamic physiological events that are monitored.

human structure is a person who happens to be a patient in need of treatment from whom samples are taken.

Sampling takes two forms. The first is dynamic, where physiological parameters are measured directly from the body, usually through electrodes. The second is static, where samples of tissue are excised or withdrawn from the living system, taken to a remote site, and analyzed.

Dynamic sampling involves an instrument system that can respond to instantaneous changes occurring in the body's systems. *Static* samples represent instrument systems that determine conditions at a particular instant of time which may or may not be related to conditions at any other instant of time.

A cardiac monitor best represents a dynamic system and a hematocrit blood sample best represents the static. The cardiac cycle may change rapidly, requiring a rapid response from both medical and nursing staffs. On the other hand, the hematocrit changes slowly, is determined from a blood sample usually taken by a finger prick into a capillary tube, and does not require constant monitoring by staff. One human system parameter is dynamic, the other is static.

1.2.1 Dynamic Sampling

Many sampling techniques may be employed to convert a dynamic physiological event into an electrical signal which may then be processed through the signal conditioner and displayed and understood by the staff. The dynamic sampling techniques may be classed into two groups: invasive and noninvasive.

Invasive sampling techniques use electrodes or transducers placed on the skin surface or inserted into the body. *Noninvasive* techniques involve using transducers that neither contact nor penetrate the body's surface.

Noninvasive techniques are always safer to use. The instrument system, however, is always more difficult to design or use. Invasive techniques present more hazard to the patient but are easier to design and more reliable in operation.

Dynamic sampling always involves some form of transducer. A *transducer* is a device that converts a physiological event from its natural mechanical or electrochemical state to an electrical signal whose amplitude, rate of change, and repetition rate are proportional to those of the physiological event.

The transducer may be as simple as a silver–silver chloride-plated surface electrode or as complex as a diaphragm-driven linear variable differential transformer.

Any electrical parameter that can be made to vary as a function of a chemical reaction, electrochemical charge, or physical variation in pressure, temperature, or mechanical displacement may be used as a transducer. This includes, but is not limited to, resistance, capacitance, and inductance. It also includes piezoelectric and photoelectric devices.

The transducer is a relatively simple electrical component driven by a mechanical or chemical source that generates a signal proportional to the source.

1.3 TRANSDUCER EFFECTS

The transducer is the only component in the instrument process that contacts both the patient and the instrumentation. It is important to understand how a transducer affects both the patient and the instruments. Table 1.1 shows the six design parameters that affect the performance of a transducer.

TABLE 1.1 The Six Design Parameters of All Dynamic Transducers

1. Sample loading
2. Output impedance
3. Damping
4. Frequency response
5. Linearity
6. Noise

1.3.1 Sample Loading

Sample loading is the effect the transducer has on the system it is trying to measure. Ideally, the transducer should not cause any change in the physiological or chemical parameter that drives the transducer. Because the driving source is mechanical or chemical in nature, the transducer must offer minimum mechanical resistance and cause minimal chemical activity.

If the mechanical resistance is high, the sensitivity of the transducer is reduced and the resulting measurement is not proportional to the change in the physical parameter but rather is a measurement of the physical parameter as modified by the resistance of the transducer—in a word, garbage—meaningless, but deceptive garbage upon which someone could make a diagnostic determination and an inappropriate treatment, placing the patient at hazard.

There is, of course, a limit to how low one can make the mechanical resistance of a transducer without having it begin to chatter and bounce from the slightest muscle tremor or twitch from the patient. There is a range of sample loading resistance which neither distorts the physiological signal nor results in erratic signals because of irrelevant signal pickup. This range varies depending on the type of transducer used and how the particular transducer is applied.

The body could be described as a large chemical processing plant. Transducers applied to or inserted in the body also have a chemical characteristic. When the chemistry of the body interacts with the chemistry of the electrode, a troublesome situation occurs. The problem may range from increased electrode resistance, resulting in a change in sample loading, to electrolysis, resulting in the release of foreign chemical elements into the human system. The latter is commonly known as "trace metal poisoning" or "organic toxicity." It is never good and may be fatal to the patient.

Electrodes and transducers that are placed on or into the body are usually a compromise between the relatively inert, chemically closely bound metal platinum and the relatively active, chemically loosely bound metal copper.

A platinum electrode does not react easily with the chemistry of the body. Its use is limited, however, because its chemical noise generation is extremely high and it is electrically inconsistent. Copper electrodes, on the other hand, react easily with the body's chemistry. Even though the electrode produces the least noise and is extremely consistent electrically, its use is limited.

The general rule is that skin surface electrodes or electrodes which are to be placed in the body for a short period of time should be made of the compound silver–silver chloride. This compound has been found to be the best compromise between noise generation, electrical stability, and chemical stability.

If an electrode is to be left inside the body of a patient for a long period of time (more than a few days), a platinum electrode should be used. The need for chemical stability overrides any other condition.

1.3.2 Output Impedance

The *output impedance* of the transducer is important in terms of the match between the transducer and the input impedance of the signal conditioner. In the majority of systems, the signal conditioner's circuit will be an amplifier of some sort.

In the study of electronics, we are taught that the best possible condition that may exist for maximum energy transfer is that condition where the output impedance of a device exactly matches the input of the device that it drives. This is a true statement for power transfer. You must remember that power equals the current times the voltage, or

$$P = IE$$

For the majority of applications in electronics, this point of equal input and output impedance is the desired method of operations. When you are connecting a surface or implanted electrode to an amplifier, *equality of electrode output impedance to amplifier input impedance is not desirable.* If the impedances are equal, a current may be established through the electrode either from the amplifier or from the patient. This undesirable electrode current will accelerate any chemical reaction and produce additional noise.

The desirable relationship is to have the transducer's output impedance small in comparison to the input impedance of the amplifier. The amplifier then senses voltage variations, but current flow is essentially prevented.

The minimum acceptable ratio of amplifier input impedance is 10 times the electrode impedance (10 : 1). At this ratio it is still possible to have some electrode current flow. The maximum ratio is on the order of a million to one (1,000,000 : 1). At this level, static noise and ambient pickup begin to occur, which offset the fact that no electrode current is possible.

For most applications where electrodes are placed onto or in the body, the ampifier input impedance is set between 20 and 50 megohms (MΩ). If the impedance is set much higher, the increase in noise signal pickup overrides the increase in transducer current limitations.

1.3.3 Damping

Damping occurs when the transducer does not follow the physiological event faithfully. There are three damping factors: underdamping, critical damping, and overdamping.

An *underdamped* transducer responds too quickly to a pulse-shaped physiological event resulting in an exaggerated, spiked, leading wave and a tendency to oscillate or ring on the trailing edge. An overdamped transducer responds too slowly to a pulse-shaped event, resulting in a delayed rise time and delayed decay slope: essentially, a rounding at both the start and the end of the pulse.

A *critically damped* transducer is the desirable state. The transducer neither overreacts or underreacts to the pulse-shaped physiological event. When the transducer is *overdamped,* its frequency response is reduced, causing the frequency response of the entire instrument system to be reduced. When the transducer is underdamped, the frequency response of the transducer is greater than the system frequency response.

When the transducer is critically damped, its frequency response matches the frequency response of the system. Damping may, with some instruments, be accomplished mechanically. A good example is the filling of the catheter between the patient's vein and the dome of a pressure transducer. If too much saline solution is used, the pressure transducer becomes overdamped and does not read the correct pressure. If too little saline solution is used, the transducer becomes underdamped and again, the system indicates an incorrect pressure. Just the right amount of saline solution, enough to completely fill the catheter and almost fill the transducer's dome, will result in critical damping, which will result in a correct indication of pressure.

Damping, with other instruments, may be accomplished electronically. Usually, a small variable capacitance is placed in with the electrode leads. The variable capacitance is adjusted until it resonates with the leads inductance. At resonance, the inductive reactance of the leads is equal to but of the opposite phase of the capacitive reactance of the trimmer capacitors. The capacitive reactance eliminates the inductive reactance and the electrode appears as a perfect resistance. The electrode is critically coupled.

1.3.4 Frequency Response

The *frequency response* of the transducer is directly related to its damping. If the frequency response of the transducer is lower than the bandwidth of the physiological event, information related to the event is lost. If the frequency response of the transducer is greater than the bandwidth of the physiological event, spurious information not related to the event is sensed and a spurious result is displayed.

The frequency response of the transducer is the band of frequencies to which it will respond. The *bandwidth* of frequencies of the physiological event are those frequencies produced by the physiological event. Table 1.2 shows a list of typical levels and frequency range of physiological signals.

TABLE 1.2 Typical Levels and Frequency Range of Physiological Signals

	Typical Signal Range	Frequency Range
Heart potential (ECG)	50 μV to 5 mV	0.05–100 Hz 3-dB points prescribed by the American Heart Association
Brain potential (EEG)	2 to 10 μV (scalp)	1–100 Hz (scalp)
Muscle potential (EMG)	20 μV to 10 mV	10 Hz to 2 kHz (needle electrode) 10 Hz to 10 kHz (gross electrode)
Electro-oculogram (EOG)	10 μV to 4 mV	0.1–100 Hz
Blood pressure pulse	5–15 m/sec	
Blood pressure indirect measurement	0–300 mmHg	0.1–500 Hz
Blood pressure direct measurement	0–40 mmHg (venous) 0–300 mmHg (arterial)	0.1–100 Hz
Blood flow	1–300 cc/sec	1–20 Hz
Heart sounds (PCG)	—	5 Hz to 4 kHz
Respiration rate	500 cc of air; 10 to 20 times per minute	0.15–1.5 Hz
Breath flow rate	3–100 liters/min	—
Untreated skin resistance	50–800 kΩ	—
Electrogastrogram (EGG)	10 μV to 80 mV	0–1 Hz

If the physiological event is rather slow or sinusoidal, as in electroencephalography, the band of frequencies generated is rather narrow. The bandwidth of the transducer should also be narrow. If the frequencies generated cover a moderately wider band such as those that occur in pressure monitoring, the transducer must have a moderately wider frequency response.

If the frequencies generated by physiological event are rather rapid, and thus contain a large band of frequencies such as in electrocardiography, the frequency response of the transducer must be large enough to accommodate them.

1.3.5 Linearity

Linearity relates to the tracking between the transducer's input caused by the physiological event and the transducer's output, which is the physiological event's electrical equivalent. Almost without exception, it is desirable to have the output proportionately track the input. That is, for a unit change in displacement, pressure, differential voltage, or temperature, there is a unit change in transducer output.

There are occasions when it is desirable to modify the linear signal generated by the transducer. This is usually accomplished in the signal conditioner, not in the transducer.

This requirement for linearity over the entire range of measurement using a particular transducer severely restricts the usable range of the transducer. This means that the transducer will respond linearly to a fairly narrow range of amplitude variations in the physiological event. If the amplitude of the mechanical signal is too small, the transducer is not sensitive enough to detect the change. If the amplitude is too large, the transducer saturates and the peaks of the signal are cut off. In either case, the transducer's output is not linearly related to its input and any resulting information is not usable.

The transducer is effective only when the amplitude of the physiological event is large enough to be sensed and small enough not to saturate the transducer. This linear window represents the range of linearity of a transducer. The range of linearity is variable depending upon the type of transducer used in a particular system.

1.3.6 Transducer Noise

Transducer noise is a major problem. There are three sources of noise. These are physiological noise, thermal noise, and ambient noise. All noise signals have one thing in common. They are not desirable signals and they degrade the quality of performance of the measurement system.

Physiological noise is usually unwanted signals generated by the transducer because it senses physiological activity either having the same frequency components or large-amplitude variations near the frequency response of the transducer. The most frequently observed physiological noise is caused by muscle tremor or limb movement. Proper placement of electrodes and positioning of the patient are the only solution.

White or *thermal noise* is generated by current flowing through a resistive circuit. It is proportional to the resistance and the square of the current. If the white noise occurs at or near the input of an amplifier, it is amplified by that amplifier and all succeeding amplifier stages. If the thermal noise is generated at or near the output of the amplifier, it is not amplified and presents less of a problem. For this reason, the input circuit of the signal processor is always of high impedance and current starved.

If excess current flows in the input circuit or the transducer, the noise of the system will rise proportionally to the square of the current times the amplification factor of the system. It does not take a great deal of thermal noise at the input to bury completely the desired physiological signal's output.

Ambient noise is also a problem. Most signal processing systems are designed to reduce or eliminate ambient noise. This type of unwanted signal is generated by the fixtures, utilities, and structure within which a physiological event is being made but totally unassociated with the event. The 60-hertz (Hz) line frequency is a prime example of ambient noise. It may be picked up riding on the patient's skin surface, from the bed the patient is laying on, at the physical junction of the transducer and the patient, or at the connection between the transducer and the signal conditioner.

The only solutions to reducing ambient noise are proper design of the instrument, shielding of all transducer cables with the shield grounded at the instrument input, proper placement of the transducer, and grounding of all electrically powered equipment to an adequate ground from the power distribution system.

If it is not apparent, all six parameters of transducers are interrelated. Sample loading affects the output impedance of the transducer; damping affects the frequency response and linearity. Noise is created if the output impedance, frequency response, and/or linearity are not correct. In practice, the transducer design and application depend on a balance between these parameters, sacrificing equalities of one to gain qualities in another. Specific transducers are discussed in later sections, coupling them directly to their applications of measurement techniques.

1.4 SIGNAL PROCESSORS

The term "signal processor" chills the hearts of the innocent. Because we live in an age of high technology, the technical terms we use have more implication than meaning.

Signal processors are simply those parts of an instrument system that are used to perform a mathematical act on the signal detected by the transducer. The mathematical act, called a *function*, is usually simple and fairly straightforward.

We could best describe a signal processor as a box with four terminals. Terminals 1 and 2 represent the input and terminals 3 and 4 represent the output. The mathematical function of the box is represented by the notation *fx*, which is read "*f* of *x*" or "function of *x*" (see Fig. 1.3).

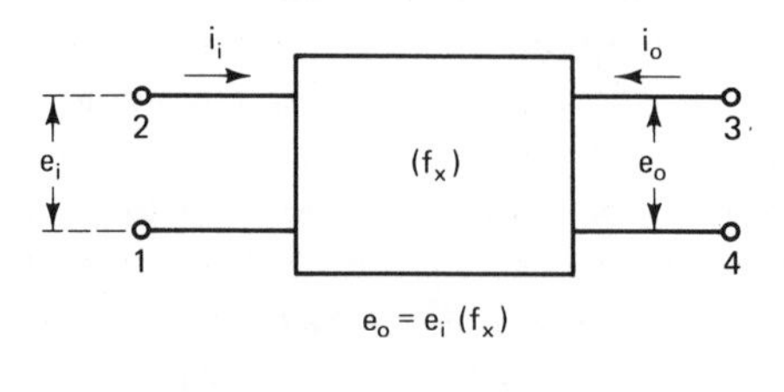

Legend: e_{in} = input voltage
e_o = output voltage
f_x = transfer function of black box

Figure 1.3 Signal processor block.

If our box represents an amplifier, the f of x would be the gain of the amplifier. If our box represented a circuit that reduced the signal by half, f of x would be the divisional factor represented by 1 over 2, or $\frac{1}{2}$.

In most instrumentation applications, our signal processor will be an amplifier especially designed for the physiological signal being processed and the transducer used to detect this signal. There are eight design parameters that affect the performance of an amplifier, as shown in Table 1.3.

TABLE 1.3 The Eight Design Parameters of Amplifiers

1. Input impedance
2. Output impedance
3. Gain
4. Power distribution
5. Noise
6. Common-mode rejection
7. Bandwidth
8. Distortion

To some extent these parameters are related to each other and related to the elements driving the amplifier as well as the elements that are driven by the amplifier.

1.4.1 Input Impedance

Most amplifiers used for instrumentation fall into a category called "electrometers." An *electrometer* is an amplifier with an extremely high input impedance. As discussed in Section 1.3.2, the input impedance of the amplifier should be at least 10 times the output impedance of the transducer, so that current drain through the electrode is kept at a minimum. The amplifier is designed to sense voltage variations at its input and the high input impedance prevents the amplifier from loading the transducer down. If the transducer is loaded down and a current is developed, the probability is that the signal voltage developed by the transducer will be effectively reduced, which is a form of distortion.

On the other hand, if the input impedance of the electrometer is too high in relation to the transducer, the input circuit will act as an antenna. It will, in effect, "pick up" signals floating in our atmosphere generated by the power lines, house wiring, and radio and television stations. Unless you live and work in the desert or in the middle of the ocean, these electromagnetic signals are constantly around you. If they are detected and amplified by your signal processor, the resulting output will be affected. The result may vary from a mild distortion to complete swamping of the desired signal by the interference.

The input impedance of the amplifier should be held somewhere between 10 and 50 MΩ, depending upon the transducer used. Shielded cable should be used to connect the transducer to the amplifier's input. The shield should be grounded at the amplifier and not at the transducer. The shielding around the wires in the cable is used to catch and bleed off the stray signals and is not intended to be used as a "ground connection" between the transducer and the amplifier.

1.4.2 Output Impedance

The output impedance of a signal processor is as important as the impedance. The output impedance must be high enough to develop an output potential related to the input signal, yet low enough so that the input impedance to the next component in the system does not load the signal processor down. Again, as in most instrument applications, we are not interested in maximum power transfer, so the output impedance need not be identical to the input impedance of the next circuit. We are also not usually concerned with maximizing efficiency in the signal processor. We are concerned with maximizing the quality of the function of the signal processor. Therefore, if the circuit we are dealing with is an amplifier, we are concerned with the degree of linearity of the amplifier but not in obtaining the maximum gain. If the signal processor is a differentiator, we are concerned with the resolution of the differentiation and not with the output-signal driving power.

In the case of the amplifier, which we are discussing, the output impedance should be adjusted so that the signal out of the processor is an exact amplified replica of the input. It must be free of distortion and noise as well as being linear. The input impedance of the following stage must be adjusted so as to load as lightly as possible the output circuit of the amplifier in question.

To accomplish these apparently opposite actions, the output impedance of the amplifier in question is adjusted to give maximum amplification with minimum distortion. The input impedance of the next stage is then adjusted so that it is at least 10 times greater than the output impedance of the amplifier in question.

If the next circuit element being driven happens to have a naturally low impedance such as would be encountered in an analog-to-digital converter circuit, hammer drive for a teletype, or an EKG stylus meter movement, all of which are driven by amplifiers, an additional stage of amplification with a high input impedance and a low output impedance would have to be added to provide isolation, called *buffering*, to prevent the output loading, which results in a distorted output signal.

It is quite easy to write the equations for both the input and output impedances of an amplifier. They are, simply stated, the relationship of input current to input voltage and output current to output voltage, represented by the equations

$$Z_{\text{in}} = \frac{E_{\text{in}}}{I_{\text{in}}}$$

and

$$Z_{\text{out}} = \frac{E_{\text{out}}}{I_{\text{out}}}$$

It is much more important to recognize the effects that the input and output impedances have on real systems.

1.4.3 Gain

The word *gain* is a generic term meaning either increase or decrease in strength or amplitude. When "gain" is used to describe an increase, it is called "positive gain," "amplification," or simply "gain." If an amplification factor is being discussed, it implies a positive gain and is stated as gain. When "gain" is used to describe a decrease in strength, it is called "negative gain," "drop," or "loss."

In instrumentation there are three gain parameters: voltage, current, and power, all relating to the amplifier's effect on the signal being processed. Every electronic stage within an instrumentation system has a voltage, current, or power gain occurring within it.

For the electrometer amplifier that we are presently discussing, the normal state is to have a positive voltage gain, a negative current gain, and a negative power gain. For a buffer amplifier used to match a high-output impedance to a low-input impedance, one would expect to find a small negative voltage gain, a large positive current gain, and a small positive power gain. The EKG's galvanometer movement would have negative power, voltage, and current gains.

It should be noted that there is no amplifier or component that has all positive gains. Usually, one of the gain factors will be positive and the other two will be negative.

The symbol for gain is the capital letter A. A subscript is used to signify voltage, current, or power gains and a minus sign is placed before the symbol indicates a 180° phase shift of the signal processed. Most signal processors possess a positive voltage gain and involve 360° of the electrical phase shift. The resulting gain expression would be

$$\frac{E_{\text{out}}}{E_{\text{in}}} = A_v$$

where E_{out} represents the output voltage, E_{in} the input voltage, and A_v the voltage gain with either 0 or 360° of phase shift.

It is possible to have a signal processing element that presents a phase shift other than 180° or 360°, but this is not the normal case. When this type of circuit is encountered, its gain and phase shift should be defined by its designer.

1.4.4 Power Distribution

The *power distribution* of an amplifier is a function of the gain and the bandwidth. Sometimes called the *gain–bandwidth product*, it defines the useful range of frequencies that may be amplified by the circuit in question.

There are three bandwidths of importance associated with an amplifier: the 1-, 3-, and 60-decibel (dB) bandwidths. The most frequently encountered bandwidth is the one defined as those frequencies that lay between the points where the amplified signal is reduced from its peak by a factor of 0.707, which corresponds to the 3-dB points (see Fig. 1.4). This bandwidth is called the *half-power point* but is in actuality only the half-power point for a one-pole

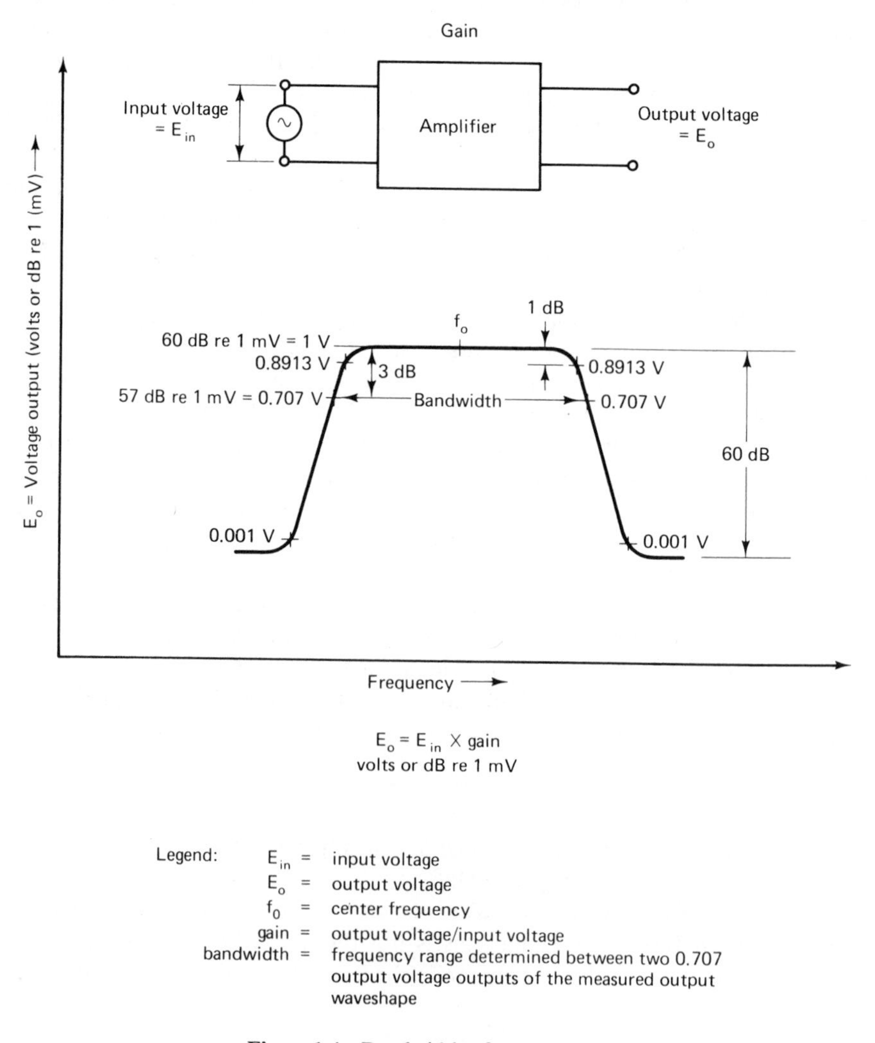

Figure 1.4 Bandwidth of an amplifier.

filter. If the stage of amplification has more than one active element or two or more filters, the half-power-point definition does not hold.

It is common practice to consider only those frequencies that lie within the 3-dB points as being usable within an amplifier. Energy outside these frequency limits is not considered to be significantly amplified to be useful. In fact, there is such a high degree of phase shift and distortion caused by the amplifier to these signals that they may substantially degrade the overall performance of a system. The ratio of the 1-dB to 60-dB bandwidths becomes important because of the degradation of performance of the amplifier.

The 1-dB bandwidth is defined as those frequencies that lay between the points where the amplifier's signal is reduced from its peak by a factor of 0.891. The 60-dB bandwidth is defined as those frequencies that lay between the points where the amplifier's signal is reduced from its peak by a factor of 0.001. From this it may be observed that the 1-dB bandwidth is always smaller than the 3-dB points and that the 60-dB bandwidth is always larger.

The slope established between the 1-dB point on the bandwidth curve and the 60-dB point define the quality of the amplifier. A single-pole filter or one active-element amplifier should have a slope of 6 dB per octave of frequency. A multiple-stage amplifier would have slopes of 12, 18, or 24 dB per octave. The higher the decibel/octave slope of the amplifier, the greater the exclusion of signals whose frequency lay outside the 3-dB bandwidth.

Gain and bandwidth are to a great extent interrelated. If the 3-dB bandwidth of an amplifier is reduced, the gain of the amplifier increases. If the 3-dB bandwidth of an amplifier is increased, the gain of the amplifier is reduced. An energy relationship occurs which may be stated: for any particular amplifier, there is a constant relationship between the amplifier's gain and its bandwidth. If the product of the gain and bandwidth are calculated, the result will be constant for a particular amplifier.

What this means is that the gain–bandwidth product of an amplifier is constant and that its gain will change inversely to a change in its bandwidth. If we have a very narrow band of frequencies that are to be amplified, we may reduce the bandwidth of the amplifier and utilize all its developed energy to amplify that narrow band of frequencies and thus produce a large gain factor. On the other hand, if the signal to be processed contains a wide band of frequencies, the bandwidth of the amplifiers must be increased, resulting in a wider distribution of the amplifier's developed energy and a reduction in the gain factor.

1.4.5 Distortion

When processing a signal, some distortion occurs regardless of the techniques used. This result can never be avoided completely. Distortion observed on a display (as opposed to being measured) has acceptable limits. For visual displays between 3 and 5%, distortion will be tolerated by most individuals with no degradation of interpretation of results. An audio distortion of from 7 to 10% is acceptable by the majority of observers. The human optical system is usually more acute than the auditory system and will not accept as high a degree of distortion.

Most noticeable distortion in a processed signal does not occur in the amplifier. It occurs at the transducers and at the display. Transducer distortion may be traced to the application technique or sampling methodology used. Display distortion may be traced to damping factors, repetition rates, and frequency responses.

Distortion from amplifiers must usually be measured and is normally below 1%. If an amplifier is distorting the signal, it is often traceable to improper loading of one of the stages of amplification. When this occurs, the distortion increases because the stage's operating point is shifted and the linearity of the stage is reduced. The corrective action is to increase the input impedance of the stage following the overloaded stage. With increased use of integrated circuits, amplifier-induced distortion has become less of a problem.

1.5 DISPLAYS

The term *display* usually brings to mind the vision of a *cathode ray tube* (CRT), which presents data in the neat tabular form we associate with computers. This automatic vision is not too far from the truth. Along with the CRT display, however, we include analog meter movements, strip chart recorders, teleprinters, digital meter displays, data listers, and data storage devices. In addition to visual display there may be audio displays ranging from simple bell tones to complex changing tones and lately to computer-generated word groupings.

If the signal that has been processed is to be displayed as a continuously changing function from which the operators are to make quick decisions, a visual display is normally used. If the processed signal is to be monitored to see that it does not exceed preset limits, audio displays are used.

There are eight design parameters related to the performance of displays and seven characteristics related to data storage, as shown in Tables 1.4 and 1.5. The following paragraphs discuss these parameters. It must be remembered

TABLE 1.4 The Eight Design Parameters of Displays

1. Input impedance
2. Linearity
3. Accuracy
4. Repeatability
5. Intensity
6. Resolution
7. Damping/Frequency response
8. Sample rate

TABLE 1.5 The Seven Design Parameters of Data Storage

1. Duration of storage
2. Total memory capacity
3. Overwrite sequence
4. Auto-listing
5. Access time
6. Buffer capacity
7. Interrupt sequence

that the perception of displays and data systems by a computer person is quite different from the perceptions of an instrumentation person, so that the definitions stated below might appear quite different from those made by a computer person.

1.5.1 Display Parameters

The display's input impedance should be substantially higher than the output impedance of the circuit driving it. A ratio of 10 : 1 should be maintained. If this is not possible, a buffer circuit should be added to the system to establish a ratio as close to 10 : 1 as possible.

If a 10 : 1 ratio is not maintained, the driven circuit will become nonlinear and the data displayed will be misleading at best.

The linearity of the display itself is important. If the display is a CRT, it has both vertical and horizontal deflections. The vertical deflections are normally used to indicate the amplitude variations of the processed signal. The horizontal deflections are normally used to indicate time relationships. On occasion, a third axis is used to indicate a spatial relationship. The horizontal deflection is always called the *x-axis*, the vertical deflection is called the *y-axis*, and the spatial deflection is always called the *z-axis*.

The actual graduation or scales on the three axes are normally ruled in millimeter divisions, with major cross lines placed at 10-millimeter (mm) [1 centimeter (cm)] spacings. The time base, calibrated in seconds, must cross the face of the CRT linearly so that it takes exactly the same time to traverse any one 10-mm segment of the grid. Similarly, the vertical deflection must represent a given magnitude of signal presented to the system from the transducer. If a physiological event as sensed by the transducer generates a potential of 5 milliovolts (mV), the vertical deflection should rise to an equivalent 5 mV. If an additional 5 mV is generated, the vertical deflection should rise again by an equal amount. To accomplish this, the CRT display and its associated circuits must operate linearly.

When the display employed is either a digital or an analog meter, the indicated output must be a linear representation of the input signal. In many cases, the metered output is not a measure of amplitude but is a measure of rate of occurrence or change. Therefore, the meter must display a linear representation of the rate.

Accuracy, repeatability, and resolution are generally confused. Of the three, repeatability is the most difficult to obtain. *Repeatability* is the ability of an instrument to display the same indication for repeated identical stimuli. A CRT display usually prossesses a high degree of repeatability, and analog meter movements have the poorest.

Accuracy and resolution are somewhat interrelated. *Accuracy* is the display's ability to indicate with precision, whereas *resolution* is the degree of gradation of the precision. For example, if an instrument's CRT display is said to have an accuracy of 1%/cm and a minor division of 1 mm, the accurate resolution would be 10%. The CRT may display the signal to an accuracy of 1%, but the observer may only be able to perceive variation in steps of 1 mm. If the major scale rulings are brought across the face of the CRT but the minors are only ruled at the edges or in the middle, the resolution is further degraded.

When an analog meter is used as a display, the accuracy depends on where the meter face is calibrated. If the meter is calibrated at its zero and maximum points, the meter is accurate only over the maximum part of the scale. If the analog meter is calibrated across the middle portion of its range, the unit is accurate only across the middle portion and becomes progressively less accurate toward either end. The resolution of an analog scale is directly dependent on the scale's gradation and the width of the meter movements stylus. The finer the graduations and the finer the stylus, the greater the resolution. A normal analog meter's accuracy is about 3% full scale, degrading to between 7 and 10% at the lower end of the scale. Analog movements used in scientific instruments normally have a calibration range accuracy of 0.1% and degrade to 2% on the extremes. Their resolution is normally one-half a minor scale division.

Digital panel meters are more accurate than analog meters because their displays are linear over the entire range of display. This results in a linear accuracy over this range. If the digital display has an expressed 1% accuracy, it should indicate within 1% over its entire scale. The digital meter's resolution is also greater, with the exception of the last digit. The last digit is always arbitrary within two units.

For precise measurement of constant signals, technicians prefer the digital meter or CRT display. For measurements of signals that change constantly, technicians prefer the analog meter or CRT display. The three most important factors of a display are its repeatability, accuracy, and resolution.

The intensity of a display is related to CRT and audio displays only. With the CRT display the intensity should be balanced so that it is just bright enough for easy observation but not so high that it burns the phosphor on the inside of the CRT's screen. If the CRT's trace is set too high, the tube's natural astigmatism and focus potential are enhanced, resulting in a loss of resolution.

Damping and frequency response become a problem with strip chart recorders. Damping is both the mechanical restriction placed on the strip chart recorder's stylus and the electronic trim of the system's amplifier. If the meter movement is overdamped, the low-frequency response of the system is degraded. Underdamping the meter movement results in chatter in the stylus at its end points, destroying the system's resolution. Overdamping reduces the system's lower-frequency response. Underdamping increases the upper-frequency response. Both states are undesirable.

Sample rate is related to digital displays, especially when a signal is stored in memory and then retrieved for display. Data are stored in memory in discrete bites of information. A signal is sampled at a predetermined rate and the information at the point of sampling is stored. The information making up the signal between sampling points *is lost.* When the memory is played back on the CRT screen, only the information captured at the sampling points is displayed.

If the sample rate is too low, the display's distortion increases and may become annoying. If the sample rate is too high, the information load into the computer memory becomes too great, requiring either an extensive core memory or pushing out older data without review as new data flow in. The sample rate for most on-line storage of physiological data should be between 40 and 200 samples per second. At about 40 samples per second, obvious visual distortion is eliminated. At about 200 samples per second, rapidly changing data may be reproduced with minimal loss of resolution.

1.5.2 Data Storage Parameters

Duration of storage and total memory capacity are interrelated. The data storage element is like a coffee can and the bits of information are like the grains of coffee dropped into the can. If you are going to take lots of samples over a long period, you will accumulate a great many bits of information, requiring a very large coffee can. A memory's *capacity* is defined in terms of groups of bits. The normal grouping of bits varies as a function of the design of the integrated circuits used to control the memory bank, but generally consists of 4, 8, or 16 bits, making up what is called a *byte.* The memory is normally structured in 4000-byte (4-kilobyte) increments.

A simple system with a relatively slow sample rate and relatively short storage duration could use a 4-kilobyte memory, with 200 to 400 bytes used for control purposes and the remainder used for data storage.

A complex system with a relatively high sample rate and a long storage duration would require from several hundred kilobytes up to more than 1 million bytes. In this case, from 4 to 16 kilobytes might be used for the internal control program and the remainder would be used for data storage.

From this it may be seen that if the duration of storage is short, such as in normal patient monitoring, the total memory capacity may be small. Where the duration of storage is long, such as encountered in the newer trend analysis and arythmia detection intensive care systems, the total memory capacity must be large. When the memory fills up with data, something must happen. Either old data are pushed out of the memory and lost or an auto-listing sequence occurs.

In the early memory-based monitoring systems, a storage duration time of from 8 to 32 seconds was used. The data were fed into the memory, pushed through by new data, and finally 8 to 32 seconds after insertion displayed on a CRT screen and dumped, never to be recovered. If the patient suffered an arythmia or some form of distress where the cardiac rate changed, setting off the alarms, a hard copy was generated by a strip chart recorder. Otherwise, all information was soft. This process of passing information through the memory, in essence delaying the information and then dumping old data as new data are accumulated, is called *overwriting*. New data are constantly overwriting or displacing older data.

On the newer intensive care monitoring systems, most large-scale laboratory systems, as well as radiologic tomographic and nuclear computer axial tomographic (CAT) scanning apparatus, the overwriting technique of data storage cannot be used. We have to use auto-listing.

In these large systems, data are accumulated at either extremely fast sampling rates or over fairly long storage duration times. A nuclear CAT scanner may take 1 million samples over a 30-minute period. The newer intensive care units may have a relatively slow sample rate but will be sampling up to three parameters on a minimum of eight patients for up to 24 hours. In either, a tremendous quantity of data is gathered, stored, organized, and then must be displayed.

Auto-listing performs this function. A portion of the memory is assigned to the control responsibility and is programmed to accept, code, and file the data. Then, either at automatic call points or when requested by the system operator, the control program retrieves the data from the file, organizes the data into a preset format, and displays them on the CRT screen for review or prints out a hard copy on a printer for either analysis or filing. In the case of tomographs, lab analysis, or CAT scans, printout is automatic upon completion of the analysis. In the case of physiological monitoring systems, printout should occur in the event of an individual patient going into distress, on a command for medical review, and when an individual patient's allocated memory space fills up.

It should be understood that in most of these larger systems, as the data are accumulated and passed into the memory, they are also normally displayed on a CRT screen, so that "real-time" monitoring of the data is possible by the system operators.

The effective speed of a computer-based system is controlled by three interrelated factors. These are access time, buffer capacity, and the interrupt sequence. Access time is the interval of time required for the "machinery" to take a byte of data and move it to its appropriate place in the memory bank. It is also the interval of time required to search for a byte of data in the memory and retrieve it for display. Buffer capacity relates to the transient storage available to hold live data while they wait for the access mechanism to store them in the memory or for the printer to display the data in either soft- or hard-copy formats.

The interrupt sequence occurs when we want to hold the accumulation of new data while we extract and display stored data. There are two interrupt sequences presently used. The first sequence takes the live data byte by byte and stores them for future use. The second takes many bytes at one time and stores the groups of bytes for reuse. The latter interrupt sequence is faster.

As discussed above, there is always an internal program used to control the action of the memory built into a system. This internal program operates on what is called a *central processor unit* (CPU), which is a specialized integrated circuit chip. It performs as the memory's director. Data are fed into and through the CPU by the buffers. The CPU controls the data flow through a system of priority interrupts which require the CPU to hold data in the buffers, move data from the buffers into the memory, store the data in appropriate memory spaces, search out previously stored data, extract the data from memory, move the data to the buffers, hold the data in the buffers, and then pass the data out of the buffers to the displays. All of this complex data handling, called *processing*, occurs faster than the blink of an eye. The only problem that occurs is when the buffers fill up and active data handling stops.

The buffers fill up when the access time becomes too large, the interrupt cycle is too frequent, or the printers cannot handle the flow of retrieved data. In these cases, the system's performance is interrupted until the filled buffers empty and then the rest of the operations recover. The only thing that is normally lost is time and the operator's temper.

1.6 CONCLUSION

This chapter has provided a discussion of instrumentation systems and the various components that make up these systems. We have discussed in general terms transducers, signal processing, data display, and data storage as they relate to each other and to the system. No element within the system may function independently of the other components. All elements affect the functions of their complementary elements.

The remaining chapters deal primarily with detailed discussion of specific elements of instrumentation and specific instrumentation systems.

1.7 REVIEW QUESTIONS

1. Draw an overall instrumentation process used in the hospital. Label each part.
2. Describe what is meant by invasive and noninvasive measurements used in bioelectronic measurements.
3. Describe the frequency response, linearity, and noise of a transducer.
4. Describe the use of amplifiers and analog and digital readout devices used in bioelectronic measurements.
5. Discuss the power distribution system of an amplifier.

1.8 REFERENCES

1. Bellville, J. W., and Weaver, C. S.: *Techniques in Clinical Physiology: A Survey of Measurements in Anesthesiology*, Macmillan Publishing Co., Inc., New York, 1969.
2. Kantrowitz, P., Kousourou, G.. and Zucker, L.: *Electronic Measurements*, Prentice-Hall, Inc., Englewood Cliffs, N.J., 1979.
3. Boylestad, R., and Nashelsky, L.: *Electricity, Electronics, and Electromagnetics: Principles and Applications*, Prentice-Hall, Inc., Englewood Cliffs, N.J., 1977.

2

MEDICAL INSTRUMENTATION SENSORS

2.1 INTRODUCTION

Medical instrumentation sensors (transducers and electrodes) have similar characteristics of conventional industrial and scientific sensors. However, these sensors act as an interface to, and may be introduced into the human body. Sensors are used to detect electrochemically generated physiological voltages. Transducers measure displacement, pressure, force, velocity, acceleration, flow, sound, and temperature variations. Electrodes are used to detect electrochemical properties in solution. This section introduces the medical technician to the principles of sensors including electrode operation.

Sensors used in medicine have to be physically compatible with both the measurement function and the physiological function of the patient. In the selection of a transducer for a particular application, there are eight important characteristics that should be considered. These are:

1. What measurement technique should be used?
2. What external voltages and/or currents must be applied to the transducer to make it function?
3. What type of electrical signal does the transducer produce in terms of its level, frequency range, and duration?
4. Are the signals produced repeatable under varying environmental or physiological conditions?
5. Is the output of the transducer stable over its operating life?

TABLE 2.1 Conversion Factor for Units of Pressure

Unit	atm	dynes/cm^2	in. H_2O	mmHg	N/m^2	psi
1 pound per square inch = 1 psi =	0.06895	68,900	27.7	51.7	6895	1
1 newton per square meter = 1 pascal =	9.8692×10^{-6}	10	4.0148×10^{-3}	7.5×10^{-3}	1	1.45×10^{-4}
1 mmHg at 0°C = 1 torr =	0.0013156	1333.3	0.535	1	13,333	0.01937
1 inch H_2O at 4°C =	0.002458	2490.82	1	1.868	249.082	0.0361
dynes/cm^2 =	9.87×10^{-7}	1	4.0148×10^{-4}	7.5×10^{-4}	0.1	1.45×10^{-5}
1 atmosphere[a] =	1	1.013×10^6	406.8	760	1.013×10^5	14.7 psi = zero gage pressure

[a]Barometric pressure above and below 1 atm is the mean sea-level pressure of 14.7 psia or 760 mmHg.

6. Is the transducer strong enough to survive the stress of its application?
7. Is the sensor linear over its operating range, and is this linearity maintained during usage?
8. What are the effects of temperature, humidity, and immersion?

These application characteristics underlie the discussion of the technical characteristics, which follows.

2.2 PRESSURE TRANSDUCERS

Over the past 10 years there has been increasing use of and concern over the measurement of pressures within the patients' cardiovascular, pulmonary, and spinal systems. The measurement of pressure is always referenced to atmospheric pressure at sea level but may be expressed in a fairly broad range of terms. Table 2.1 lists these terms and indicates conversions.

The majority of pressure-measuring instruments use transducers that operate on the principle that the deformation of an elastic metal membrane accompanying the establishment of a balance between opposing forces as sensed by an active element driven by a diaphragm may be used as a measure of pressure. A typical pressure transducer and its specification are shown in Fig. 2.1. The active element in the unit shown is an unbonded strain gauge. Of equal popularity are transducers that use linear variable transformers. We shall discuss both types.

2.2.1 Strain Gages

The strain gage is based on the fact that the resistance of a wire is directly proportional to its length and resistivity and inversely proportional to its cross-sectional area. It may be observed that if we stress a wire, its resistance will change proportionally to the stress. See Fig. 2.2 for resistance as a function of wire length or tension.

If we stretch a conductor so that it is stressed halfway between its length at rest and its length where its tensile strength is exceeded, we have a simple strain gate. A reduction of energy on the stretching axis results in a reduction of the length, an increase in cross-sectional area, and a decrease in the overall resistance of the wire. If we now stretch the wire by adding more force than is required for the original condition, we increase the length of the wire, decrease the diameter, amd obtain an increase in overall resistance of the wire. Thus, the strain gage measures the relative change of displacement around a particular reference point.

The types of resistance strain gages currently used include the unbonded metallic-filament strain gage, the bonded metallic-foil gage, and the bonded piezoresistive or semiconductor gage.

Cleaning and Sterilization

Waterproof construction and corrosion resistance permit the disinfection and sterilization of the P23 ID Transducer by immersion in the appropriate chemical agent. Ethylene oxide gas sterilization is recommended. (Detailed instructions for cleaning and sterilizing the P23 ID are given in Product Bulletins MP260A and MP260B.)

Specifications

Pressure range	−50 to +300 mmHg
Maximum over-pressure	5000 mmHg
Sensitivity*	50µ V/V/cm ±1%
Volume displacement	0.04 mm³/100 mmHg pressure, approximately
Bridge resistance	350Ω, nominal
Non-linearity and hysteresis*	±1.5 mmHg at 0 to 300 mmHg ±0.1 mmHg at 0 to 10 mmHg
Zero balance*	±15 mmHg
Thermal coefficient of sensitivity (typical)*	0.015%/°F
Thermal coefficient of zero (typical)*	0.12 mmHg/°F
Operating temperature	−65° to +175°F
Electrical leakage	
AC	2 µA maximum at 115V rms 60 Hz
DC	100 MΩ minimum at 50V DC
Insulation	Withstands 10,000V DC
Rated excitation voltage	7.5V DC or AC through carrier frequency
Maximum excitation voltage	10V DC or AC through carrier frequency
Weight	1.6 oz (45.4 grams)
Size	
Length	2.21 inches (56 mm)
Maximum diameter of base	.710 inch (18 mm)

*For 7.5V excitation

Domes

TA1011D	Disposable Diaphragm Dome with Linden fittings
TA1010D	Disposable Diaphragm Dome with Luer Loks
TA1011	Reusable Dome with Linden fittings
TA1010	Reusable Dome with Luer Loks

Principles of Isolation, Model P23 ID

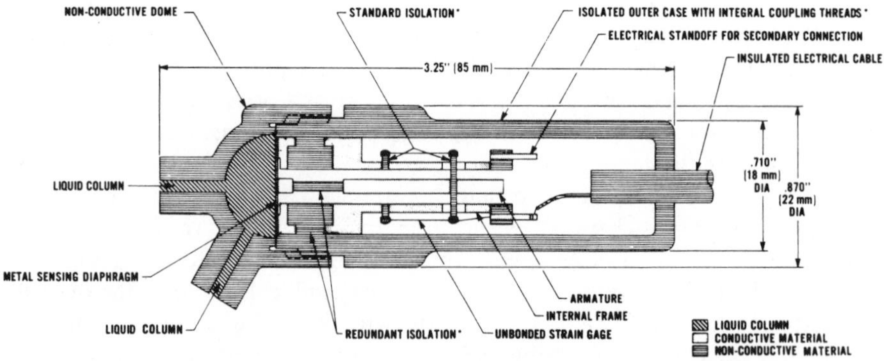

*Three modes of isolation are provided: (1) **External** isolation of the case with a plastic sheath which provides protection from extraneous voltages; (2) **Standard internal** isolation of the sensing (bridge) elements from the inside of the transducer case and from the frame; and (3) **Additional internal** isolation of the frame from the case and the diaphragm in case of wire breakage. Thus, isolation of the patient/fluid column from electrical excitation voltages is assured, even in the event of failure of the standard internal isolation.

Figure 2.1 Internal structure of the Gould/Statham P23 ID transducer. (Courtesy of Gould Inc., Measurement Systems Division, Oxnard, Calif.)

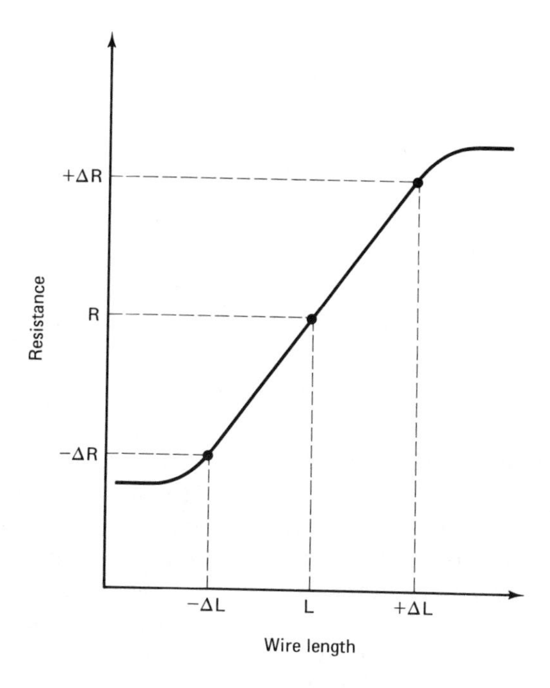

At (R, L) $\quad R = \dfrac{\rho l}{a}$

At ($-\Delta R, -\Delta L$) $\quad R = \dfrac{\rho l}{a} - \dfrac{\Delta\rho\,\Delta l}{\Delta a}$ where $\Delta\rho\uparrow$, $\Delta l\downarrow$, $\Delta a\uparrow$

At ($+\Delta R, +\Delta L$) $\quad R = \dfrac{\rho l}{a} + \dfrac{\Delta\rho\,\Delta l}{\Delta a}$ where $\Delta\rho\downarrow$, $\Delta l\uparrow$, $\Delta a\uparrow$

Figure 2.2 Resistance as a function of wire length or tension.

Unbonded strain gage elements are made of one or more filaments of resistance wire stretched between supporting insulators. The supports are either attached directly to the diaphragm used as a sensing element or are fastened independently, with a rigid insulator coupling the elastic member to the taut filaments, as in the Gould unit shown in Fig. 2.1. The displacement of the diaphragm causes a change in the taut filament length, with a resulting change in resistance.

The resistance of a wire used in a strain gage may be calculated by using the expression

$$R = \frac{\rho l}{a} \pm \frac{\Delta\rho\,\Delta l}{\Delta a}$$

where R is the wire's resistance, ρ the resistivity (Ω/cir mil^2), l the length (millimeters), a the cross-sectional area (cir mil), and Δ the small change caused by compression or stress.

At the rest position, Δ is zero and the second part of the equation drops out. When the gage is stressed, the Δ factor becomes positive and the expression becomes

$$R = \frac{\rho l}{a} + \frac{\Delta\rho\,\Delta l}{\Delta a}$$

When the gage is compressed, the Δ factor becomes negative and the expression becomes

$$R = \frac{\rho l}{a} - \frac{\Delta\rho\,\Delta l}{\Delta a}$$

The strain gage transducer is extremely linear over its operating range of pressures. When it is connected in a Wheatstone bridge arrangement, it also becomes extremely sensitive.

The Wheatstone bridge is used extensively within instrumentation. This circuit which consists of four elements connected into a bridge, is shown in Fig. 2.3. At rest all four arms of the bridge are balanced. An input driving voltage is placed across the bridge at points marked *A* and *C*. Currents are developed in the circuit, one current flowing through R_1 and R_2 and a second current flowing through R_3 and R_4. As long as the four arms are equal, the two circulating currents will be equal and the voltage developed between points *B* and *D* will be balanced out or null. The bridge may then be described as balanced.

This may be expressed mathematically. If the four elements making up the bridge happen to be strain gage resistance wires with a value of 500 Ω in the unstrained (at rest) case and the driving voltage were 10 V, then

$$I_1 = I_2 = \frac{E_D}{R_A}$$

where I_1 and I_2 are the two circulating currents, E_D the applied driving voltage, and R_A the total bridge arm resistance.

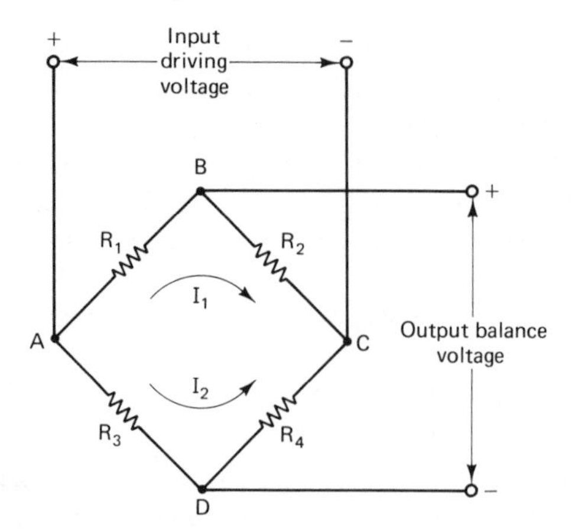

Figure 2.3 Wheatstone bridge.

The resulting calculation would be

$$I_1 = I_2 = \frac{E_D}{R_A} = \frac{10\text{ V}}{1\text{ k}\Omega} = 10 \text{ milliamperes (mA)}$$

Then the voltage drop across each resistor would be

$$V_{R_1} = V_{R_2} = V_{R_3} = V_{R_4} = IR$$

$$V_{R_1} = V_{R_2} = V_{R_3} = V_{R_4} = (500\ \Omega)(10\text{ mA})$$

$$V_{R_1} = V_{R_2} = V_{R_3} = V_{R_4} = 5\text{V}$$

The voltage at point B is found by

$$V_B = E_D - V_{R_1}$$

$$V_B = 10\text{ V} - 5\text{ V}$$

$$V_B = 5\text{ V}$$

The voltage at point D is found by the same expression:

$$V_D = E_D - V_{R_2}$$

$$V_D = 10\text{ V} - 5\text{ V}$$

$$V_D = 5\text{ V}$$

The output voltage from the bridge is the differential of the voltage found at points B and D, which may be expressed as

$$V_0 = V_B - V_D$$

$$V_0 = 5\text{ V} - 5\text{ V}$$

$$V_0 = 0\text{ V}$$

The result indicates that when at rest the bridge's output is zero.

Now, if we place a strain on the bridge, such that the value of the arm R_1 is increased to 600 Ω, the bridge becomes unbalanced. The two currents become unequal and the calculation becomes

$$I_1 \neq I_2$$

$$I_1 = \frac{E_D}{R_1 + R_2} = \frac{10\text{ V}}{(500 + 600)\ \Omega} = \frac{10\text{ V}}{1100\ \Omega} = 9.09\text{ mA}$$

$$I_2 = \frac{E_D}{R_3 + R_4} = \frac{10\text{ V}}{(500 + 500)\ \Omega}\ \frac{10\text{ V}}{1\text{ k}\Omega} = 10\text{ mA}$$

But the voltage drops can no longer all be equal:

$$V_{R_1} \neq V_{R_2} \neq V_{R_3} = V_{R_4}$$

The voltage drops across each resistor become

$$V_{R_1} = I_1R_1 = 9.09\text{ mA} \times 600\ \Omega = 5.46\text{ V}$$

$$V_{R_2} = I_1R_2 = 9.09\text{ mA} \times 500\ \Omega = 4.54\text{ V}$$

$$V_{R_3} = V_{R_4} = I_2R_3 = 10\text{ mA} \times 500\ \Omega = 5\text{ V}$$

The voltage at point B becomes

$$V_B = E_D - V_{R_1} = V_{R_2} = 4.54 \text{ V}$$

The voltage at point D remains as it was before at 5.0 V.

The output voltage becomes

$$V_0 = V_B - V_D$$

$$V_0 = (4.54 - 5.0) \text{ V}$$

$$V_0 = -0.46 \text{ V}$$

If two active arms, on opposite sides of the bridge, are used, the effect is to increase the effective bridge output by a factor of 2: in essence doubling the sensitivity of the bridge.

A strain gage Wheatstone bridge is made up by using four equal strain gage wires. Two of the wires are fixed and used as the two unchanging arms of the bridge. Two are attached to the diaphragm and thus not fixed and act as the variable arms of the bridge. When the diaphragm is compressed due to an increase in pressure, the strain gage elements are stretched, resulting in an output signal that is amplified and displayed as a level of pressure.

2.2.2 Linear Variable Differential Transformers

The *linear variable differential transformer* at first appears complex, but is relatively simple in actual practice.

The strain gage is frequency insensitive. This means that the strain gage's driving signal may be at any reasonable frequency from dc or 0 Hz on up. The linear variable differential transformer (LVDT) is different. It operates at a specific frequency.

As shown in Fig. 2.4, the LVDT has its primary winding driven by an ac signal from an oscillator usually operating between 15 and 200 kHz. The smaller the transformer, the higher the frequency. When a force is applied to the magnetic core of the LVDT, it is displaced, shifting the magnetic coupling between the primary winding and the two secondaries.

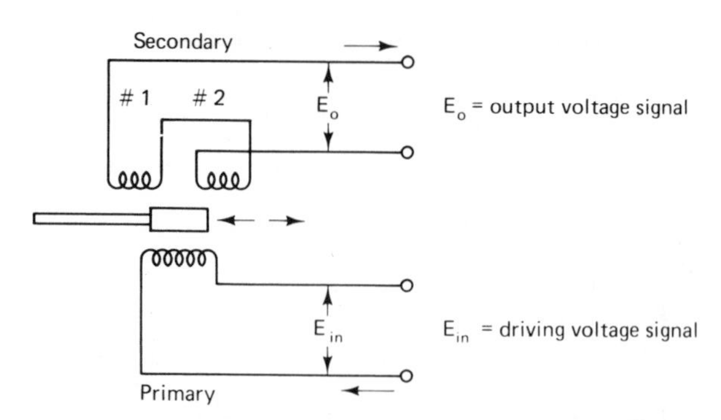

Figure 2.4 Linear variable differential transformer circuit.

At rest, represented by zero millimeters of mercury (0 mmHg), the core is at dead center and the coupling between the primary and two secondaries is equal. When negative pressure is applied, the core moves out and the primary coupling to secondary 1 is greater than the coupling to secondary 2. The result is that the driving signal coupled to secondary 1 is greater than the driving signal coupled to secondary 2. When positive pressure is applied, the core moves in and the primary signal coupled to secondary 2 is larger than that coupled to secondary 1.

The signal from the differential transformer is fed out to a ratio detector, which in effect measures the differences in driving signal as coupled to the secondaries. The principal advantage of the LVDT is that it may be used over a wide range of displacement while remaining linear. The secondary windings tend to offset slight variations or inconsistencies, improving the linearity.

The primary disadvantage of LVDTs is that an impulse of pressure may tear the diaphragm, ruining the device.

2.3 ELECTRODES

The sensors that detect electrochemical reactions in a living organism are called *electrodes*. Biopotential electrodes may be separated into three categories based on their physical construction. These are skin surface electrodes, needle electrodes, and microelectrodes.

If the signal from a sensor is distorted or obscured by unwanted noise called *artifacts*, it becomes difficult either to improve or to reconstruct the signal. To ensure a good signal-to-noise ratio signal, care must be taken in the selection, placement, and praparation of electrodes.

Electrodes are conductors that provide a current path between the desired signal (voltage and impedance changes that occur within the body) and the signal conditioner (amplifier, modulator, etc.). Electrodes must be made of conducting material that perimits stable current transfer between the subject and the signal conditioner. The material *must* be inert to the chemicals on or in the body. Active metals such as zinc and nickel replace hydrogen when in contact with the skin. This causes ions to gather on the electrodes, producing *polarization* that blocks the signal. A less active metal, such as silver, produces fewer ions and less polarization. Thus, silver-plated electrodes with a silver chloride coating (silver–silver chloride) provide good current transfer with minimum polarization.

The physical size and shape of the electrode should minimize artifacts (noise) caused by motion between the subject and the electrodes. Many types and sizes of electrodes are used.

The metal plates or disks are attached with tape, straps, or adhesives. Wire meshes, circular or square, are filled with conductive jelly to obtain good contact for long periods of time. Another electrode type uses a metal cup with the skin pulled into the cup by a vacuum.

2.3.1 Skin Surface Electrodes

Electrodes used to detect signals created within the human body are more than simplistic probes. They are electrochemical transducers. Biopotential electrodes (see Fig. 2.5) are constructed of silver–silver chloride stud, mounted on an adhesive pad. The stud projects downward into a pool of salt-saturated gel (KCl). The contact to the skin surface is through the KCl gel. A snap connector is attached to the top of the stud and the connector's lead wire is connected to the cable going to the instrument being used.

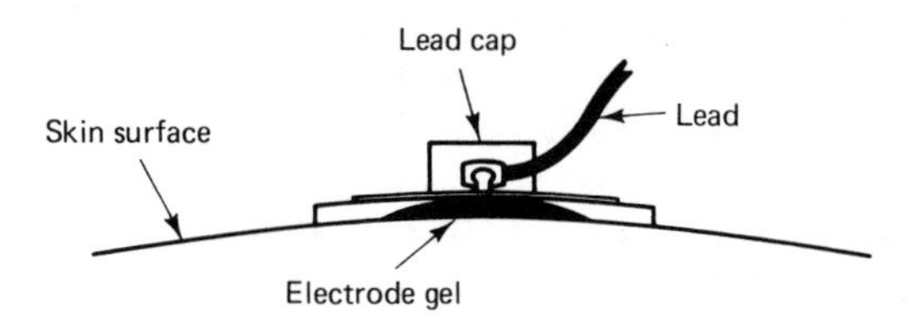

Figure 2.5 Biopotential electrode.

The biopotential electrode surface and the skin surface form a half-cell between a metal and an electrolyte. An electrical potential may be developed across this half-cell which is proportional to the ionic exchange between the metal and the electrolytes of the body. A double layer of charge may be established at the interface, causing the electrode–skin half-cell to appear as a capacitance. The single biopotential electrode may thus be represented by the equivalent circuit shown in Fig. 2.6.

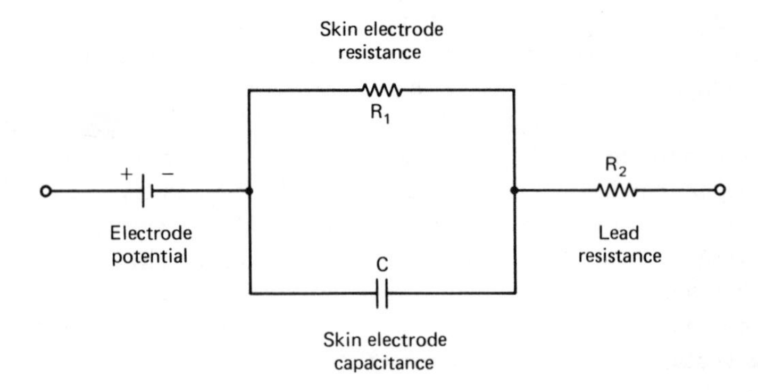

Figure 2.6 Equivalent circuit of a biopotential electrode.

An actual measurement of skin suface biopotentials requires the placement of at least two identical electrodes. The small instantaneous difference between the individual electrode potentials depends on the real difference of ionic potential between the two points of placement.

2.3.2 Needle Electrodes

Small, stainless steel or platinum needle electrodes are used for animal research and some electroencephalographic measurements. To reduce noise and unwanted pickup, wires from each electrode are usually run close to the body up to a single point, where they are all brought out together and run to the instrumentation.

2.3.3 Microelectrodes

Microelectrodes are elements originally constructed to make measurements within a single cell. They were pipettes whose tips were heated and drawn out to a fine point in the 1-micrometer (μm) range. This fine point could then be filled with an electrolyte compatible with the fluid in the cell and a platium wire fixed into the top of the pipette, extending down into the electrolyte. The measurement cell is formed by the creation of the two interfaces: the wire–electrolyte and the electrolyte–cell fluid.

With the advent of microsurgery involving brain tissue, a metal electrode developed. Metal microelectrodes are formed by etching a stainless steel wire to a fine point. Either chemical or electrolysis etching may be employed. After etching, the fine wire is coated with a layer of insulation. The metal microelectrode may be used to make measurements or, by sleaving the insulated electrode through another conductor which is used as a ground, for electrosurgery.

2.3.4 Comments on Electrodes

If the two electrodes used for a biopotential measurement are of different materials, a significant dc voltage may be produced, which can result in a current through both the electrodes and the amplifier to which the electrodes are connected. This dc potential, called the *electrode offset voltage*, may be mistaken for a true physiological event.

Significant noise currents are also common from biopotential electrodes. These currents can be caused by chemical activity within the electrode, muscle or skin flex, and/or charges developed by foreign material (such as nylon) that come into contact with the patient. It has been found that the silver–silver chloride electrode is the most stable biopotential electrode material available.

The size and shape of biopotential electrodes are important to determination of the actual electrode impedance. The larger the physical size of the electrode, the smaller its inherent impedances. Surface electrodes generally have interface impedances of from 2 to 10 kΩ, whereas needle electrodes, with their extremely small surface areas, can have impedances of several hundred kilohms.

Skin surface electrodes are generally large in contact area and are used to measure ECG, EEG, and EMG potentials from the surface of the skin.

2.4 SEMIPERMEABLE MEMBRANE ELECTRODES

Semipermeable membrane electrodes can be classified into two categories: reference and measurement electrodes. By international agreement, the hydrogen electrode has been chosen as the *reference standard* and has arbitrarily been assigned an electrode potential of 0 V. All *measurement* electrodes are referenced with respect to the hydrogen electrode.

Semipermeable membrane electrodes are used to measure the concentration of an ion or a particular gas dissolved in blood, urine, or spinal fluid. Outside the hospital, this general type of electrode is widely used to monitor or control chemical processes ranging from water purification to the manufacture of glue.

As stated above, the hydrogen gas–hydrogen ion has been designated as the reference interface and has been arbitrarily assigned an electrode potential of 0 V. The reference electrodes are based on the principle that an inert metal, such as platinum, readily absorbs hydrogen ions. If a properly treated piece of platinum is partially immersed in a solution containing hydrogen ions and is also exposed to hydrogen gas which is passed through the electrode, an electrode potential is formed. The reference electrode lead is attached to the platinum and carried out to the meter. Unfortunately, the hydrogen electrode is not stable and serves as a relatively poor reference. To overcome this lack of stability, two types of electrodes are currently used as reference: the silver–silver chloride and calomel electrodes.

With the *silver–silver chloride electrode*, the interface is connected to the solution via a dilute potassium chloride (KCl) salt bridge. The stability and level of the end electrode potential depends on the concentration of the KCl.

The *calomel* (mercurous chloride) *electrode* is probably the most popular reference electrode. In this case the interface between mercury and mercurous chloride generates the electrode potential. By placing the calomel side of the interface in a potassium chloride (KCl) solution, an electrolytic bridge is formed to the solution from which the measurement is to be made. The calomel electrode is stable and, like the silver–silver chloride electrode, its electrode potential depends on the concentration of the KCl solution.

Measurement electrodes are normally formed by the establishment of ionic exchange across a membrane such as glass, Teflon, or other material with a specific permeability. Measurement probes for specific gases, dissolved in blood or other fluids, can be constructed by using a platinum wire, insulated down to its tip. This electrode is placed into an electrolyte with its reference electrode and a dc bias potential is placed between the two electrodes. The electrolyte is usually separated from the solution to be measured by a semipermeable membrane across which the ions of the specific gas can diffuse. Measurement techniques that use this type of electrode are called *polarographic methods*.

The major causative factor in pH meter and polarographic instrument failure is the failure of the semipermeable membrane electrode. Even the slightest scratch on the surface of these electrodes will seriously reduce the effectiveness of the instrument. They must be kept clean and free of salt (KCl) buildup, properly conditioned by storage in pH 7.0 distilled water, and properly filled with fresh KCl where required. Most membrane electrodes cannot be repaired after they have failed.

2.5 THE THERMOCOUPLE AND THERMISTER TRANSDUCER

The *thermocouple* (TC) is a temperature transducer that develops an EMF that is a function of the temperature difference between its hot and cold junctions. TCs made of base or noble metals are commonly used to measure temperatures from near absolute zero to about +3200°F, and special units are available for temperatures to +5600°F. Shielded and unshielded (bare) junction varieties with a wide choice of sheath materials, such as Inconel, stainless steel, and noble metals, are standard items that can be obtained in a number of different configurations. Thermocouples may be used in bioelectronic measurements to measure temperature. The thermocouple is not popular as a medical transducer because of the Peltier effect, which occurs when one junction is warmed and the other cooled. A current error is introduced which must be minimized in the thermocouple's measurement.

Thermistors are semiconductors whose resistance changes with temperature. They are extremely popular medical transducers. Although their existence has been known for about 150 years, they did not experience extensive application until about 1940. They are commonly made of sintered oxides of manganese, nickel, copper, or cobalt, and are available in disk, wafer, rod, bead, washer, and flake form, with power-handling capabilities from a few microwatts up to 25 watts (W). Standard units have a high temperature coefficient of resistance and are produced with resistances ranging from a few ohms to 100 MΩ.

Thermistors can be connected in series–parallel arrangements, for applications requiring increased power-handling capability. High-resistance units find application in measurements that employ long lead wires or cables. Thermistors are chemically stable and can be used in nuclear environments. Their wide range of characteristics also permits them to be used in limiting and regulation circuits, as time delays, for the integration of power pulses, and as memory units.

2.6 PIEZOELECTRIC TRANSDUCERS

Although the piezoelectric effect has been known for practically 75 years, it is during the last 30 years that practical piezoelectric transducers have become common. Piezoelectricity means simply "pressure" electricity; that is, if particular types of crystals are squeezed along specified directions, an electric charge is developed by the crystal.

Crystals for bend, shear, or compression modes can be designed for a particular medical application. Examples include piezoelectric microphones for detecting heart sounds and blood pressure.

Significant advances in performance characteristics of motion, force, and pressure transducers have been made in the past decade by use of new piezo materials. New transducers are available that measure high amplitudes and provide high outputs at small amplitudes. This is achieved by using improved sensing elements, including newer piezoelectric ceramics, piezoresistive materials, and special designs by comparison. Other transducers have limited dynamic ranges, and are more susceptible to shock, vibration, temperature, and humidity.

The charge developed in piezoelectric transducers is proportional to the piezoelectric constant of the material and to the applied stress. The constant depends on the mode of operation employed. Although quartz crystals are used in some units, the manufactured ceramics are now popular, since they exhibit higher piezoelectric constants, provide higher output, and are less susceptible to environmental effects, such as case strains and transverse forces or motions. Lead–zirconate–titanate ceramics are used extensively, and other proprietary ceramics are employed in medical bioelectronic measurement devices. Electret (ceramic) microphones are also applicable for medical applications.

Recently, a new rugged and lightweight physiological pressure transducer (Fig. 2.7) which uses quartz crystal sensors to measure partients' blood, gastrointestinal, intrauterine, intracranial, and other physiological pressures has been announced by Hewlett-Packard.

Several distinct advantages are offered the hospital staff by the quartz crystal technology used in the HP 1290A transducer. It is more rugged and durable than conventional transducers that use thin-metal diaphragm sensors. It can withstand hard day-to-day usage and handling, can be gas sterilized, and even scrubbed with brush and detergent—all without affecting the transducer's accuracy.

With its flat form, rather than the cylindrical body of earlier transducers, the HP 1290A can easily be strapped to a patient's limbs to permit shorter tubing for more accurate dynamic measurement. The compact transducer is 88.9 mm long and 31.3 mm wide (3.5 in. × 1.25 in.) and weighs only 30 grams (g).

A disposable dome, which attaches with a simple twist-and-lock motion, is available for the transducer. The circular dome is constructed of transparent plastic, which makes it easy to see and dispel unwanted bubbles that could distort the accuracy of the measurement.

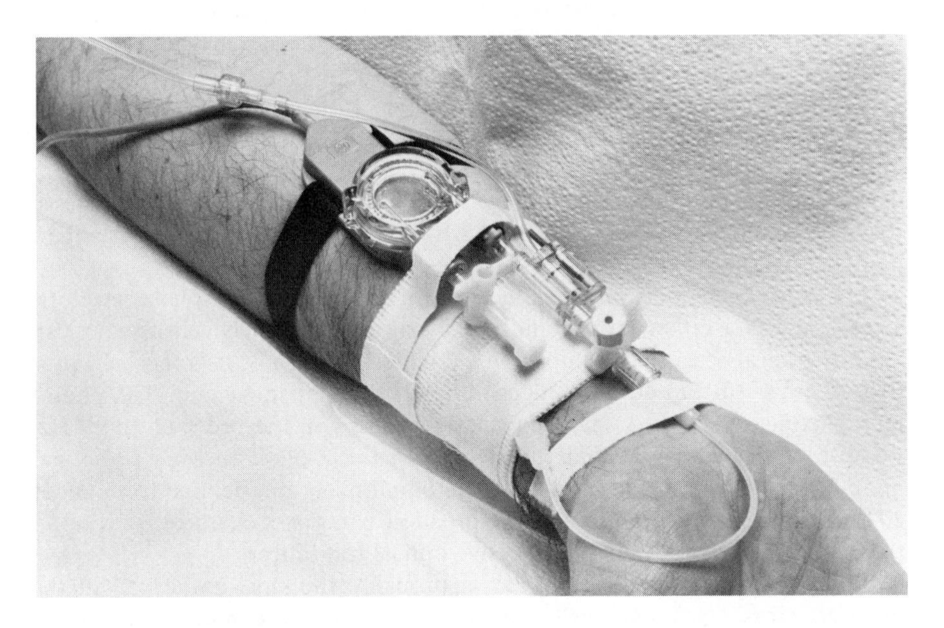

Figure 2.7 Hewlett-Packard 1290A. (Courtesy of Hewlett-Packard Co., Palo Alto, Calif.)

The HP 1290A may be used with any Hewlett-Packard pressure amplifier used in critical care areas, operating rooms, catheterization laboratories, and research laboratories.

2.7 OPTICAL TRANSDUCERS

In the field of bioelectronic measurements there is a great emphasis on optics used for measurement. All optical instruments must use an optical transducer to convert the radiated light (visible, ultraviolet, or infrared) into an electrical signal that can be calibrated and recorded. There are many types, sizes, and shapes of optical sensors, ranging from sophisticated phôtomultipliers to rather simple solid-state diodes.

A comparison of the characteristics of various photosensitive devices should be made at this point. Each device has particular characteristics that determine its use in a particular system. The vacuum *photodiode* is relatively insensitive but has an extremely high frequency response, on the order of 1 megacycle, and is therefore useful for measuring high-intensity, rapidly varying light energy. Its spectral sensitivity approximately 10^{-3} to 10^{-6} W of light energy of 10^{-3} to 10^{-6} lumen. Below this intensity the energy cannot be measured. The band-pass or spectral response is determined by the cathode material and is generally limited to the visible-light region above 4000 angstrom units (Å)

[400 nanometers (nm)] and below 6400 Å (640 nm). Because of the low potentials involved, there is little white or thermal noise generated, but its overall usefulness is limited.

The gas *phototube* is like the vacuum diode in that its spectral sensitivity is low, 10^{-3} to 10^{-5} W. But in all other respects it is vastly different. It passes a much lower frequency response, on the order of 10 kHz, and has a higher device current, resulting in higher thermal noise. This device also possesses a nonlinear output and thus is not suited for analytical measurement. It is best used as a trigger device.

The *photomultiplier* has a distinct advantage over all other photodetectors. That is, it has a rather large gain, which makes it extremely sensitive. It can have a spectral sensitivity of up to 10^{-13} lumen (10^{-13} W). Its range of sensitivity is from 10^{-4} to 10^{-13} W, depending on the individual unit. The photomultiplier also has a very fast frequency response, on the order of 100 MHz. Because of the high potentials involved with this device, there is a thermal noise factor that must be considered. Photomultipliers can be used from about 200 to 1500 nm, which represents the broadest band-pass characteristics of all the devices. It is also the most expensive optical transducer.

The cadmium sulfide *photocell* is probably the most limited of all the photodetectors. This photocell closely approximates the human eye and it proves that we are really partially blind. The spectral sensitivity is 1 to 10^{-5} W, depending on the wavelength. Its frequency response, about 100 Hz is very slow. Its optical frequency range is limited to the visible spectrum. Its chief advantages are its simple, rugged construction and low power requirements. It is small and easily packaged. Because of this, it is used in small, portable equipment design.

The photodiode and photovoltaic cell are similar to the photocell in many respects but are vastly improved over it in others. For one, they have a frequency response of 10 kHz, a tenfold improvement, and a sensitivity range of 10^{-0} to 10^{-10} W, depending on the wavelength.

One type of photodetector that we have neglected to this point is the *phototransistor*. This device consists of a base-to-emitter photodiode and a base-to-collector diode. The photodiode responds to light energy just as a simple photodiode does, but the transistor arrangement of the base–collector junction provides gain, causing the phototransistor to have the propensity of increased sensitivity.

2.8 INFRARED DETECTORS

Only recently it has been demonstrated that there also exist temperature elevations too minute to assess tactilely and that these may indicate areas of excessive cellular or metabolic activity. With sensitive apparatus called a *thermograph*, such subtle elevations can now be "seen" or thermographed, and the temperature "maps" or thermograms so obtained can be qualitatively and quantitatively analyzed. Cancer, a process involved in increased vascularization and metabolic activity different from that of normal tissues, can thus be detected. The knowledge revealed by thermography may therefore have far-reaching diagnostic application to this pathology and to many other problems of medicine and biology.

The input of an infrared detector is radiant energy from the human skin. The output is an electrical signal or visible picture. Infrared-sensitive elements accomplish this detection of incident radiation by virtue of several fundamental properties: photoconduction, the phosphor effect, and the measurement of several thermal effects. Strokes and occlusive diseases of the cartoid artery are demonstrable on thermograms. Thermograms may also be used to aid the selection of patients for angiography and to evaluate the effectiveness of therapy.

Since arthritis is an inflammatory process, it is not surprising that its presence is often detected by increased heat. The fluctuations in temperature as they are seen thermographically can provide a measure of the severity of the process in the joints involved, and serial thermograms can help to appraise the value of therapy.

The depth and degree of burns and frostbite injuries can also be shown thermographically and can help the surgeon to judge the necessity and optimal time for skin grafting, as well as displaying the demarcation between areas of vital and dead tissue.

2.9 THE LASER

The *laser* (light amplification stimulation by emission of radiation) is being used in medicine today for such things as retinal detachment operations, where a part of the eye's retinal tissue can be welded back into place with the intense beam of light energy. Primarily, the argon laser is used for photocogulation of the retinal vessels not tueking or welding. Lasers are presently being evaluated for uses in dentistry, where they may make drilling and filling a tooth almost painless because of the rapidity of the process.

The carbon dioxide laser, as a surgical tool, is suitable for primary treatment of small carcinomas of the vocal cord. The cure rate has been reported to be about the same with laser surgery as has been attained with radiation therapy. With the CO_2 laser, odynophagia, dysphagia, sore throat, loss of hair, radiation laryngitis, and radiation sickness are avoided. The only reported aftereffect is a mild sore throat.

2.10 FIBER-OPTIC TRANSDUCERS

Fiber optics deals with the transmission or guidance of light rays along transparent fibers of glass or plastic material. Like the laser, this space-age technology has found its way into medicine. Fiber-optic technology is useful for endoscopy in the gastrointestinal system. A physician may view the internal organs with minimal effect on the patient.

2.11 REVIEW QUESTIONS

1. Describe five characteristics of a medical sensor.
2. Discuss the use of the strain gage sensor in medical work.
3. Discuss the use of a linear variable differential transformer in medical work.
4. Discuss three types of electrodes used in bioelectronic measurements.
5. Discuss the use of the following in biolectronic measurements:
   (a) Lasers
   (b) Fiber-optic devices
   (c) Thermocouples
   (d) Thermistors
   (e) Piezoelectrical devices

2.12 REFERENCES

1. Welkowitz, W., and Deutsch, S.: *Biomedical Instruments: Theory and Design*, Academic Press, Inc., New York, 1976.
2. Cobbold, R. S. C.: *Transducers for Biomedical Measurements*, John Wiley & Sons, Inc., New York, 1974.
3. Kantrowitz, P.: Transducer Development for the Artificial Heart Assist Devices, *Journal of the Audio Engineering Society*, Vol. 17, No. 5, October 1969.
4. Arthur, K.: *Transducer Measurements*, 2nd ed., Tektronix, Inc., Beaverton, Ore., 1971.
5. Hahn, G. A.: The Carbon Dioxide Laser in Gynecology, *Philadelphia Medicine*, Vol. 74, No. 9, September 1978.

3

AMPLIFIERS USED IN INSTRUMENTATION

3.1 INTRODUCTION

The electrochemical signals generated by a patient's physiological systems are usually small and require amplification. A simplified block diagram of a typical physiological monitor is shown in Fig. 3.1.

The transducer detects and converts the electrochemical signal from the barrier created by the patient's skin surface into an electronic signal, which is then used to drive the amplifier circuits, which increase the signal active elements, such as transistors or integrated circuits, used in the design of the

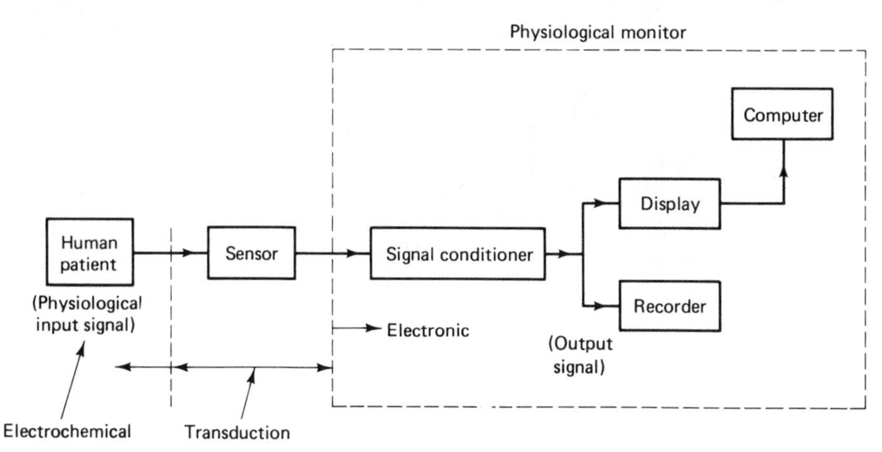

Figure 3.1 Idealized block diagram of a physiological monitor.

amplifier's circuits. Amplifiers used within physiological monitors have the following characteristics:

1. High input impedance, generally over 10 MΩ
2. High-gain open loop, generally over 1000
3. Moderate output impedance
4. Low white noise
5. Low harmonic distortion
6. Bandwidths defined by the frequency content of the signal
7. High linearity

3.2 INPUT ISOLATION

In the case of physiological monitors, the signal-conditioning amplifier circuits consist of preamplifier input isolation circuits, electrode selection switching, voltage amplifiers, and the power amplifiers used to drive the display.

Preamplifier isolation circuits are used to increase the input impedance of the monitoring system in order to isolate the patient from the instrument. The patient must be isolated because most monitors are line operated. The stray ac circuits that may leak through the unit's power supply to the patient through the transducer electrodes must be reduced to a minimum. The isolation amplifiers, by raising the system's input impedance to about 50 MΩ, achieve this objective. A typical isolation amplifier circuit is shown in Fig. 3.2.

In this circuit, two identical transistors (Q_1 and Q_2) are connected in cascade emitter-follower circuits. The input impedance as seen by the electrode

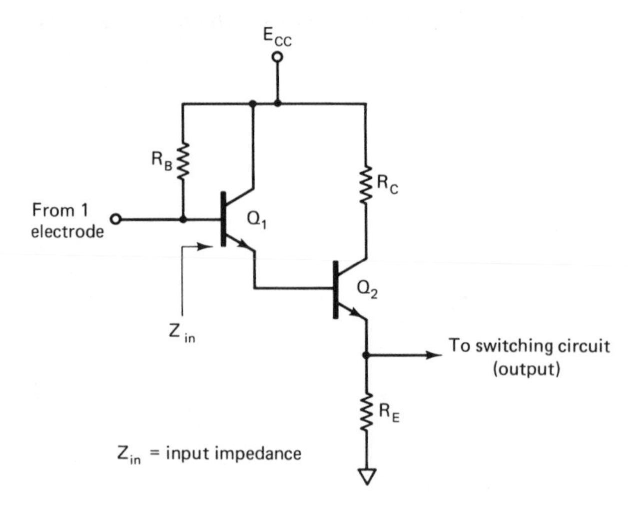

Figure 3.2 Darlington pair used as an isolation amplifier with the output taken across the emitter.

transducers is found by multiplying the common-emitter current gain beta (β) by the output impedance. Because the transistors are identical, the expression for the input impedance may be written as

$$Z_i = \beta^2 Z_0$$

The emitter resistance R_1 is used to maintain the bias of Q_1 and does not affect the output Z to the degree that the emitter resistor R_2 does. The output impedance becomes the equivalent parallel resistance formed by R_2 and the input impedance of the next stage, in this case the switching circuits. The normal switching circuit impedances are on the order of 10 kΩ. Normal R_2 values are on the order of 1 kΩ or lower. The 10 : 1 ratio results in the value of R_2 being the determining factor in establishing the output impedance of the Darlington pair. The beta of typical transistors used in Darlington pairs is on the order of 98 to 99, which results in an input impedance of about 10 kΩ. This is an extremely high impedance, but not high enough. To increase the impedance even further, a feedback network may be connected between the emitter of the second transistor and the collector of the first. This feedback loop, called a *bootstrap*, is shown in Fig. 3.3.

The feedback voltage created by the bootstrap voltage-dividing network, R_1 and R_2, is injected into the collector circuit of the first transistor. An increase in signal level at the input of the circuit causes an increase in signal through the divider at the collector, changing the transistor's bias point, in effect increasing the input impedance. The feedback developed is proportional to the values of R_1 and R_2 in series and R_3 in series. The input impedance is normally increased to about 50 MΩ using this method.

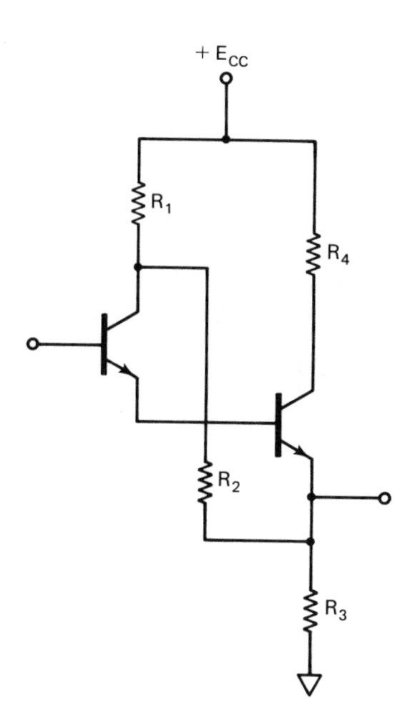

Figure 3.3 Cascaded emitter follower with bootstrapping.

It should be noted that R_4 in Fig. 3.3 and R_3 in Fig. 3.2 are used to limit the current flowing through Q_2. Whereas the voltage gain of these circuits is approximately 0.9, the current and power gains are quite high. The second transistor would burn out if its current were not limited.

From three to five electrode transducers are used with a physiological monitor. The "waves" detected by these transducers are actually the electrochemical voltage differences between any two of the electrodes. Each electrode requires an isolation amplifier between it and the switching circuits.

Signal amplifiers may be classified as either dc or ac circuits.

3.3 DC AMPLIFIERS

The dc amplifier operates from zero frequency up into the audio frequencies without appreciable loss of gain. To illustrate the dc amplifier, we chose to use three metal-oxide semiconductor field-effect transistors (MOSFETs) connected in cascade (see Fig. 3.4). The MOSFETs used have nominal drain currents of 3 mA. The drain potential is set at 10 V with respect to ground. The input signal is applied to the gate of the first MOSFET, amplified, and passed on to the next transistor's gate from the first drain. The output voltage is taken from the center tap of the voltage dividers formed by the 100 −kΩ resistors at the last transistor's drain. The lower resistor is returned to −10 V and because the drains are biased at +10 V, the quiescent output voltage rests at zero. When an ac source drives this dc amplifier, an amplified output voltage is achieved.

The major problem with direct-coupled amplifiers is thermal drift. The first stage of amplification must be compensated so that a change in temperature does not cause a change in bias levels. Bias-level changes will be seen by the amplifier as if they were regular signals. They will be amplified by the first

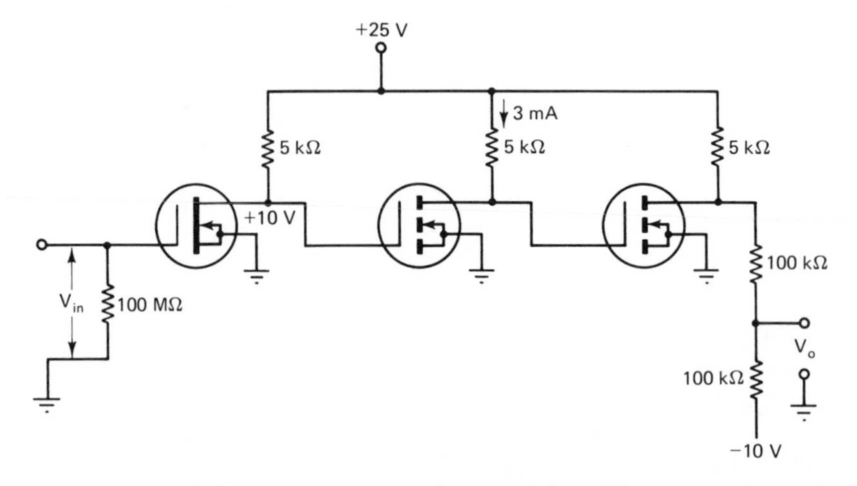

Figure 3.4 Three-stage direct-coupled amplifier.

stage and each stage following the first stage. The accumulative effect will be a large output swing caused by a relatively small thermal change. Compensation corrects this effect.

An amplifier's dc drift is stated in terms of change in output voltage as a function of a change in temperature stated in degrees Celsius. The temperature compensation used must be exactly opposite this drift and of the same magnitude.

Although we have discussed the MOSFET dc amplifier, there are many other circuits that fall into this class. They will all have similarities regardless of the form they take. All dc amplifiers have bias interaction between stages, an absence of coupling components, and stabilization attained by feedback circuits.

A high-gain signal amplified may be constructed by connecting three *RC* coupled amplifier stages in cascade, as shown in Fig. 3.5. The input to the first stage is placed between the base of the transistor and the power supply's ground reference. The output of the first stage is taken between the transistor's collectors and the power supply ground. The signal voltage, V, at the output of the first stage is coupled to the next stage via the coupling capacitors C_{C_3}. Each stage consists of a single bipolar-junction transistor in the common-emitter configuration, using a universal bias circuit.

The conventional NPN transistor as used in each stage are basically off-switch devices and must be biased "on" to their operating point. For the current to flow, each NPN transistor in the three stages must have its base made more positive than the emitter. To increase the voltage gain of each stage, the emitter resistors have been shunted to ground by an emitter bypass capacitor.

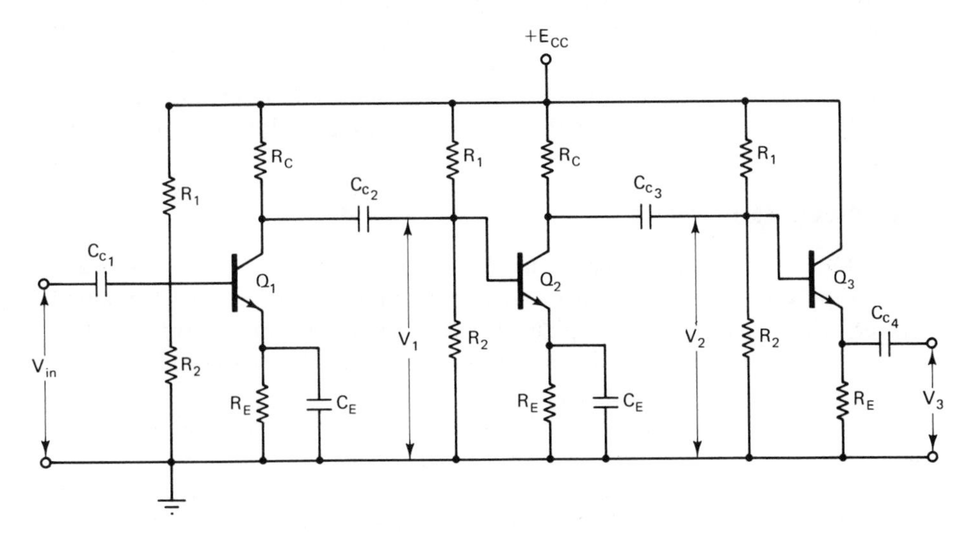

Figure 3.5 Three-stage *RC*-coupled amplifier. The three common emitter-follower stages result in a phase shift of 180°.

The second-stage coupling capacitor, C_{C_2}, connects the collector of Q_1 to the base of Q_2. As far as the dc voltages are concerned, C_C behaves as an open circuit so that the dc collector voltage of Q_1 does not affect the bias voltage at the base of Q_2. The quiescent operating voltage for each stage can be set separately. The input voltage, V_{in}, capacitively coupled to the base of Q_1, appears in amplified form with a 180° phase shift as AV_{in} or V_1. The output of the second stage, V_2, is A^2V_{in}, and in the third stage the output is V_3 or A^3V_{in}.

To make a quick input resistance calculation for the three-stage *RC*-coupled amplifier of Fig. 3.5, calculate the input resistances of Q_1, Q_2, and Q_3 and calculate the emitter currents. Since the three universal bias circuits are identical, calculation of the dc emitter current will apply to the three transistors. The dc base voltage, V_B, of the first stage is determined by multiplying the parallel combination of R_1 and R_2 by the collector voltage. As an example, for Fig. 3.5 if

$$R_1 = 60\ \text{k}\Omega$$

$$R_2 = 30\ \text{k}\Omega$$

$$E_{C_c} = 12\ \text{V}$$

then

$$V_B = \frac{30}{30 + 60}(12) \text{ or } 4.0\ \text{V}$$

The small forward voltage drop across the emitter junction is neglected so that the emitter voltage V_E is also 4.0 V. The emitter resistor for each stage is R_E and is assumed to be 4.0 kΩ. The emitter current I_E is then 4 V/4 kΩ, or 1 mA for each stage. The three transistors operate under identical dc conditions so that the input resistance of each stage is approximately $50\beta/I_E$ (mA). If $\beta = 100$ for each stage, the input resistance is 50(100)/1, or 5 kΩ. The actual input resistance of the second stage is the parallel combination of R_1, R_2, and R_C. If R_C is 5 kΩ, the input resistance of the second stage is (20)(5)(1000)/(20 + 5), or 4 kΩ. The actual input resistance of the first, second, and third stages is 4 kΩ.

3.4 POWER AMPLIFIERS

The primary function of *power amplifiers*, also called large-signal amplifiers, is to supply driving power into a load with minimum distortion. Maximum power gain is achieved. An amplifier biased class A is the most common linear circuit because the bias circuits are relatively complex but stable. The class A circuit is, however, the least power efficient.

The class AB push-pull amplifier is the most desirable where fair efficiency, large amounts of power, and low distortion are essential.

Class B amplifiers produce a larger distortion characteristic and are not suitable in audio amplifiers unless well-balanced push-pull circuitry is used. Class C amplifiers are the most efficient bias mode and are used in amplitude-insensitive systems.

Power amplifiers are used extensively in clinical applications. Some applications include electrosurgical output circuits, EKG stylus driving circuits, and relay driving and control circuits.

3.5 PHASE SPLITTERS

A transformless push-pull emitter-follower amplifier, as shown in Fig. 3.6, creates two outputs which are shifted 180° out of phase on an input voltage with respect to each. This circuit is called a *phase splitter* or phase shifter.

The basic advantage of the circuit is the inherent 180° phase shift between the transistor emitter–base junction and the base–collector junction. The outputs are taken from the collector and emitter of Q_1 and are exact replicas of the input signal which is introduced to the base of Q_1. Both output signals should be about 75% of the amplitude of the input signal.

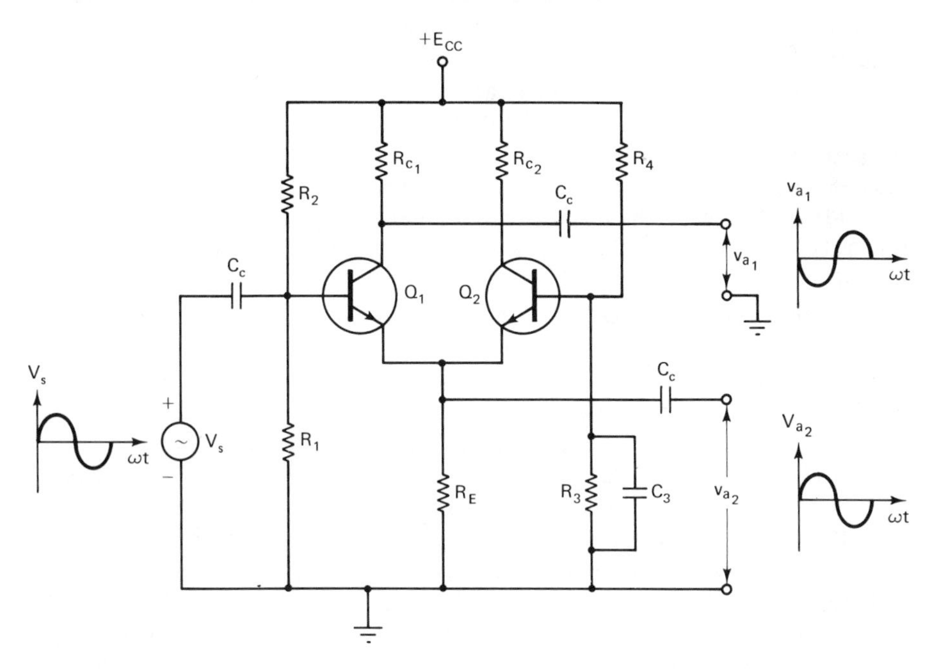

Figure 3.6 Emitter-coupled phase-splitter circuit. With the additional (emitter-follower) circuitry, the two output impedances can be equalized.

3.6 DIFFERENTIAL AMPLIFIERS

The amplifiers used for bipolar measurement such as encountered with electrocardiogram (ECG), electroencephalogram (EEG), and electromyogram (EMG) instruments are generally classified as *differential amplifiers*.

The differential amplifier amplifies the signal that appears between two (bipolar) input terminals. Under ideal conditions this signal will be the energy from a physiological source; however, due to practical design limitations, some undesirable signal will appear across the inputs. This undesirable information is called a *common-mode signal.* A common-mode signal may come from a variety of sources, including 60-Hz power distribution pickup signals, poor grounding, power supplies leakage, or it can be inherent in the type of physiological monitoring being done. A high-quality differential amplifier can reject most common-mode signals, depending on the balance and matching of the two transistors used.

The reduction of common-mode signal is called *common-mode rejection* (CMR). Common-mode rejection of 100 to 120 dB is commonly specified for ECG, EEG, and EMG measurements. The common-mode rejection at 60 and 120 Hz is of major concern and CMR measurements should be made.

The approximate input voltage level to be sensed for physiological monitoring is from 1 mV for EEG measurements to near 10 μV for EMG measurements.

The frequency of these voltages are all in the dc to audio-frequency range, as ECG signals lay between 0.05 and 100 Hz, EEG signals between 5 and 100 Hz for scalp measurements and for the cortical section of the brain's 0.5 to 100-msec pulses, and EMG signals between 10 Hz and 10 kHz.

Figure 3.7 shows the schematic of an input stage to a physiological monitor. This differential amplifier has a voltage gain of 15. A CMR of 120 dB is typical.

Transistors Q_1 and Q_2 are connected in cascade on the positive supply side, and transistors Q_3 and Q_4 are in cascade on the negative supply side. Combined, the two sides operate at an essentially constant current controlled by Q_5, which is a high-impedance current source.

The input signal when applied to the FET gates causes the balance of the current through the two halfs of the differential amplifier to shift, which produces a difference of voltages developed cross R_5 and R_6.

If the input signal driving Q_2 is made more positive than the input signal driving Q_3, the collector current through Q_2 is increased and the corresponding collector current through Q_3 is decreased. The result is that the collector of Q_1 is driven negative and the collector of Q_4 is driven positive. A larger output signal is developed because the difference between the input signals is larger.

A single physiological monitor may be used to measure ECG, EEG, and EMG signals by following the input differential amplifier with a special differential amplifier called a gain set (or selector). Such an amplifier is shown in Fig. 3.8.

In this circuit, Q_1 and Q_2 form a differential amplifier. The dc bias current for each amplifier is established by R_2 for Q_1 and R_4 and Q_2. With the two sides of the circuit balanced, no current will flow through R_3. When signals are presented to the bases of Q_1 and Q_2, the currents developed by the bias network

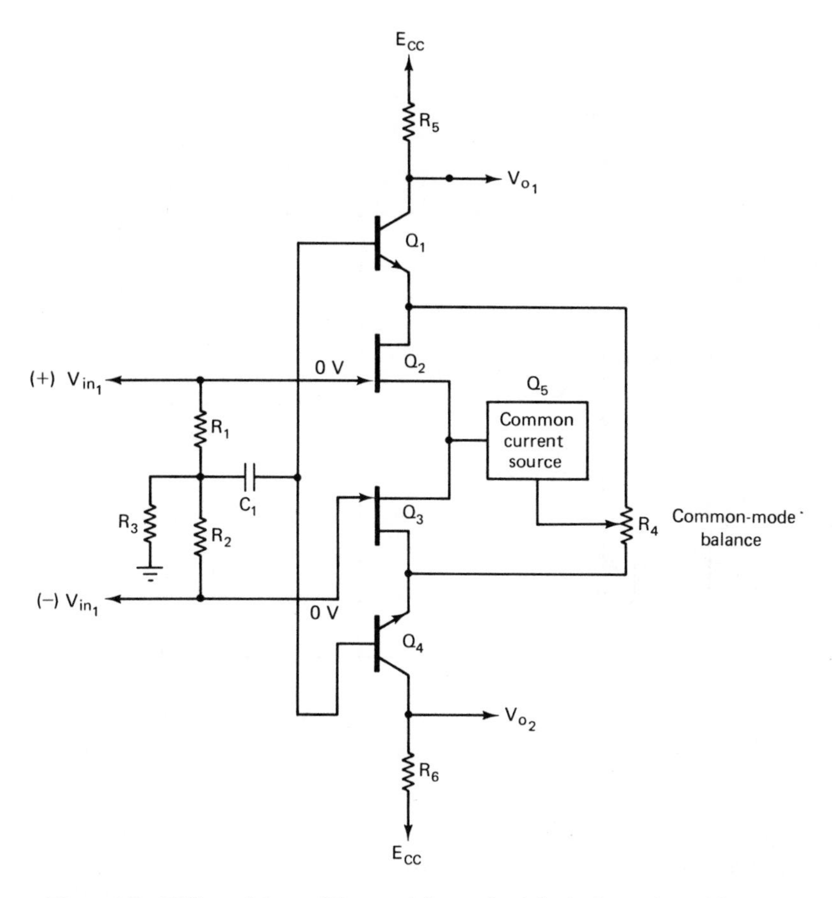

Figure 3.7 Differential amplifier used in a physiological monitor. (Courtesy Tektronix, Inc. © 1967. All rights reserved.)

formed by R_3, R_2, and R_4 shift, causing the currents through the two transistors to change by offsetting amounts. This results in an amplified but inverted output signal as seen from the collectors of Q_1 and Q_2. The base-to-emitter diodes prevent reverse base-to-emitter junction breakdown when the stage's input signal are larger than the transistors can handle.

Gain selection for EKG, EEG, or EMG operation is accomplished by the shunt selector switch, which changes the shunt load impedance between the collector circuit of Q_1 and Q_2. Each increase in position increases the effective gain of the circuit by a factor of 10.

C_1, which forms a capacitive shunt between the collectors of Q_1 and Q_2, establishes the bandwidth. Because the shunt impedance changes by a factor of 10 with each increase in selector switch position, the bandwidth also changes by a factor of 10. It is limited to 100 Hz for ECG, about 1 kHz in the EEG position, and 10 kHz in the EMG position.

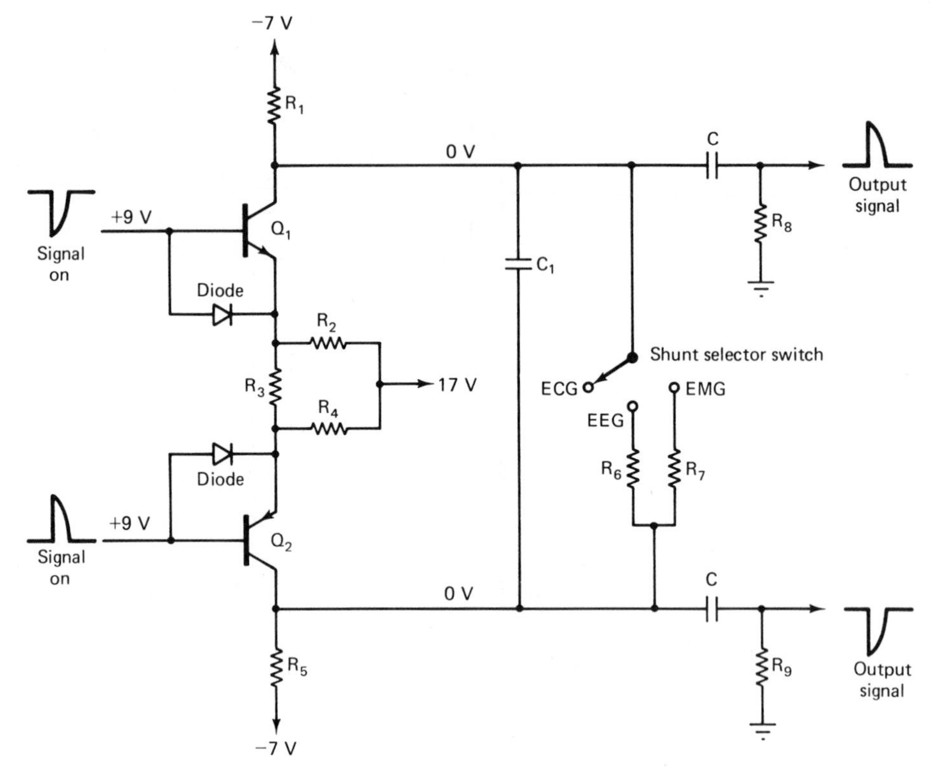

Figure 3.8 Gain selection stage following a differential amplifier. (Courtesy Tektronix, Inc. © 1967. All rights reserved.)

3.7 FEEDBACK

Feedback means sending back to the input a voltage derived from the output which is proportional to either the output voltage or current. The objective is to cause the amplifier to maintain a stable output linearly related to the input but unaffected by other factors.

The output circuit of a constant-voltage system will have a low impedance. A constant-current system will have a high impedance.

Negative feedback occurs when the signal fed back from the output subtracts from the input. Positive feedback occurs when the signal fed back adds to the input. Unless properly designed, a system with positive feedback is inherently unstable. Except for oscillators, positive feedback is usually undesirable. Negative feedback stabilizes a circuit and is usually desirable. Negative feedback causes the following effects on a system:

1. Gain stability
2. Increased bandwidth

3. Reduction in self-generated noise and distortion
4. Improved phase response
5. Increased input impedance levels

An amplifier using a negative feedback network is shown in Fig. 3.9. A portion of the output voltage determined by the ratio of R_1 to R_2 is fed back to the input terminals of the operational amplifier. With the switch open, the voltage gain, G, is found to be the output voltage, V_o, divided by the input voltage, V_s. When the switch is closed, the input voltage, V'_{in}, is the addition of $V_s + \beta V'_o$. We know that the output voltage, V_o, with the switch open is GV_{in}. Therefore,

$$V_{in} = V_s + \beta V'_o \qquad \text{with feedback} \tag{3-1}$$

$$V_o = GV_{in} \qquad \text{without feedback} \tag{3-2}$$

$$V_{in} = \frac{V_o}{G} = V_s + \beta V'_o \tag{3-3}$$

$$V'_o = GV_s + \beta V'_o \tag{3-4}$$

$$\frac{V'_o}{V_s} = \text{overall gain} = \frac{G}{1 - \beta G} \qquad \text{with feedback} \tag{3-5}$$

The overall gain of the amplifier with feedback is a function of the internal feedback factor, βG.

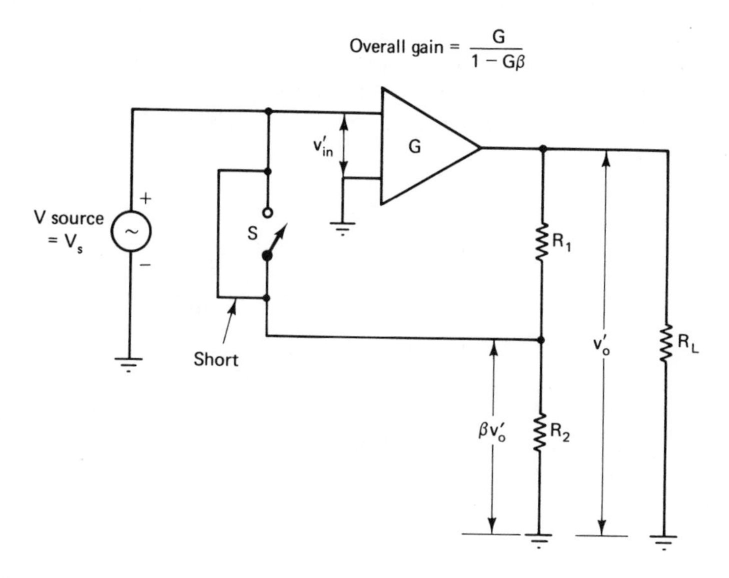

Figure 3.9 Schematic diagram using a feedback network. In this circuit, the switch is closed. Remove the short to open the switch.

3.8 OPERATIONAL AMPLIFIERS

Operational amplifiers have widespread use in bioelectronic equipment. Their use will increase steadily because of ever-improving techniques in integrated-circuit fabrication, increased improvement in reliability, and reduction in size.

The *operational amplifier* is a stable, high-gain, dc amplifier which is normally used with a large amount of negative feedback. In this manner the functional amplifier circuit is made relatively insensitive to circuit loading and the effects of temperature and time on amplifier parameters. To an excellent approximation, the characteristics of the amplifier in a given circuit are the characteristics of the external feedback elements.

The ideal operational amplifier has the following characteristics:

1. Large input and low output impedance
2. Large voltage gain and bandwidth
3. Minimum drift with temperature
4. Minimal dc offset voltage

Because of its high incidence of use, the operational differential amplifier is the one selected for discussion. We assume operation in the low or middle audio-frequency range. A balanced operational differential amplifier is shown in Fig. 3.10.

In the circuit shown in Fig. 3.10, the output is proportional to the difference between the signals applied to the inverting (—) and noninverting (+) terminals of the operational amplifier.

The ratio of the resistance value of R_1 to R_2 is set equal to the ratio of R_3 to R_4 for maximum common-mode rejection. The common-mode voltage E_{cm} in Fig. 3.10 is extremely small and should approach zero.

If $R_1 = R_3$ and $R_2 = R_4$, the output signal developed with input signals at both inputs should be

$$V_o = \frac{e_1 - e_2}{R_1} R_2$$

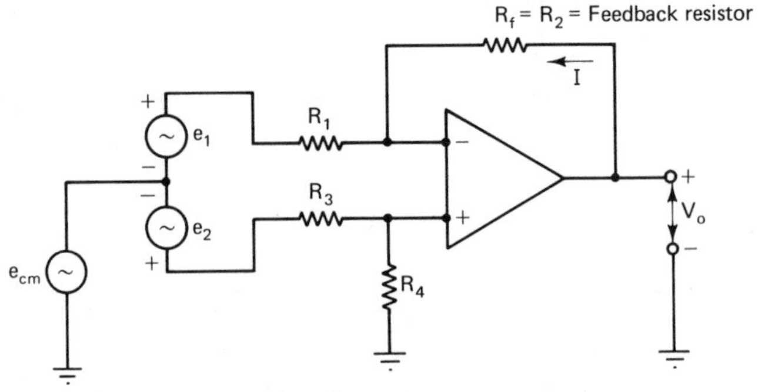

Figure 3.10 Operational differential amplifier.

If $e_1 - e_2$ equals the input vnltage, V_i, the gain of the stage may be found by

$$V_o = V_i \frac{R_2}{R_1}$$

$$\frac{V_o}{V_i} = \text{gain} = \frac{R_2}{R_1}$$

3.9 ELECTROMETER AMPLIFIERS

The *electrometer amplifier* has an extremely high input impedance and high gain. An input impedance of greater than $10^{12}\ \Omega$ and a gain in excess of 1000 is required. Electrometers are used almost exclusively for laboratory or scientific instruments. It is essential, because of this amplifier's high gain, to keep the thermal and bias voltage drifts as low as possible.

Originally, special electrometer tubes were used in these circuits. With the introduction of junction field-effect transistor (JFET) and complementary-symmetry/metal-oxide semiconductor (CMOS) integrated circuits, the electrometer has entered the modern solid-state era.

Electrometers must be used where the transducer's operation will be affected by an amplifier load impedance of less than $10^{10}\ \Omega$. This includes such devices as spectrophotometers, flame photometers, blood gas analyzers, and pH meters. The spectrophotometers and flame photometers use photodetectors and the blood gas analyzers and pH meters use semipermeable electrodes. In both cases, an extremely small signal is generated by an extremely high impedance transducer. The signal would be lost and the transducer reduced in sensitivity if the transducer is loaded by a lowered amplifier input impedance.

3.10 CARRIER AMPLIFIERS

In biomedical instrumentation, it is often desirable or necessary to use transducers such as the strain gage, which may be excited by an ac signal to measure slowly changing events. Devices known as *carrier amplifiers* permit the amplification of slowly changing (low-frequency) events as well as those containing static or dc components by means of stable, high-gain ac amplifiers. Carrier amplifiers possess the advantage of operating over a very narrow band of frequencies.

By limiting the bandwidth requirements of the amplifier, both internal and external noise which falls outside the passband of the amplifier will not appear in the output. The signal-to-noise ratio is improved dramatically.

The principle underlying the carrier amplifier consists of causing the lower-frequency event to modulate a high-frequency (carrier) signal used to energize the transducer. After amplification the modulated carrier is stripped of its modulation signal, which is then displayed.

The carrier amplifier circuit shown in Fig. 3.11 uses a Wheatstone bridge. One leg of the bridge is made with the strain gage transducers.

The carrier signal is impressed across the bridge through the secondary of the transformer T_1. A second secondary of T_1 may be used to provide the driving voltage to a floating power supply if compelete power-line isolation is required within the system.

At the null or the zero point of the strain gage, a constant-amplitude ac voltage is passed to the amplifier. At deflection of the strain gauge, the amplitude of the carrier is changed. It is this *amplitude modulation* that becomes the important information within the system.

The amplitude-modulated carrier is fed into the amplifier. From the amplifier the signal is passed to the detector, where the modulation is rectified off the carrier and displayed.

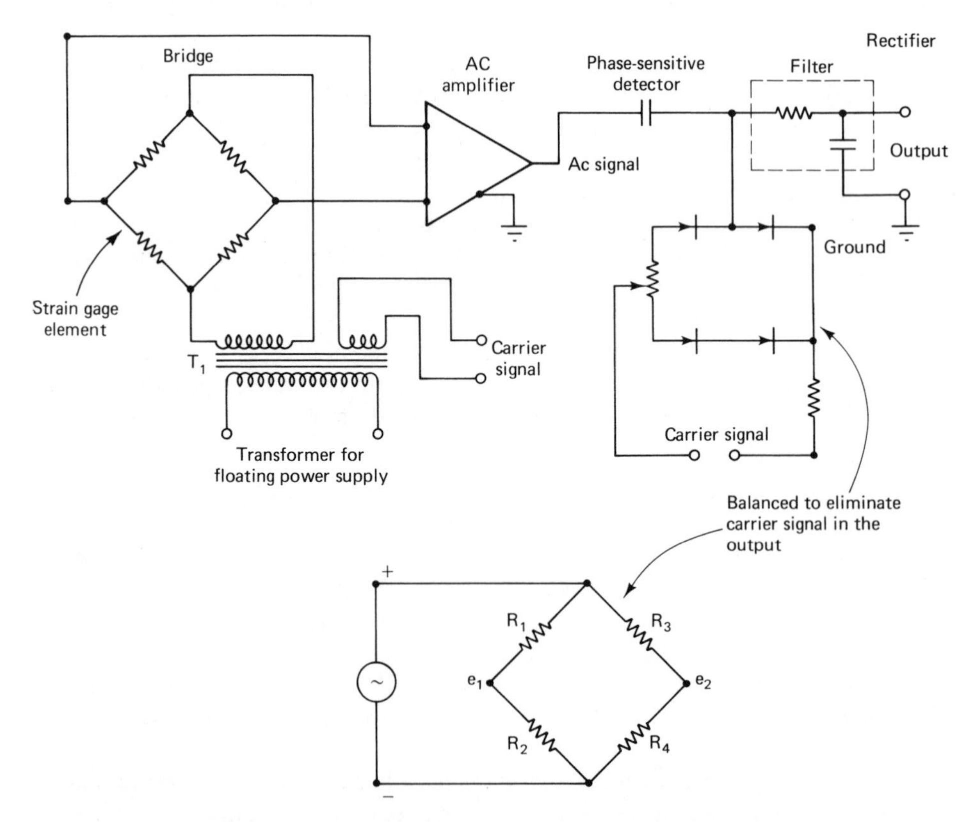

Figure 3.11 Carrier amplifier configuration. Balanced condition: $e_1 = e_2$ or $e_1 - e_2 = 0$; unbalanced condition: $e_1 \neq e_2$ and depending on the resistance ratio, $e_1 - e_2$ may be + or − at any instant when the driving polarity is such that the top is + with respect to the bottom. Thus, the phase of the driving generator of the output is 0° or 180° for pure resistive elements.

3.11 INSTRUMENTATION POWER SUPPLIES

Power supplies used within instrumentation usually must have good line regulation, load regulation, low ripple content, and good thermal stability. They must also have the capacity to deliver 1.5 times their load-rated current. Most instrument power supplies will be designed for the instrument with which they are used. There is little, if any, universal power supply utilization.

Instrument power supplies tend to be full-wave rectifier/series shunt regulator/constant-voltage units. For this reason this is the type of supply that we discuss. Line regulation is the ability of a power supply to maintain a constant output voltage with variation in the 115-V ac line. For medical applications an instrument's power supply should have a line regulation of $\pm 0.1\%$ with an input variation between 105 and 125 V. Its line regulation should be at least $\pm 1.0\%$ with an input variation between 95 and 135 V.

A power supply's load regulation is the ability to maintain a constant output voltage with a changing load current. A good power supply will maintain load regulation of $\pm 0.1\%$ with a load-current variation of 50% over design-rated load.

Ripple is the ac component riding on the rectified and filtered dc output voltage. The ripple varies according to filter capacity and load current. A high-impedance load draws little current and produces very little ripple. The supply is termed *lightly loaded.* A low-impedance load draws a great deal of current and reduces the effectiveness of the filter. This increases the ripple content in the output. The supply is termed *heavily loaded.* A ripple factor of 0.01% of the output voltage level is considered fair for an instrument power supply.

The power supply shown in Fig. 3.12 is typical of those used with biomedical instrumentation. In this case, the 115-V ac line voltage is fed into the supply through SW_1 and F_1 into the primary of T_1. Through transformer action, the line voltage (usually stepped down) is brought to the diodes D_1 and D_2, which make up a full-wave rectifier. The ac voltage is rectified into a positive pulsating dc potential, which is then smoothed out by the filter network, consisting of R_1 and C_1 and C_2. A low-ripple unregulated dc voltage is then fed into the pass or series regulator Q_1. At the junction of C_2 and the collector of Q_1, a sample current is extracted through R_2 to back bias the zener diode D_3 to establish an internal reference voltage. The variable resistor R_3 is connected in parallel with the zener D_3 so that the stable zener voltage may be made adjustable. Adjusting R_3 changes the reference voltage, which will cause the regulated output voltage to change.

The emitter of Q_1 connected directly to the output and the top of R_4. The resistors R_4, R_5, and R_6 form a voltage divider that samples the output-voltage level.

We now have two sample voltages. One, referenced to the output, is load-sensitive and is tapped off the variable resistor R_5. The second, referenced to the input through the network R_2, D_3, and R_3, is line sensitive.

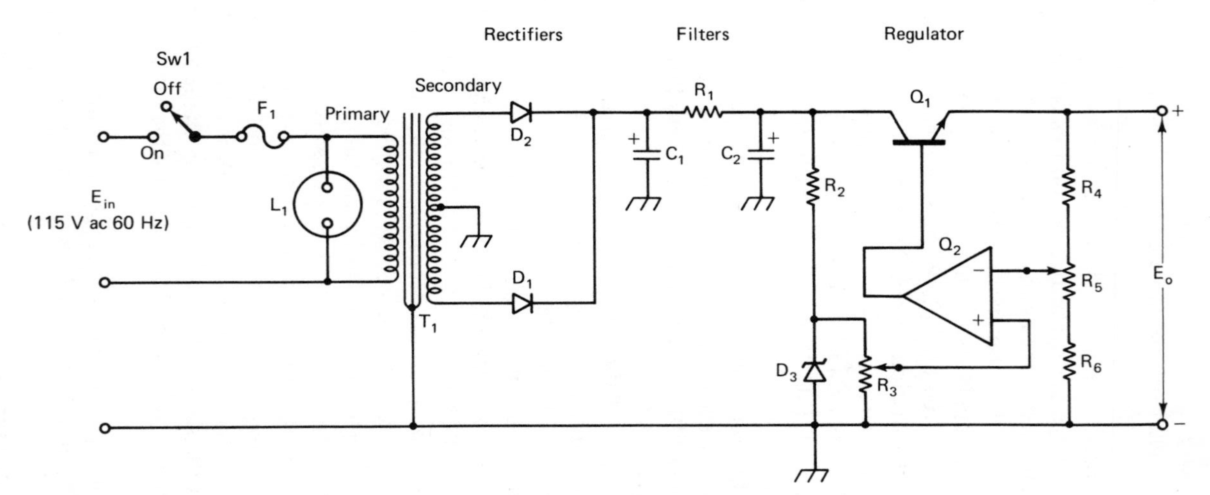

Figure 3.12 Typical power supply used in biomedical instrumentation.

The two sample voltages are passed to the reference amplifier Q_2, which detects the difference. If the output voltage rises, the potential on the wiper of R_5 rises; the differential signal increases positively, causing the output of the reference amplifier to decrease, causing the potential on the base of Q_1 to decrease, which reduces the emitter voltage, which is the output voltage.

This constant balancing between the two references maintains the regulation of the power supply.

3.12 REVIEW QUESTIONS

1. List five characteristics of a physiological monitor.
2. Draw a medical dc amplifier. Discuss each part.
3. What is the purpose of a medical power amplifier?
4. Discuss the purpose of differential amplifiers in bioelectronic measurements.
5. Discuss the effects of negative feedback of a medical amplifier.
6. Discuss the purpose of an operational amplifier in bioelectronic measurements.
7. Discuss the purpose of electrometer amplifiers in bioelectronic measurements.
8. Discuss the purpose of a carrier amplifier in bioelectronic measurements.
9. Discuss the purpose of JFETs and CMOSs for medical applications.
10. Draw and label the following:
    (a) Differential operational amplifier
    (b) Electrometer amplifier
    (c) Carrier amplifier

3.13 REFERENCES

1. *COS/MOS Integrated Circuits Manual*, Technical Series CMS 271, RCA/Solid-State Division, Somerville, N.J., © 1972.
2. Geddes, L. A., and Baker, L. E.: *Principles of Applied Biomedical Instrumentation*, 2nd ed., John Wiley & Sons, Inc., New York, 1975.
3. Kantrowitz, P., Kousourou, G., and Zucker, L.: *Electronic Measurements*, Prentice-Hall, Inc., Englewood Cliffs, N.J., 1979, Chapter 10.
4. *Motorola Application Notes on Transistor Theory*, Motorola Semiconductor Products, Inc., Phoenix, Ariz.
5. Bilboon, F.: *Medical Equipment Service Manual: Theory and Maintenance Procedures*, Prentice-Hall, Inc., Englewood Cliffs, N.J., 1978.

4

RECORDERS AND DISPLAY DEVICES

4.1 INTRODUCTION

Bioelectronic signals usually supply useful physiological data. To be meaningful in the sense of medical interpretation, these instantaneous data must be displayed in a usable format and recorded so that it may be analyzed in a static presentation. Display and recording devices used in medical and clinical instruments include analog and digital meters, special dedicated oscilloscopes as well as strip chart and X-Y recorders.

Displays present transitory information, whereas *recorders* usually generate permanent records. A recorder that traces out the continuously changing physiological event is called a *graphic* or *oscillographic recorder*.

4.2 OSCILLOGRAPHIC RECORDERS

The oscillographic recorder consists of the following basic components:

1. An electromechanical device to convert an electrical input signal to a proportional mechanical movement.
2. A stylus[1] which leaves a trace record on the chart paper as it moves across the paper.
3. A chart paper assembly consisting of a chart paper supply roll with an

[1] A stylus may be ink or thermal in nature.

associated drive mechanism to move the chart paper across the writing table.

The heart of any recording system is, therefore, the stylus, the stylus movement, and the paper and ink.

Many recorders have internal signal conditioning units (amplifiers) to enlarge the signal so that the excursion of the stylus will be large enough to provide a usable permanent written record.

The basic factors in selection and use of a recorder are:

1. Frequency response
2. Sensitivity (damping and power)
3. Range
4. Accuracy
5. Type of presentation

4.3 GALVANOMETRIC RECORDERS

The major device used to convert the electrical signal to a mechanical signal is the *galvanometer*. In galvanometric recorders, the stylus is attached to a coil in the field of a permanent magnet. The stylus or pen records the bioelectronic signal on graph paper which is moved across a sharp platen at an established speed (Fig. 4.1). For physiological measurements, the speed is set at 12.5, 25, 50, or 100 mm/sec. For chemical or biological recording, the speed is usually set at 0.1 in/sec.

Figure 4.2 shows a *light-beam galvanometer*. Since the light-beam galvanometer moves only the small mass of the mirror, the frequency response is increased to about 10,000 Hz. By comparison, a thermal stylus strip chart recorder's bandwidth is limited from dc to about 100 Hz. The beam of light generated by the light bulb is reflected off a target mirror and traces a line on photosensitive chart paper.

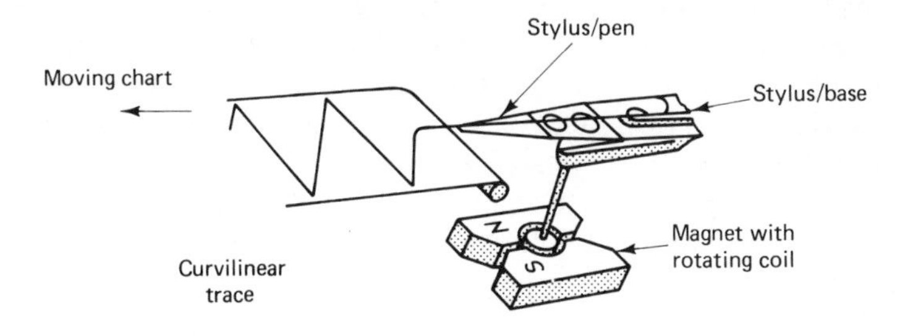

Figure 4.1 Basic galvanometric recorder.

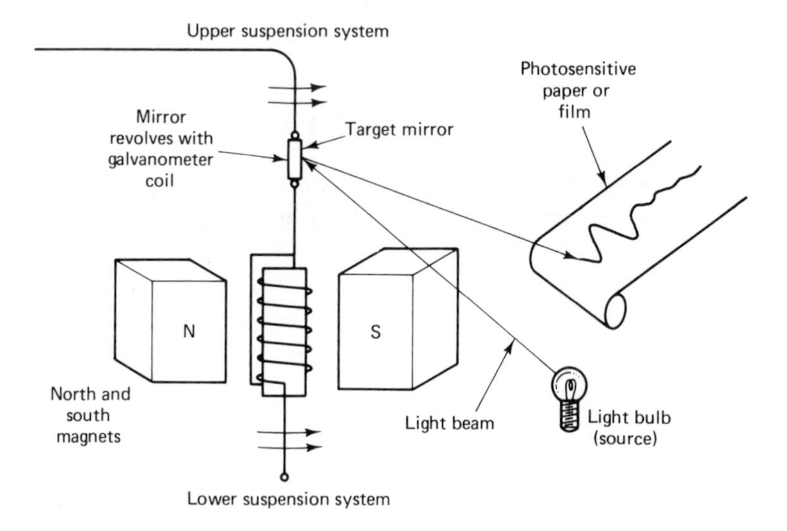

Figure 4.2 Galvanometer light-beam recorder.

4.4 POTENTIOMETRIC RECORDERS

The *potentiometric recorder* operates on a servomechanism principle, with the position of the stylus arm being deflected by a mechanism mechanically attached to the stylus arm which is powered by a motor that is controlled by a slide-wire potentiometer. Because of the slow speed of the servo system and the mass and frictional forces associated with the mechanism, potentiometric recorders are low-frequency devices with a frequency-response limit of dc to about 10 Hz. They may also be referred to as a *null balance recorder.*

In the circuit shown in Fig. 4.3, the input signal is placed across the entire resistive attenuator. The level is set through wiper *B* and passed to the slide-wire potentiometers from resistor *C*. The error voltage detected by the lower slide wire is fed into the negative-going input of the servoamplifier at point *D*. The error voltage is amplified by the servoamplifier marked *E*. This amplifier has a high power gain, developing an output between 2 and 10 W. The amplifier's output is used to drive the servomotor *M*. If the input signal is larger than the reference voltage, the error voltage is negative and the motor will be driven in a direction that drives the potentiometer wiper and pen upscale until the error voltage becomes zero. If the reference voltage is larger than the input signal, the error voltage is positive, the amplifier output reverses itself, and the motor will be driven in a direction that drives the potentiometer's slide wire and pen downscale until the zero error voltage or null point is again reached.

Potentiometric recorders are used in clinical laboratories in conjunction with most complex laboratory instruments.

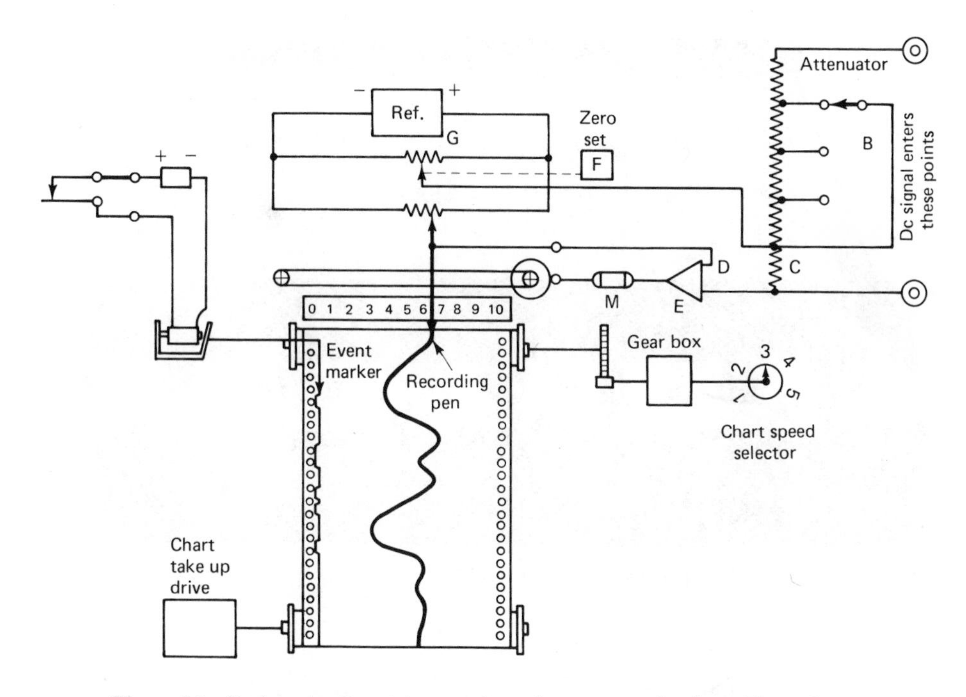

Figure 4.3 Basic potentiometric recorder using servomechanism. (From *Basic Electronic Instrument Handbook*, Clyde F. Coombs, Jr., Editor-In-Chief. © 1972. Used with permission of McGraw Hill Book Co.)

4.5 THE X-Y RECORDER

In the *X-Y recorder*, two related signals are used to drive one pen. One signal, called the dependent variable, is used to drive a slide-wire galvanometer for y-axis displacement. The second signal, called the independent variable, drives a second slide-wire potentiometer for x-axis displacement. The two slide wires work independently and simultaneously. Rapid plotting of dependent variable against independent variables is possible. A typical use of this recorder is the plotting of chemical reactions as a function of time.

Most X-Y recorders use self-balancing potentiometers whose pens trace on chart paper placed on a calibrated flat bed. Figure 4.4 shows a block diagram of a typical X-Y recorder. In each channel, a signal enters an input attenuator used by the operator to adjust the recorder's signal level within a range of from 0 to 5 V/cm. The signals are then passed to the slide wire, where they are compared with internal reference voltages. The error voltage is fed to the servomotors, which the error voltage drives to null on both axes simultaneously. As either input signal changes, the servos move so as to maintain the null balance.

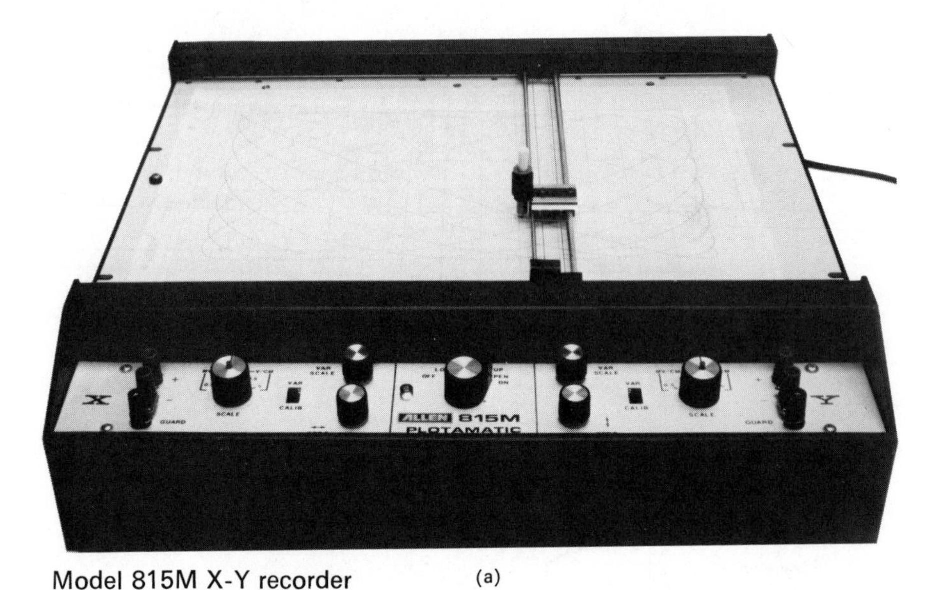

Model 815M X-Y recorder (a)

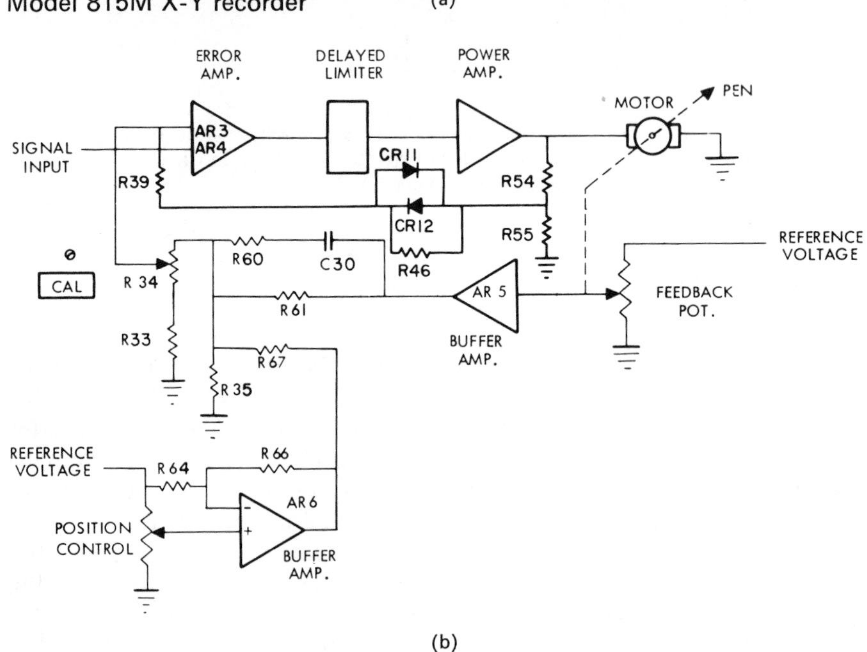

(b)

Figure 4.4 Basic X-Y recorder. Allen Datagraph servo amplifier and buffer amplifiers (both axes similar). (Courtesy of Allen Datagraph, Inc., Salem, N.H.)

4.6 MAGNETIC RECORDERS

A *magnetic tape recorder* is either an analog or a digital information storage device. Tape recorders permit data storage for long periods of time with retrieval at a later time. By recording information at low speed and reproducing at higher speed, the data may be transferred in a short time. This compression is important when a great deal of data is being handled.

In the block diagram shown in Fig. 4.5, the input signal is fed through an operational amplifier to an electromagnetic recording head, where a magnetic track is impressed on the tape as it is moved across the face of the head. The tape moves forward and crosses the face of the reproduction head. The magnetic tracks on the tape cut the magnetic field set up in the head and the information is transferred from the tape onto the head. The signal is then amplified and passed onto the display.

It is normal to use either the recording head or the reproduction head. It is possible to use them simultaneously, using the recorder as a signal-time delay.

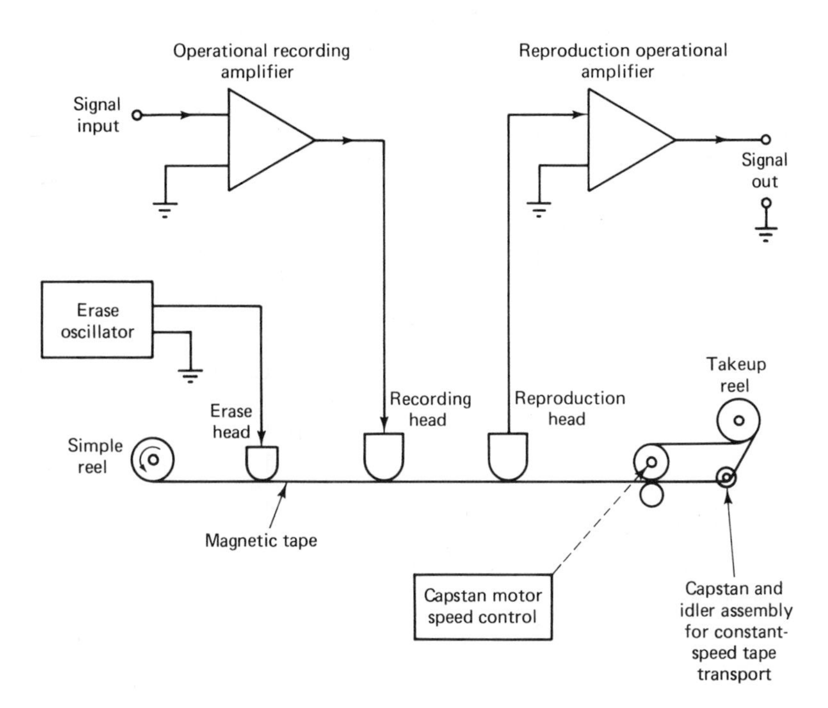

Figure 4.5 Block diagram of a magnetic tape recorder.

4.7 DEDICATED OSCILLOSCOPES

A *dedicated oscilloscope* is a cathode ray tube (CRT) set into a circuit designed for a particular purpose. For example, a 5-in. CRT for physiological monitoring would be set into a horizontal sweep circuit with 3- and 6-second sweep times. Its vertical amplifiers would be differential and have 0 to 10 mV/cm of scope deflection gain. An oscilloscope dedicated to a computer operation with an alphanumeric display would be much like your home television set, with an extremely high articulated horizontal sweep of 525 lines and possibly color guns for vertical display.

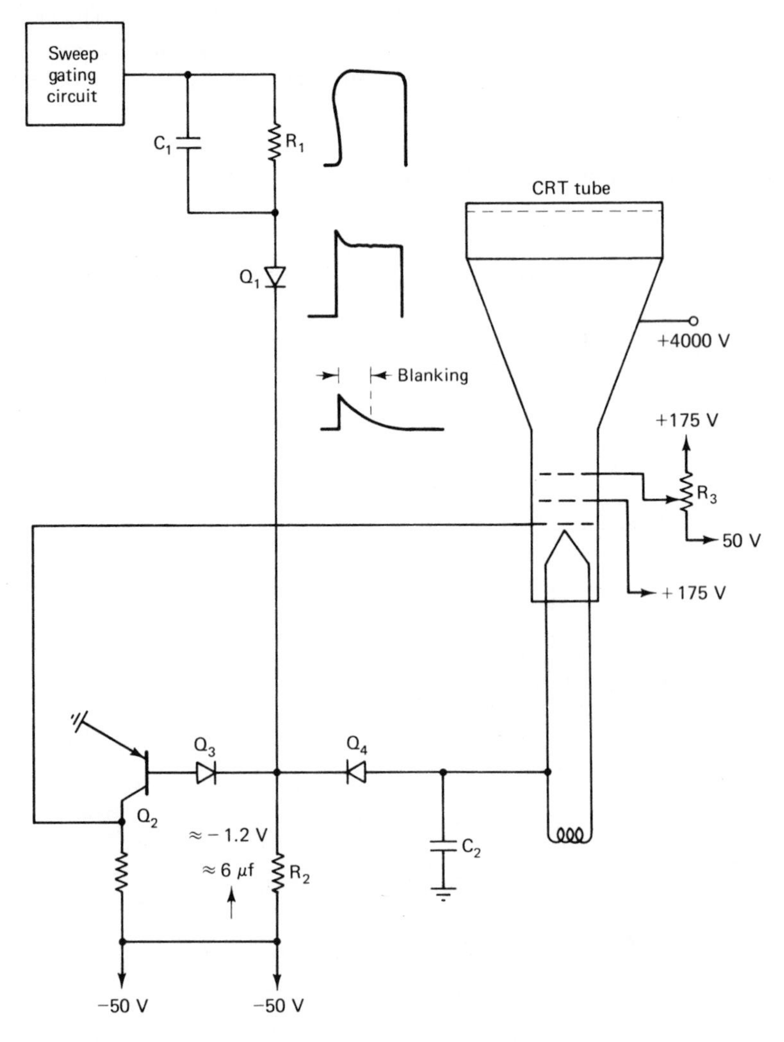

Figure 4.6 Simplified schematic diagram of a CRT circuit. (Courtesy Tektronix, Inc. © 1967. All rights reserved.)

Generally, a CRT display used in medical electronics will employ electrostatic focus and electromagnetic deflection. Because of constant beam current when the CRT is turned on, the CRT has a fixed intensity. The focus CRT adjustment resistor R_3 provides the beam focusing. The cathode can be directly heated for good efficiency and fast warm-up. The CRT cutoff voltage is about -35 V. Figure 4.6 shows a simplified schematic diagram of a CRT circuit. Q_2 and the cathode ray tube form a negative feedback circuit, which sets the CRT grid voltage to the required level so that the CRT is held close to ground potential. The fixed cathode voltage results in a constant beam current of about 6 μA through the CRT.

Because of the base-to-emitter voltage of Q_2 and the junction voltage of the diode Q_3, the voltage at the Q_3 and Q_4 junction is set at about -1.2 V. A current of about 6 μA is established through R_2 with the majority of this current following through Q_4 into the CRT. The negative feedback is used to stabilize the 6-μA bias current through the CRT. The collector potential of Q_2 and thus CRT bias is -12 to -20 V under the quiescent condition just described.

It is desirable to turn the CRT off during sweep retrace. This is accomplished by a 30- to 40-msec blanking pulse from a sweep gating circuit. The 30- to 40-msec blanking pulse is of sufficient duration for the sweep to retrace and the Miller effect capacitance to stabilize within the CRT.

The blanking pulse deteriorates after the *RC* controlled time of 30 or 40 msec, allowing Q_2 and Q_3 to turn on and raise the CRT control grid to cutoff. The CRT turns on and the negative feedback stabilizes, reestablishing the original quiescent operating condition.

Figure 4.7 shows a sine-wave form and an 8×10 cm (5 in). graticule used in industrial and medical applications.

Typical input impedances of an oscilloscope range from 1 to 10 MΩ shunted by about 20 picofarads (pF) of capacitance. CRTs may have frequency responses in excess of 10 MHz.

4.8 THE STORAGE OSCILLOSCOPE

The *storage oscilloscope* has become an important instrument in that its digital sampling techniques allows measurement of extremely fast physiological events. With memory, these scopes also allow the storage of repetitive information over a period of time. These oscilloscopes may sense and display frequencies up to 400 MHz. This represents events that occur in the 2.5×10^{-9} sec range.

In addition to displaying high-frequency waveforms, storage oscilloscopes have the following functions:

1. Storing several waveforms as data for comparison over a long period.
2. Transferring stored data from one part of an automated test system to another part.

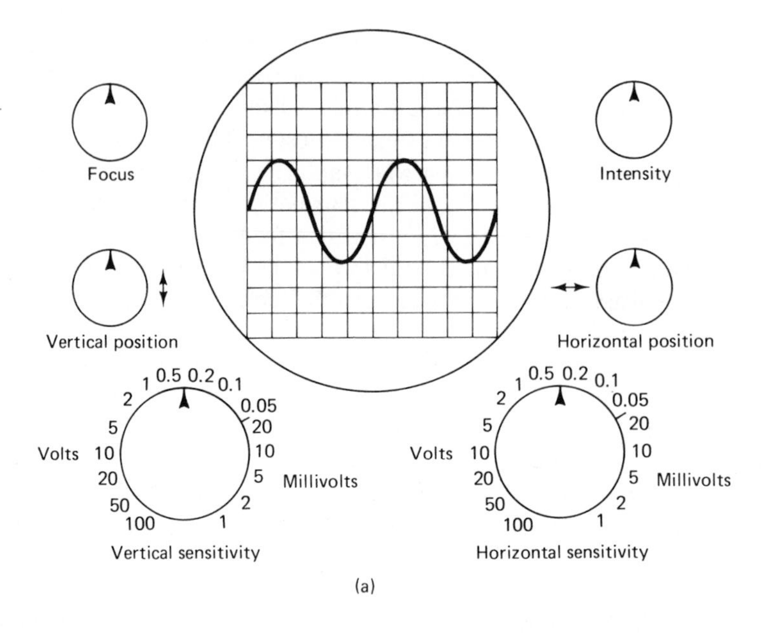

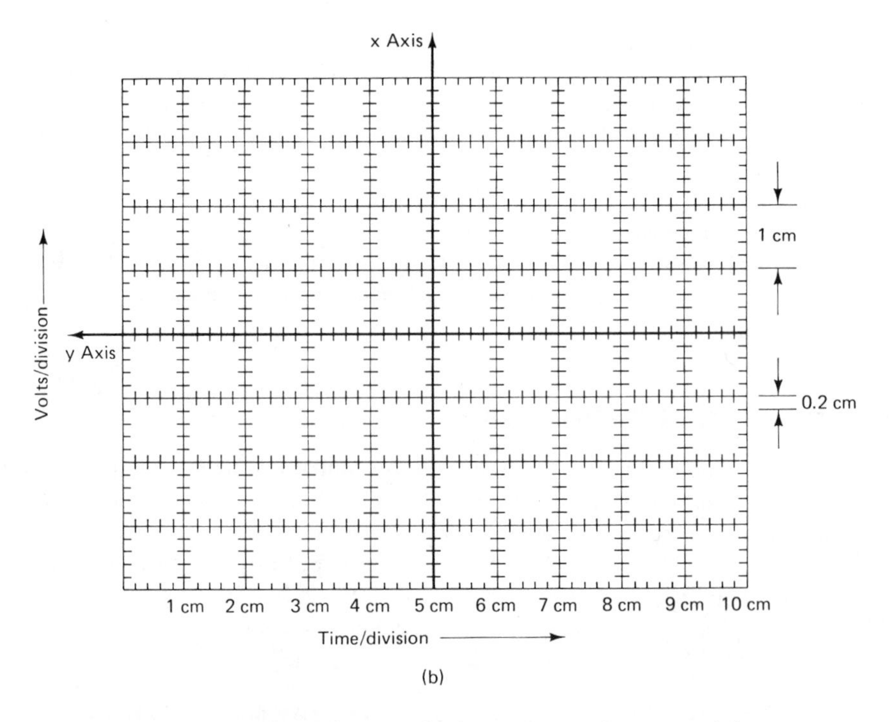

Figure 4.7 (a) Waveform of CRT, with instructions on how to read the scope face. (b) 8 × 10 cm graticule of CRT. (Courtesy of Hewlett-Packard Co., Palo Alto, Calif.)

3. Specific data processing to preset (and limited) programs.
4. Interfacing between an instrument system and a separate computer.

Analog storage oscilloscopes use a CRT tube with a special screen phosphor. Digital oscilloscopes sample the signal under measurement and store these sampled data in a semiconductor memory. When the stored information is to be displayed, it is extracted from memory as a function of the storage sequence, passed through a digital-to-analog converter, and displayed. There are three forms of displays. The first appears as individual dots, the second as dots joined by straight lines, and the third as waveforms developed as interpretations of the stored data.

4.9 REVIEW QUESTIONS

1. Discuss the use of oscillographic recorders in bioelectronic measurements.
2. Discuss the use of galvanometric recorders in bioelectronic measurements.
3. Discuss the use of X-Y recorders in bioelectronic measurements.
4. Discuss the use of the CRT in bioelectronic measurements.
5. Discuss the use of the storage oscilloscope in bioelectronic measurements.

4.10 REFERENCES

1. Strong, P.: *Biophysical Measurements*, 1st ed., 3rd printing, Tektronix, Inc., Beaverton, Oreg., 1973.
2. Kantrowitz P., Kousourou, G., and Zucker, L.: *Electronic Measurements*, Prentice-Hall, Inc., Englewood Cliffs, N.J., 1979.

5

ELECTROCARDIOGRAPHY AND HEART SOUND MEASUREMENTS

5.1 INTRODUCTION TO CARDIAC MUSCLE PHYSIOLOGY

The *heart* is a hollow muscular organ shaped like a human fist, situated in the thorax between the lungs and above the diaphragm. In the adult, it averages 4.7 in. in length, 3.5 in. in width and 2.3 in. in thickness. It weighs 0.61 to 0.75 pound in the male adult, 0.51 to 0.61 pound in the female adult, and 0.044 pound in the newborn child. Its shape is roughly that of an inverted cone and it is suspended by the great vessels so that the broader end, or base, is directed upward, backward, and to the right. The pointed end, or apex, points downward, forward, and to the left. It is rotated so that the right side is almost in front of the left.

In Fig. 5.1 the heart is divided into right and left halves, called the *right heart* and the *left heart*. The *atria* are separated by the *atrial septum* and the *ventricles* using a muscular partition called the *ventricular septum*, which extends from the base of the ventricles to the apex of the heart.

The right side of the heart contains venous blood. The right atrium, which has thin walls, receives blood. The right ventricle, which has thick walls, expels blood.

The left side of the heart contains oxygenated arterial blood. The left atrium, which has thin walls, receives blood. The left ventricle, which has thick walls, expels blood.

The main cardiac muscle is called the *myocardium*. This tissue includes muscle bundles of the atrium and ventricular and the atrioventricular *bundle of His* located in the septum of the right and left sides of the heart.

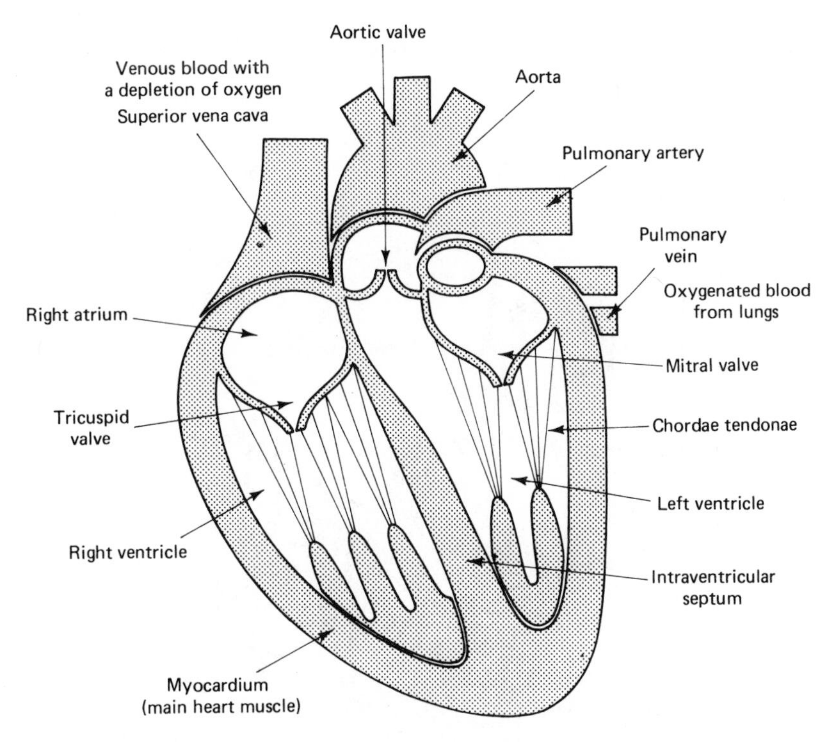

Figure 5.1 Normal heart as a pump and muscular structure. (Courtesy of Roche Laboratories, Division of Hoffmann-La Roche Inc., Nutley, N.J.)

The heart, a muscular pump, self-generates periodic electrical impulses and stimulates the cardiac muscle fibers to contract themselves in a rhythmic fashion, thereby circulating the blood through the cardiovascular system.

For the heart to function in its capacity as a pump and to circulate the blood throughout the body, it must stimulate the muscles of the various areas to contract at specific time intervals and with a specific amount of force. This is accomplished by electrical discharges that follow a specific pattern. The conduction mechanism of the heart is shown in Fig. 5.2.

The electrocardiogram (ECG) is a graphic representation of the electrical depolarization and repolarization of the heart. Under normal conditions, as shown in Fig. 5.2, the depolarization of the heart originates in the *sinoatrial node*, also known as the *pacemaker* of the heart. The depolarization then spreads through atrial muscle and the internodal atrial tracts, arriving at the atrioventricular node, where it continues through the bundle of His, forward via bundle branches, and terminating through *Purkinje fibers* onto the ventricular myocardium.

The atrial and ventricular muscles, the working muscle fibers of the heart, contract in a manner similar to that of a skeletal muscle. They form two separate and distinct functional syncythia, the atrial syncytium and the ventricular

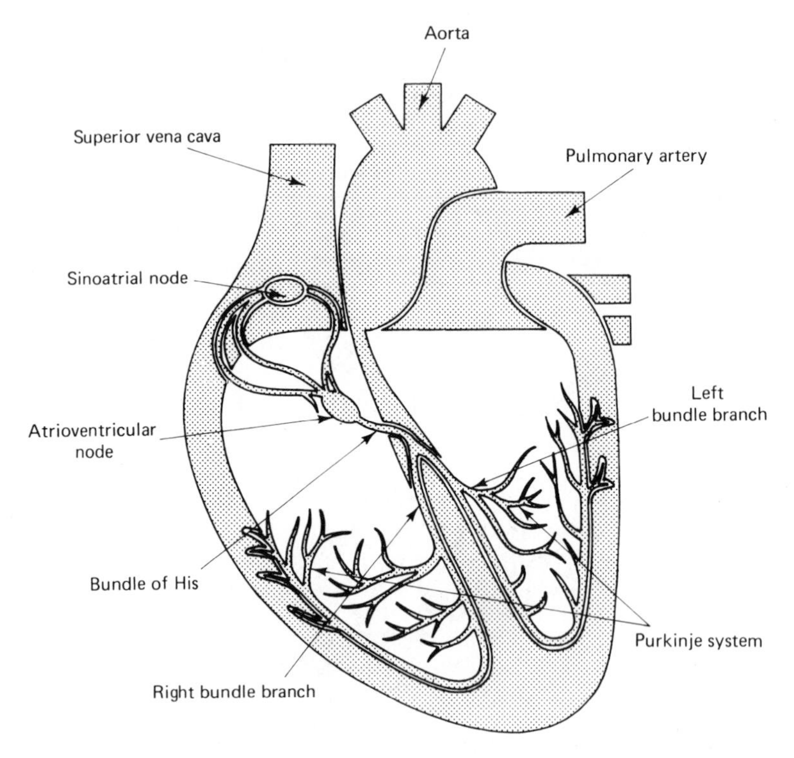

Figure 5.2 Conduction mechanism of the heart. (Courtesy of Roche Laboratories, Division of Hoffmann-La Roche Inc., Nutley, N.J.)

syncytium, which are normally interconnected only via the specialized conducting system muscle fibers of the His–Purkinje system.

The specialized conducting system muscle fibers can contract only weakly, since they contain a few contractile fibrils. Their main function is to provide a means of cardiac impulse initiation and propagation throughout the atrial and ventricular syncytium of the heart, in order to excite the cells of the working fibers and maintain cardiac function and rhythm.

Cardiac muscle fibers are composed of individual cylindrical cardiac muscle cells connected in series, with adjacent fibers laterally interconnected, via frequent branching and recombining, to form a latticework.

In Fig. 5.3, to measure the resting membrane potential of cells of the cardiac conducting system, we can submerge in a salt solution an isolated fiber preparation from the cardiac specialized conducting system. With both electrodes of the pair of recording electrodes submerged in the salt solution [Fig. 5.3(a)], a potential difference is recorded. A negligibly small potential difference between the two recording electrodes may not result in a true zero reading. At the instant that the micropipette recording electrode penetrates the cell membrane of a cell in a superficial fiber of the isolated preparation

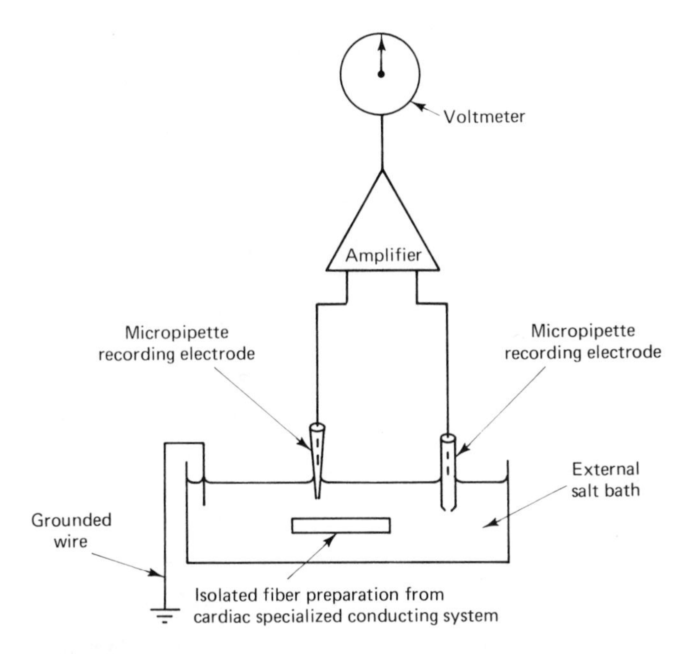

Figure 5.3 Measurement of the resting membrane potential with recording electrodes.

from the cardiac specialized conducting system, a sudden negative deflection should take place on the recording voltmeter, resulting in a reading ranging from approximately −65 mV for cells from the atrioventricular node to approximately −95 mV for cells from the Purkinje system, relative to the potential of the external salt bath, which is grounded [Fig. 5.3(b)]. These potential differences are the *resting membrane potentials* of the corresponding cells of the cardiac specialized conducting system.

The atria first contracts pumping blood to the ventricles, then the ventricles contract pumping blood to the lungs and peripheral tissues and organs.

The normal ECG trace is generated by the depolarization and repolarization vector generated by the heart muscles. An enlarged section of an ECG chart record is found in Fig. 5.4.

By using combinations of recording electrodes on the body surface, the time variations in the net cardiac vector can be recorded. This recording is called a *vector cardiogram* (VCG).

The spatial VCG is a three-dimensional representation of the locus of the "head" of the net cardiac vector throughout a complete cardiac cycle as shown in Fig. 5.5. It cannot be recorded as such with two-dimensional recording equipment, but its projection onto the frontal plane (frontal VCG), onto the transverse plane (transverse VCG), and onto the sagittal plane (sagittal VCG) can be recorded by utilization of special lead systems.

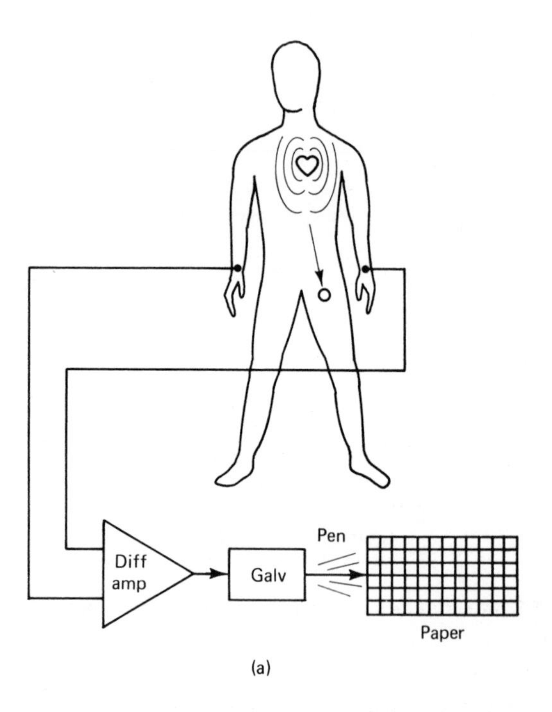

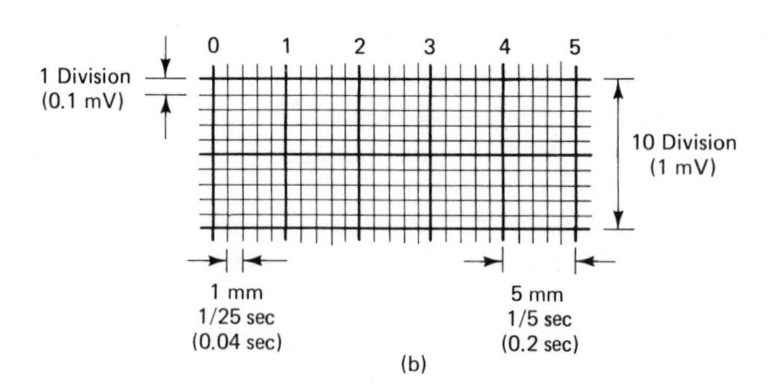

Figure 5.4 (a) Pictorial showing that the action of the composite myocardium results in the ECG, which can be represented by the vector 0. The vector cardiogram permits amplitude-phase analysis of this vector. (b) Enlarged section of the ECG chart record, showing the tracing that results for each heart "beat." Times given are for 25-mm/sec chart speed; for 50 mm/sec, divide times given by 2. (Courtesy of Hewlett-Packard Co., Palo Alto, Calif.)

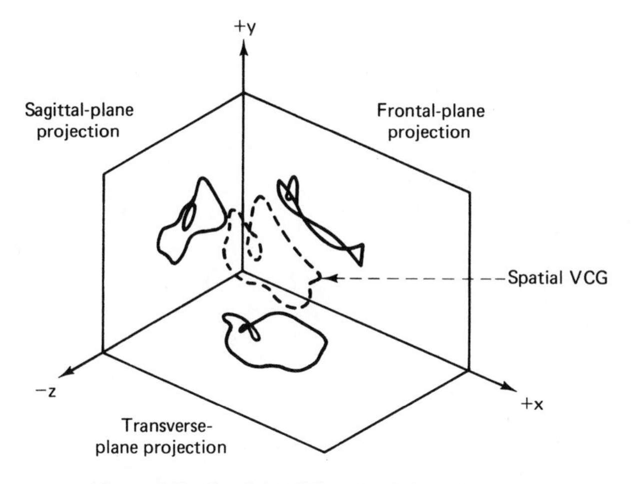

Figure 5.5 Spatial VCG and VCG projections.

5.2 INTRODUCTION TO THE ELECTROCARDIOGRAM

The *electrocardiogram* (ECG or EKG) is a graph of the voltage variations resulting from the cardiac cycle and plotted against time. The cardiac cycle is the depolarization and repolarization of the cardiac muscle, which are electric field functions that are measurable at the surface of the body where the electrodes are located.

Through ECG analysis, the physician can detect the possibility of:

1. Heart arrhythmia
2. Premature contractions
3. Myocardial infarction
4. Ectopic beats
5. The timing of other physiological events in the cardiac cycle
6. Hypertension
7. Congenital heart defects
8. The patient's metabolic rate
9. Other cardiac syndromes

A definite diagnosis usually requires secondary techniques.

The ECG block diagram shown in Fig. 5.6 shows ECG electrodes placed on arms, legs, and chest. These leads are fed to a defibrillation protective device and through buffer amplifiers to a lead switch selector. The lead switch selector feeds the differential amplifier and further circuitry to drive a galvanometer recorder as well as an oscilloscope readout device.

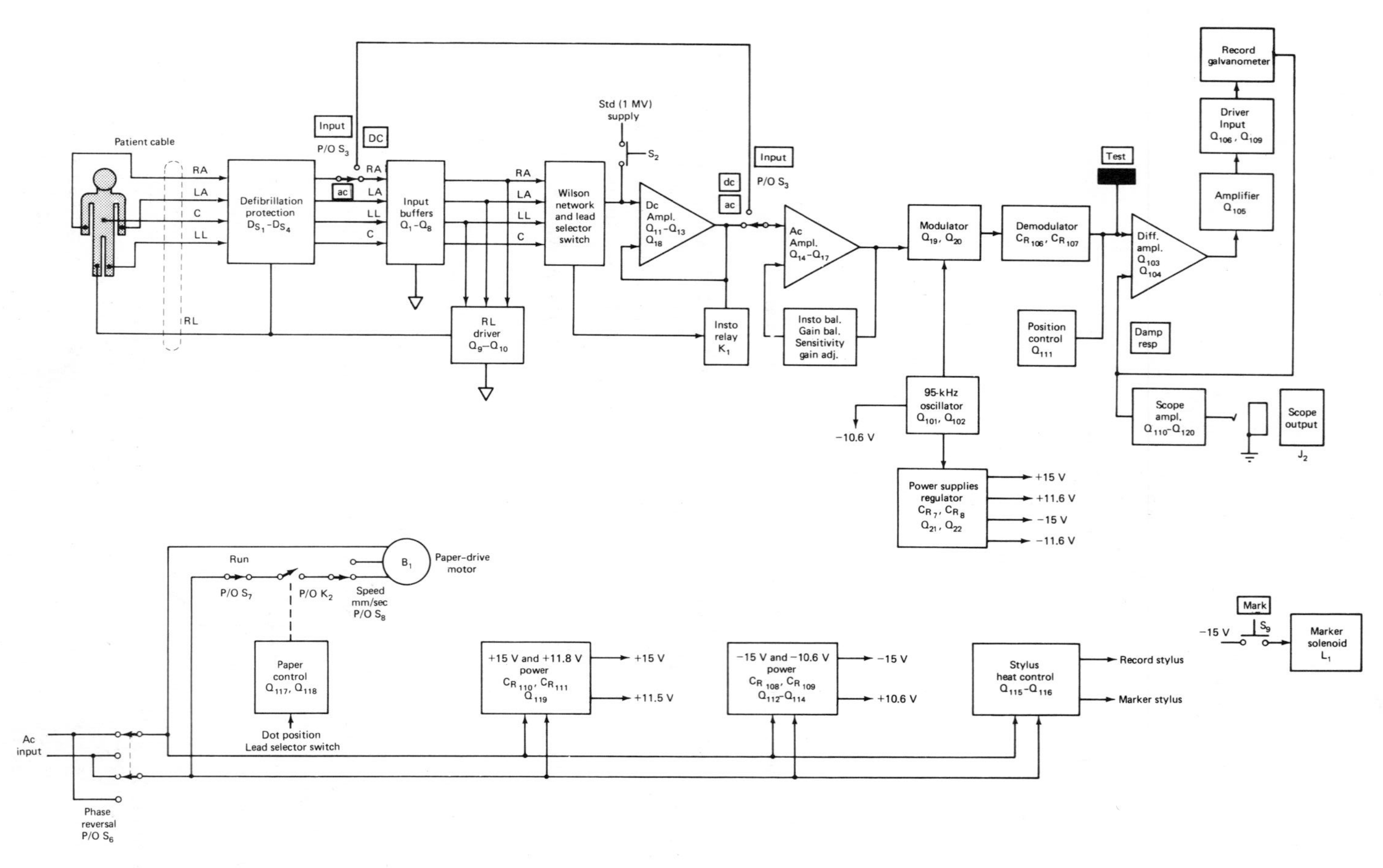

Figure 5.6 Electrocardiogram. (Courtesy of Hewlett-Packard Co., Palo Alto, Calif.)

5.3 ECG LEAD AND WAVEFORM CONFIGURATIONS

William Einthoven devised the original ECG lead system. He assumed that at any given instant in the cardiac cycle, the frontal-plane representation of the net electrical activity of the heart is a two-dimensional vector, as shown in Fig. 5.5. The length of the arrow representing the vector is proportional to the instantaneous net depolarization or repolarization voltage or potential difference, and its direction is the corresponding net direction of depolarization or repolarization in the heart. He further assumed that the origin of the vector (i.e., the heart) is located at the center of an equilateral triangle, the apices of which are the shoulders and the groin area. By assuming that the arms are extensions of the shoulders and that the legs are extensions of the groin area (the ions in the interstitial fluids of the body allow for good electrical conduction between the apices and their extensions), the apices of the triangle effectively represent the locations of the three limb electrodes.

The ECG measured from any one pair of the three pairs of bipolar limb leads I, II, and III is a time-variant single-dimensional projection of that vector onto the corresponding side of the equilateral triangle, as illustrated in Fig. 5.7. Einthoven's conventions for "+" and "−" are also depicted in Fig. 5.7.

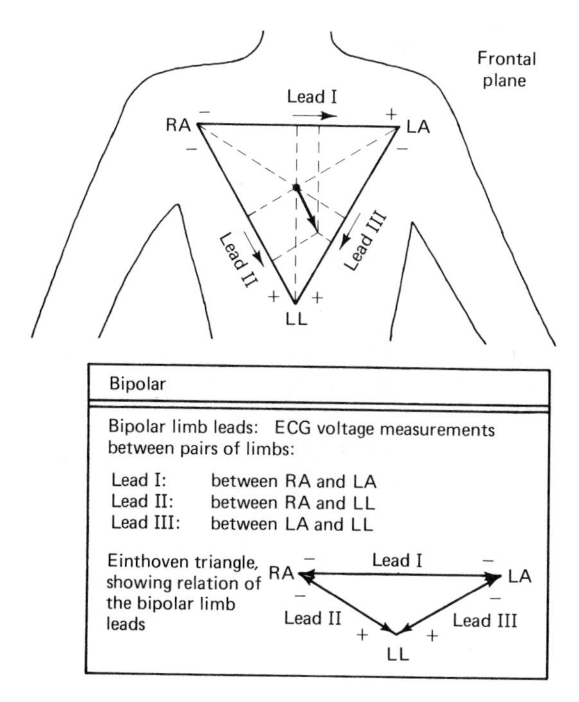

Figure 5.7 Einthoven's triangle and the net electrical heart vector. (Courtesy of Hewlett-Packard Co., Palo Alto, Calif.)

In Fig. 5.8, the bipolar and unipolar limb leads as well as the unipolar chest leads are shown in ECG techniques.

ECG leads I, II, and III are the standard bipolar limb leads. They record potential differences between two apices of Einthoven's triangle, as illustrated in Fig. 5.7 and explained further below, as follows:

Standard Bipolar Limb Lead	Potential Difference Recorded
I	LA − RA
II	LL − RA
III	LL − LA

ECG leads aVR, aVL, and aVF are called *augmented unipolar limb leads.* They record the potential of a single limb (or a single apex of Einthoven's triangle) relative to the average potential of the other two limbs or two other pieces of the Einthoven's triangle (Fig. 5.7).

ECG leads V_1 to V_6 are the *precordial chest leads.* They record the potential of a point on the chest wall relative to the potential of the midthoracic cavity, which is approximated by the average of the potentials of the three limbs (or the three apices of Einthoven's triangle) and explained further below, as follows:

Precordial Chest Lead	Potential Difference Recorded
V_1	$V_1 - \frac{1}{3}(LA + RA + LL)$
V_2	$V_2 - \frac{1}{3}(LA + RA + LL)$
V_3	$V_3 - \frac{1}{3}(LA + RA + LL)$
V_4	$V_4 - \frac{1}{3}(LA + RA + LL)$
V_5	$V_5 - \frac{1}{3}(LA + RA + LL)$
V_6	$V_6 - \frac{1}{3}(LA + RA + LL)$

The components of the normal electrocardiogram (Fig. 5.9) include the P wave, P-R interval, QRS complex, the S-T segment, and the T wave. Briefly, the electrical activity components of the ECG can be described as follows:

1. The *P wave* represents electrical activity associated with the original impulse from the sinoatrial node and its subsequent spread through the atria.
2. The *P-R interval* represents a period from the start of the P wave to the beginning of QRS complex. This represents the time taken for the original impulse to reach the ventricles and the initial ventricular depolarization. The impulse has traversed the atria and atrioventricular node.
3. The *QRS complex* represents the depolarization of the ventricular muscle. Together, this complex reflects the time necessary for the impulse to

BIPOLAR LIMB LEADS

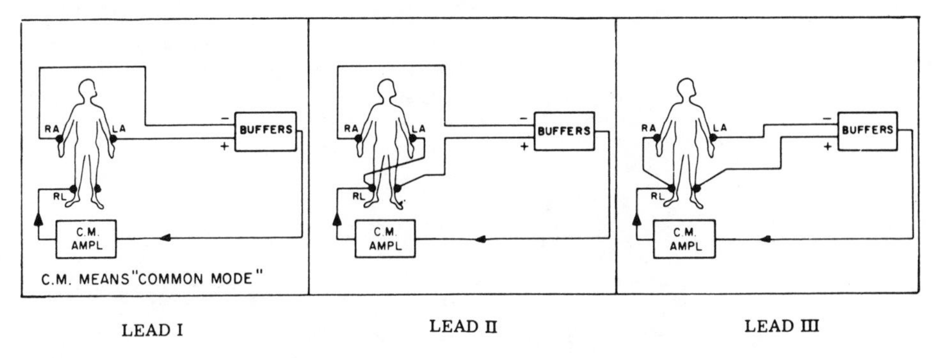

UNIPOLAR LIMB LEADS

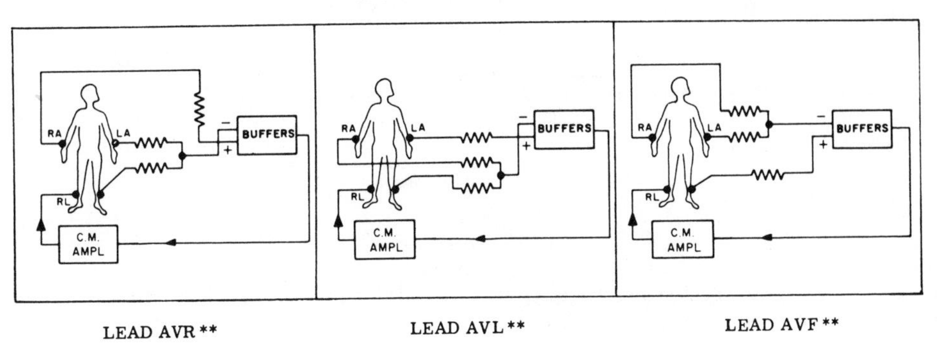

** Also known as "augmented" leads

UNIPOLAR CHEST LEADS

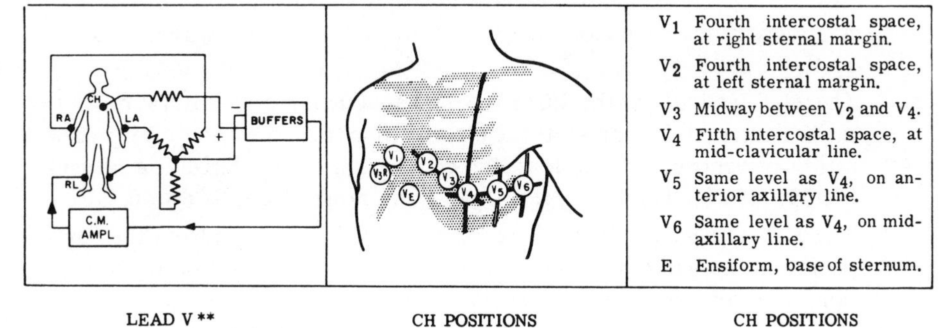

Figure 5.8 ECG techniques. (Courtesy of Hewlett-Packard Co., Palo Alto, Calif., © by Hewlett-Packard.)

ECG intervals

	Normal duration (sec) Average	Range
PR interval	0.18	0.12-0.20
QRS duration	0.08	0.07-0.10
QT interval	0.40	0.33-0.43
ST interval (QT minus QRS)	0.32	

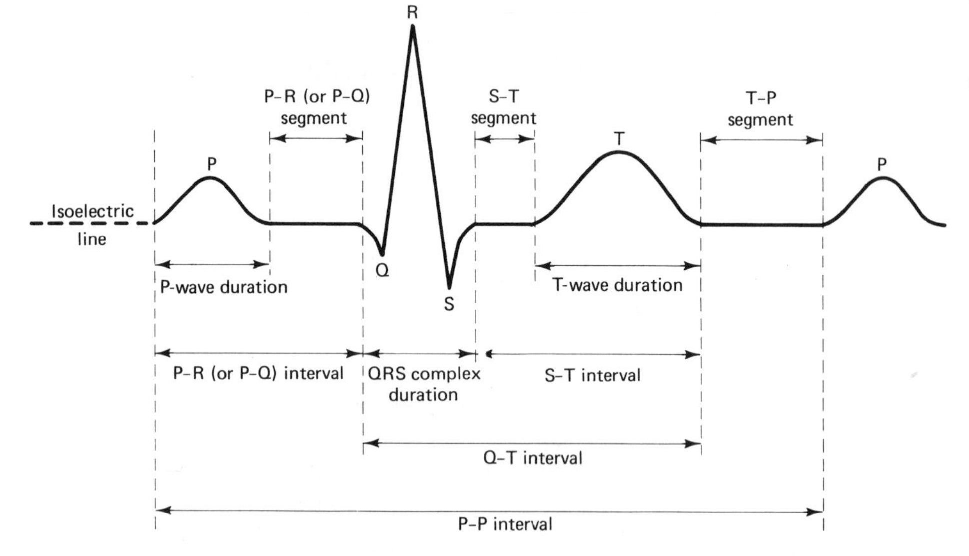

Figure 5.9 Typical normal ECG cycle recorded with a bipolar limb lead II. (Courtesy of Hewlett-Packard Co., Palo Alto, Calif.)

spread through the bundle of His and its branches to complete ventricular activation.

4. The *S-T segment* describes the period between completion of the depolarization and repolarization (recovery) of the ventricular muscle.
5. The *T wave* represents the recovery phase or contraction.
6. A *U wave* is discernible on the ECG following the T wave with a duration up to 0.24 sec.
7. The *Q-T interval* on the ECG is equal to QRS complex duration plus the S-T interval. It represents the elapsed time from the onset of depolarization of ventricular muscle to the completion of repolarization of ventricular muscle. The Q-T interval normally ranges from 0.26 to 0.49 sec.
8. The *T-P segment* on the ECG represents the elapsed time from the completion of repolarization of ventricular muscle to the onset of depolarization of atrial muscle in the next ECG cycle.
9. The *cycle duration*, or period of an ECG cycle, is the elapsed time from any point in an ECG cycle to the corresponding point in the next ECG cycle; for example, the P-P interval, which normally is equal to the R-R interval (elapsed time between two successive R waves). It normally ranges

from 0.60 sec (corresponding to a heart rate of 60 beats/min) to 1 sec (corresponding to a heart rate of 100 beats/min).

Figure 5.10 shows ECG strips of a normal patient and a 1-mV peak to calibrate the voltage measurement. A good ECG is sharp, has stable baseline, and is clear and exact. The ECG uses limb leads, augmented limb leads, and precordial chest leads to achieve a complete diagnosis of the electrical activity of the heart. The *ventricular rate* is approximately 60 divided by the R-R

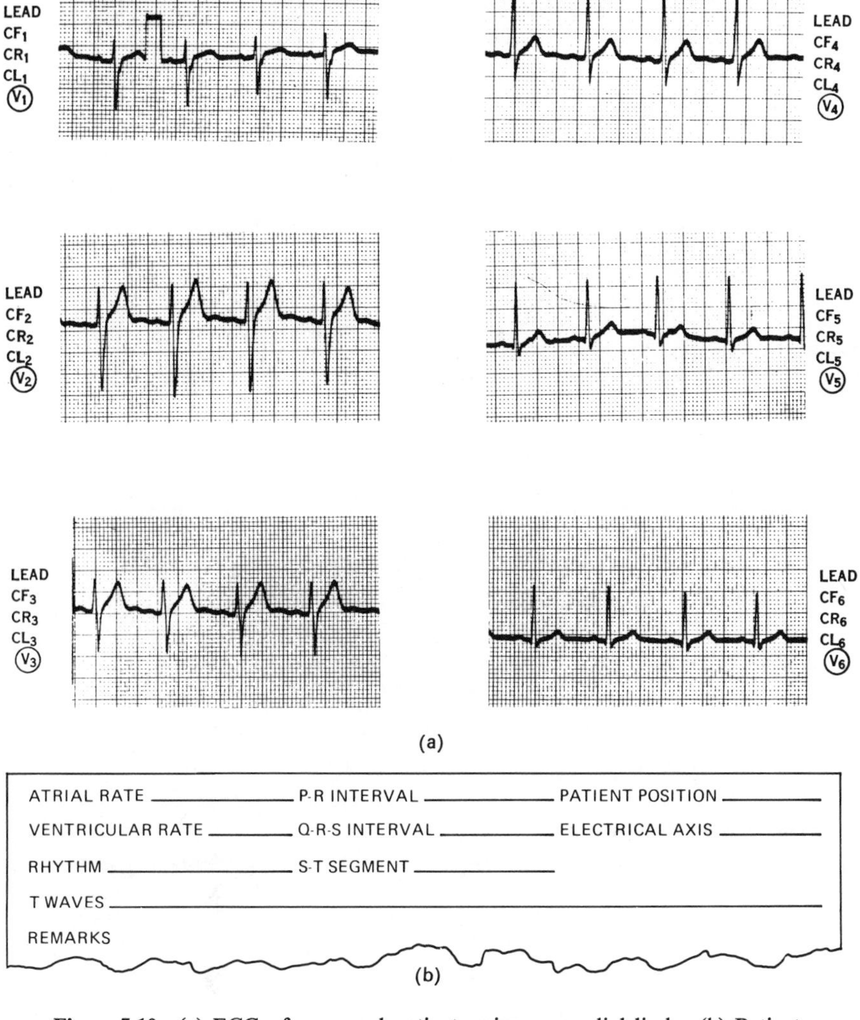

ATRIAL RATE ________ P-R INTERVAL ________ PATIENT POSITION ________

VENTRICULAR RATE ________ Q-R-S INTERVAL ________ ELECTRICAL AXIS ________

RHYTHM ________ S-T SEGMENT ________

T WAVES ________

REMARKS

(b)

Figure 5.10 (a) ECG of a normal patient, using precordial limbs. (b) Patient record. (Courtesy of Hewlett-Packard Co., Palo Alto, Calif.)

interval in beats per minute; for lead 1 of Fig. 5.11 the ventricular rate is 60/18 $\times$ 40 msec or 85 beats/min. The *atrial rate* is inversely proportional to the P-P rate. The ECG provides this information.

Tables 5.1 through 5.3 represent the bioelectronic measurement "bench marks" used in conjunction with the ECG.

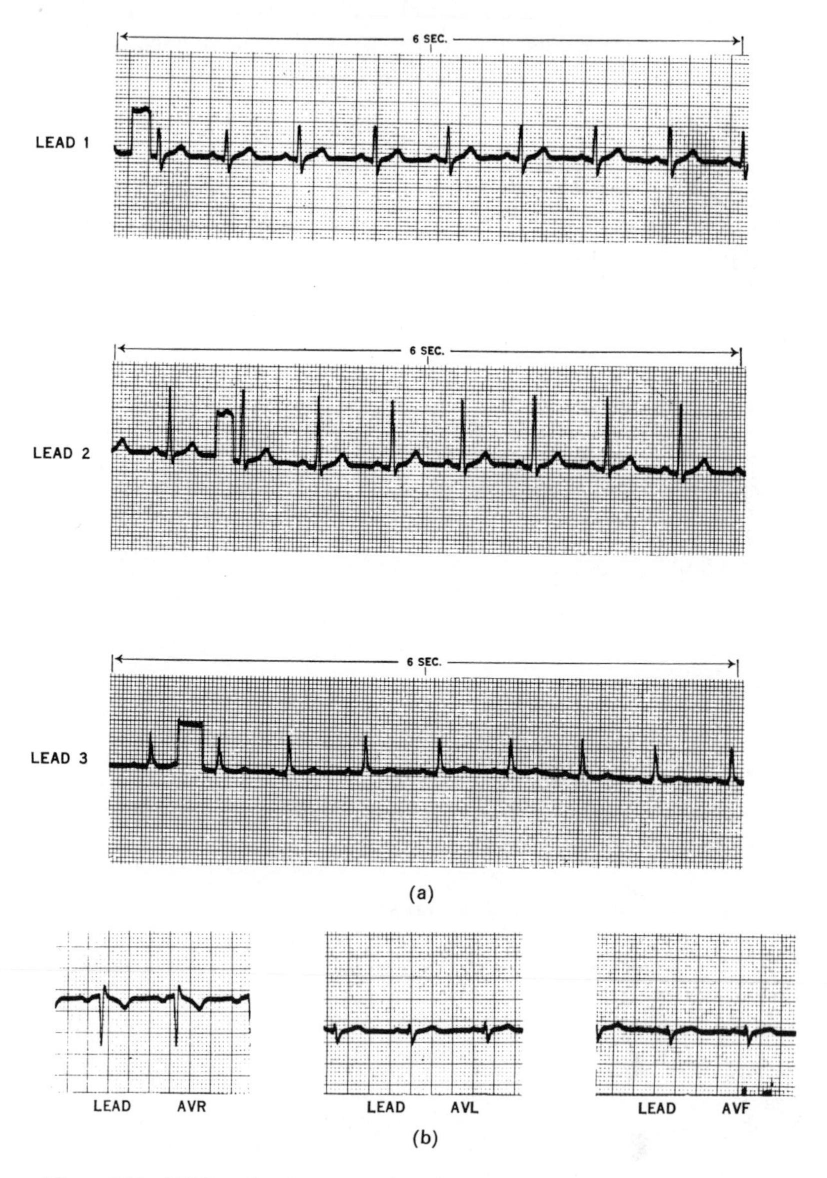

Figure 5.11 ECG strip of a normal patient: (a) using limb leads; (b) using augmented limb leads. (Courtesy of Hewlett-Packard Co., Palo Alto, Calif.)

TABLE 5.1 Nominal Range of Amplitudes of Electrocardiographic Wave Values (mV)

Wave	Amplitude Range		
	Lead I Potential	Lead II Potential	Lead III Potential
P	0.015 to 0.12	0.00 to 0.19	−0.073 to 0.13
Q	0.0 to 0.16	0.0 to 0.18	0.0 to 0.28
R	0.02 to 1.13	0.18 to 1.68	0.03 to 1.31
S	0.0 to 0.36	0.0 to 0.49	0.0 to 0.55
T	0.06 to 0.42	0.06 to 0.55	0.06 to 0.3
	aVR Potential	aVL Potential	aVF Potential
P	−0.179 to 0.01	−0.085 to 0.140	−0.06 to 0.160
Q	0.0 to 0.90	0.0 to 0.22	0.0 to 0.19
R	0.0 to 0.33	0.0 to 0.75	0.02 to 0.15
S	0.0 to 0.15	0.0 to 0.90	0.0 to 0.71
T	−0.54 to 0.0	−0.16 to 0.27	0.04 to 0.46
	V_1 Potential	V_2 Potential	V_3 Potential
P	−0.08 to 0.18	0.15 to 0.16	0.0 to 0.18
Q	—	—	0 to 0.05
R	0.0 to 0.94	0.04 to 1.52	0.06 to 2.24
S	0.08 to 2.13	0.19 to 2.74	0.09 to 2.22
T	0.03 to 1.22	−0.14 to 1.44	0.0 to 1.60
	V_4 Potential	V_5 Potential	V_6 Potential
P	0.01 to 0.23	0.0 to 0.24	0 to 0.19
Q	0.0 to 0.16	0.0 to 0.21	0 to 0.27
R	0.18 to 3.20	0.42 to 2.42	0.25 to 2.60
S	0.02 to 2.09	0.0 to 0.97	0.0 to 0.84
T	0.05 to 1.31	0.0 to 0.96	0.0 to 0.67

TABLE 5.2 Nominal ECG Recorder Characteristics

Characteristic	Parameter Range or Type
Chart speed	25 and 50 ± 1% mm/sec
Chart scale	1-cm major divisions
	1-mm minor divisions
Chart type	Thermal-sensitive paper
Sensitivities	$\frac{1}{4}$ cm/mV, ± 1%
	$\frac{1}{2}$ cm/mV, ± 1%
	1 cm/mV, ± 1%
	2 cm/mV, ± 1%
Standardization	1 mV positive dc
−3-dB Bandwidth	
High or diagnostic	0.05–100 Hz, ± 1%
Low or monitor	0.5–70 Hz, ± 1%
Linearity	0.5% over 40-mm chart
Common-mode rejection	100 dB or better
Cable leads	5 leads on single-channel
	10–14 leads on multichannel
Stylus pressure	15–25 g (20 g nominal)

TABLE 5.3 Cardiac-Related Physiological Function Range

Function to Be Measured	Parameter Range	Frequency
Electrocardiogram	10 μV to 5 mV	0.05 to 85 Hz, $\pm$ 5%
Arterial blood pressure (direct cannulation)	30 to 300 mmHg	dc to 100 Hz
Venous blood pressure (direct cannulation)	0 to 20 mmHg	dc to 20 Hz
Systolic blood pressure (indirect)	50 to 300 mmHg	dc to 50 Hz
Diastolic blood pressure (indirect)	20 to 60 mmHg	dc to 50 Hz

In medical centers, the output of many patient monitoring systems can be fed into a central computer that analyzes, stores, and gives a printout of the patient's vital signs, including the ECG.

Since over 400,000 people in the United States are potential heart patients, a computer ECG analysis system is a necessity. To determine the ECG, electrodes are externally connected to chest and fed to an ECG amplifier, which can be digitized and sent to a computer for analysis as well as being fed over telephone lines to a cardiac physician's office.

Exercise ECG testing is a recent diagnostic tool used in evaluating heart patients. Sophisticated computers are used in many hospitals to store patient's heart information obtained by means of exercise testing.

5.4 INTRODUCTION TO CONTINUOUS MONITORING OF THE ECG IN THE CORONARY CARE UNIT[1]

Patients critically ill as a result of heart attacks have their ECGs monitored continuously day and night in a coronary intensive care unit (CICU). This is done to detect the onset of ventricular fibrillation, in order that immediate lifesaving action can be taken. This monitoring of the ECG must be continuous for many days.

Automatic monitoring of the ECG and automatic detection of abnormal ECG waves with appropriate alarms is necessitated if a dangerous trend appears. Immediate lifesaving action must be taken by the medical personnel when a valid alarm is sounded. The automatic monitoring and detection instrument must be extremely reliable and have negligible rates of missed ECG events. Furthemore, since an alarm will cause a good deal of activity, even if a false alarm, and since any undue activity that excites heart patients is

[1]Portions of Section 5.4 are based on W. H. Guier, "Cardiovascular Sensors in the Clinical Environment," *IEEE Transactions on Industrial Electronics and Control Instrumentation*, Vol. IECI-17, No. 2, pp. 158–160, April 1970, © 1970, IEEE.

inadvisable, automated ECG monitoring instruments should have a very low false-alarm rate.

Figure 5.12 presents a schematic of electrode placements on the body with the typical signals generated from the ECG for continuous monitor.

The electric potential difference sensed by the electrodes is typically about 1 mV, so that every effort must be taken to minimize noise pickup. There is one more annoying clinical problem. Within a few hours to one day, the skin can build up a "horny" layer, which tends to greatly increase the skin–electrode resistance, thereby decreasing the signal-no-noise ratio. Usually, a special conducting electrode paste is used under the electrodes to minimize the effect of this horny layer, and every care is taken by the CCU nurse to maintain good electrical contact.

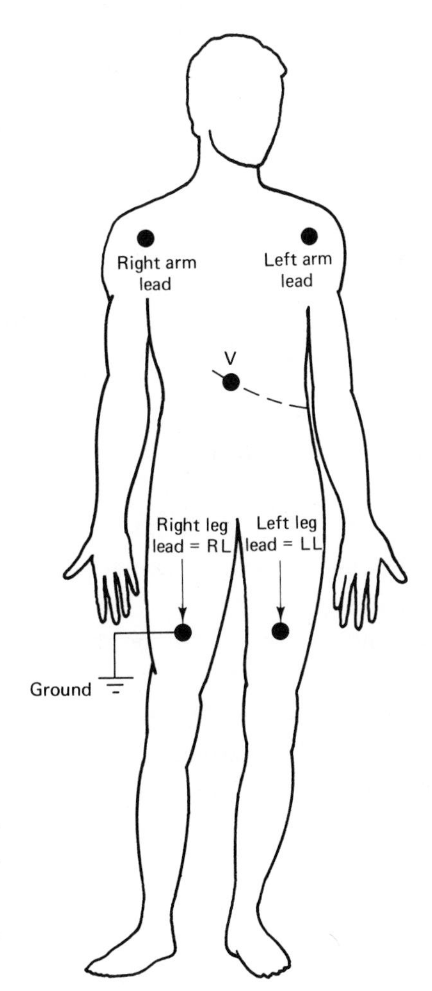

Figure 5.12 Typical ECG leads for continuous monitoring. The potential differences—I′, RA − LA; II″, RA − LL; III′, LA − LL; 1/3 (RA + LA + LL) − V, chest—approximate diagnostic limb leads I, II, and III.

The ECG leads are designed to efficiently pick up electrical activity of the heart muscle and, therefore, efficiently pick up any other muscle electrical activity. One cannot ask a patient to hold still for 5 days, so that every effort must be taken to place the electrodes to minimize the "noise" from skeletal muscle activity. For this reason the electrodes are placed toward the trunk of the body rather than on the more accessible arms and legs. There are no electrodes on the back of the patient, for obvious reasons. Also, you can easily imagine the problems associated with making the electrodes small and of low profile and of attaching them in such a way that they and the associated wire leads are not uncomfortable to the patient. Miniature wireless transmitting ECG electrodes would help solve the problem.

It is convenient to place the electrodes so that the resulting signals approximate those which are normally used for the diagnostic ECG. In Fig. 5.12, the electrode positions are labeled RA for right arm, LA for left arm, RL for right leg, and LL for left leg, since they approximate the potentials obtained from the right arm, left arm, and so on, used in diagnostic ECGs. Similarly, the potential differences given are labeled I′, II′, and III′, since they approximate the standard diagnostic limb leads, I, II, and III.

High-quality isolation amplifiers are required so that any electrical faults in the signal conditioning or other circuits cannot result in an electrical shock to the patient. The electrodes, their wires, and the interfacing electronics must be able to withstand the therapeutic actions taken in case of heart stoppage. One such action consists of providing a special high-voltage electrical shock (typically several thousand volts) to the patient's body to resynchronize the electrical activity of the heart. This is called a *defibrillating shock*.

Most of the problems associated with ECG surface electrodes can be prevented by correct skin preparation techniques and a thorough understanding of the importance of proper electrode application.

Some of the ECG problems are:

1. A baseline or straight line but no ECG tracing
2. An intermittent trace
3. A wandering baseline or instability of the baseline
4. A weak signal
5. Ac interference
6. Cable movement artifact
7. Poor electrode adhesion

Such problems should be brought to the attention of the clinical engineer or technician on call.

The necessary equipment for calibration of the ECG includes:

1. An oscilloscope with a bandwidth of dc to 500 kHz or more, minimum sensitivity of 1 mV, and a sweep rate of about 5 to 10 msec/division

2. A digital multimeter
3. A time mark generator
4. A function generator
5. A universal counter
6. Test fixtures:
   (a) Lead test jib
   (b) CMRR-leakage test fixture
   (c) Precision attenuator
   (d) Low-pass filter and attenuator
   (e) Signal rejection network
7. A calibration system

The primary purpose of an ECG patient monitor is to provide a display of the bioelectric impulses that activate heart muscles, in their proper timing and sequence.

Electrodes attached to the skin are used to pick up the weak cardiac signals generated by the patient and deliver them to a differential amplifier. These signals, on the order of 1-mV amplitude, could be buried under other low-frequency noise signals or artifacts picked up by the leads and the patient. Careful shielding of input leads, coupled with a professional differential-amplifier technique, provides usable signals. Calibration procedural checking includes the following:

1. Check the cable resistance to the system and the leakage resistance between wires.
2. Check the common-mode rejection ratio. A signal is applied to all three inputs simultaneously, and the ability of the circuit to reject the common signal is observed on the monitor screen.
3. Isolation leakage should be calibrated and checked every 6 months or whenever the ECG is dropped or abused.
4. A calibration signal is fed into the ECG to check the system gain, frequency response, trigger circuit, sweep rate, and alarm function.

These calibrations are required of *all* physiological monitors. Some requirements for the ECG leads are given in Table 5.4. Figure 5.13 shows leads I′, II′, and the chest lead for a patient who has serious cardiac electrical disturbances. If you look carefully, you can find four distinctly different electrical patterns, with three of these indicating abnormal or ectopic beats. ECGs of this sort indicate both a present medical emergency and a poor prognosis for the patient. Figure 5.13 indicates the complexity of the ECG signals that can occur and where the structure of the signal is medically important. Any amplifiers in the signal path as well as any recording equipment must have a bandwidth

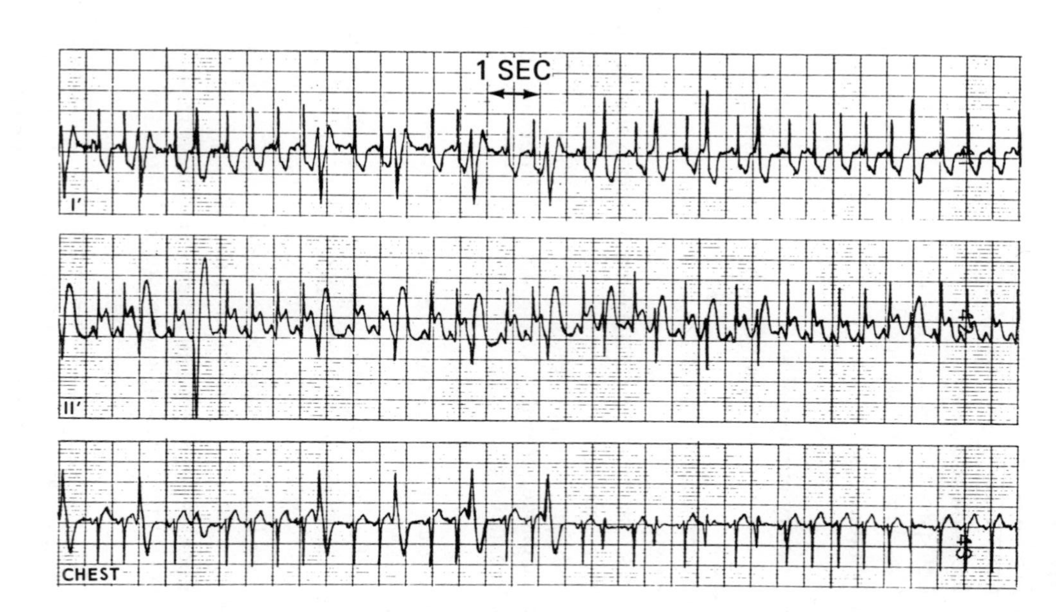

Figure 5.13 Seriously ill cardiac patient (note four different types of beats). (Reprinted from Fig. 2, p. 159, courtesy of the *IEEE Transactions on Industrial Electronics and Control Instrumentation*, Vol. IECI-17, April 1970, © 1970, IEEE.)

beyond 60 Hz. This is a most annoying fact when attempting to eliminate noise pickup due to increased electrode–skin resistance.

TABLE 5.4 Requirements for ECG Leads

1. Low skin resistance.
2. Minimum muscle noise.
3. Approximate standard leads.
4. Electrodes and wires: comfortable and minimum interference with the motion of the cardiac patient.
5. Electrically safe to the cardiac patient.
6. Capable of withstanding defibrillating shock.

As the electrodes and leads form part of the input to the differential amplifier, elimination of 60-Hz interference on the ECG monitor calls for balancing the electrodes and leads. One must make sure that the electrodes and leads are as nearly equal as possible. Clean electrodes are important. Dirty electrodes, electrodes with dried ECG paste, and tarnished electrodes have widely variable resistances and allow widely variable amounts of current through them. In short, they serve only to unbalance the inputs to the differential amplifier.

Placement of the electrodes is also important. Both electrodes should be placed on smooth skin, and it is recommended that the electrodes be placed near the shoulders and away from the hands and arms. In the operating room, electrodes with paste can be slipped under the back at the shoulders. Tape is not necessary as the patient's weight against the mattress or cushion will press the electrodes firmly against the skin.

Stray 60-Hz pickup signals may also enter the leads, especially if they are draped all over the operating table. In this case one lead may be closer to a 60-Hz pickup source than the other lead, thereby unbalancing the signals in the leads, even though the electrodes may be properly attached. To minimize rather than equalize this lead pickup, the leads should be twisted together, ensuring equal 60-Hz inductive energy pickup by both leads.

In most patient care units and the operating room, generators of stray 60-Hz energy, and therefore sources of interference signals, are everywhere: power cords draped over the floors, large lights (both incandescent and fluorescent), numerous power mains in the wall, and even the chassis of improperly grounded instruments. It is unlikely therefore that any two people in the same room are receiving the same amount of 60-Hz energy. In fact, it is entirely probable that some parts of the body are receiving more 60-Hz energy than other parts. The right arm may be in contact with or closer to a 60-Hz source than the left arm is, thereby unbalancing the inputs to the differential amplifier. For this reason it has been suggested that the electrodes be placed near the shoulders, or away from the hands and arms, which act as antennae bringing the 60-Hz signal to the body. With shoulder electrode placement, the effects of one arm being closer to a 60-Hz transmitter than the other arm is minimized.

One solution to the 60-Hz interference problem in the operating room is to use a rejection or a notch filter with active elements. Commercial units are available with a falloff of approximately 96 dB/octave at 60 Hz. This filter passes all frequencies except 60 Hz and can be placed into the circuit after the signal has been amplified. Make sure that you have the manufacturer's permission to insert the filter into the instrument's circuit prior to its insertion. If you do not have this "approval," you take the risk of voiding all warranties.

Figure 5.14 indicates patient movement and some muscle noise. An interesting new design introduced by Hewlett-Packard has circuit that automatically detects loose electrodes or poor patient leads but does not alarm on noise present on a relatively normal ECG.

5.5 HEART SOUNDS

In 1816, René T. H. Laennec devised the *stethoscope* and published his report in 1819. Prior to this, the physician placed his ear against the chest wall to hear heart sounds.

Laennec had made a consistent practice of direct auscultation for several years before he devised the stethoscope. In 1816, Laennec, when he was examining a girl for heart disease, took a quire of paper, rolled it very tightly and applied one end of the roll to the precordium and the other end to his ear. He was surprised to hear the beating of the heart. Laennec described his stethoscope as simply consisting of a cylinder of wood, perforated in its center longitudinally and formed so as to come apart in the middle, for the benefit of being easily carried.

Since Laennec, the evolution of the stethoscope finally led to today's diaphragm (or membrane) chest piece, patented by R. C. M. Bowles. When it was recognized that both the bell and the diaphragm have merit and in effect complement each other, a composite chest piece permitting rapid change from one to the other by means of a valve was the next advance. In 1926, H. B. Sprague introduced the combination chest piece now in general use. Stethoscopes also pick up chest and breath sounds.

Even today, over 150 years after its introduction, the acoustical stethoscope is the most used instrument in the physician's bag, although it is still a very simple device. It tells the physician a great deal about the mechanical functioning of the heart. This is important because it is the heart that pumps blood throughout the body to maintain life. An acoustical stethoscope is shown in Fig. 5.15.

Heart sounds are valuable in giving the physician a clue to the heart's physiological status. Heart sounds range in frequency from 5 to 2000 Hz. However, for special applications, frequencies to 4 kHz are detectable.

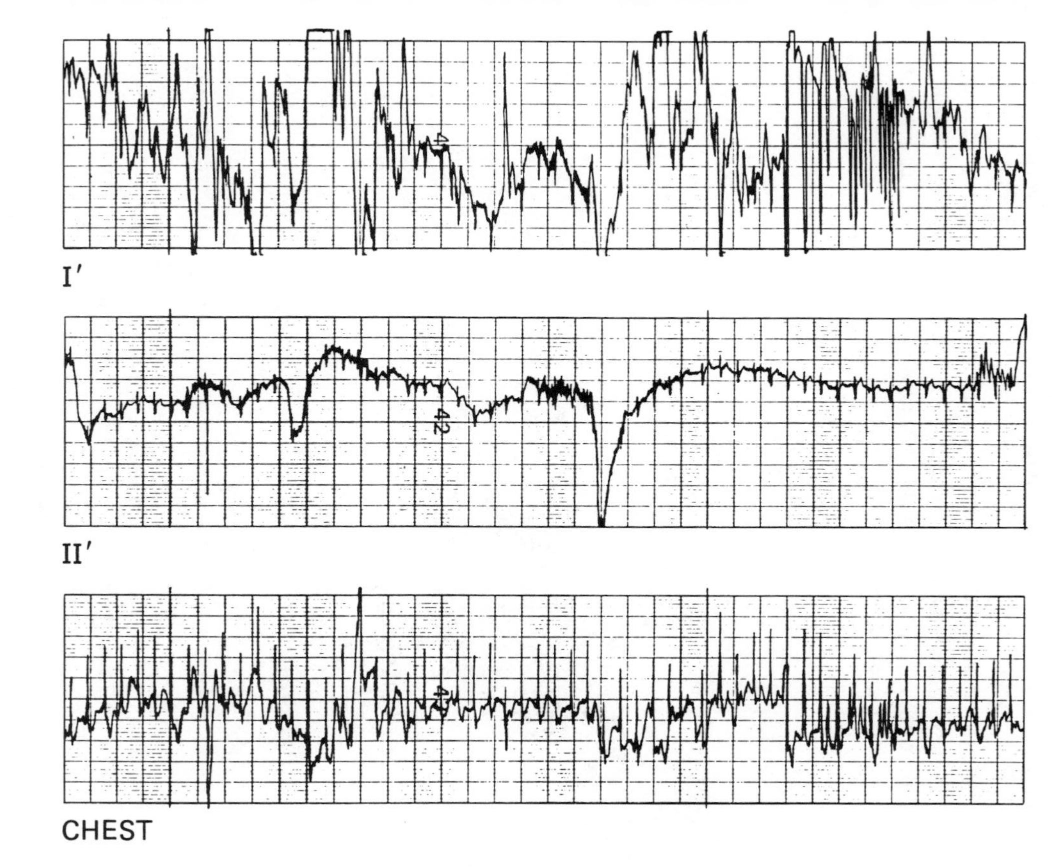

Figure 5.14 Patient motion with a loose electrode. (Reprinted from Fig. 5, p. 161, courtesy of the *IEEE Transactions on Industrial Electronics and Control Instrumentation*, Vol. IECI-17, April 1979, © 1970, IEEE.)

An electronic stethoscope is a microphone whose signals are amplified and recorded or fed into the system for analysis. This microphone is placed on the chest in the same manner as a stethoscope. If the semilunar valves do not close properly, for instance, blood leaks back into the ventricles from the arteries, producing a hushing sound. The sounds may now be simulated by "lush-sh." This condition is called a *heart murmur.*

In Fig. 5.15, the double-headed stethoscope provides a wide frequency range for general application and for use in cardiology. Slight pressure variations on the head "tunes" the diaphragm to "bring in" the frequencies desired.

The bell chest piece is particularly useful in detecting faint low-frequency sounds and murmurs. It fits conveniently in smaller areas, such as between ribs or in the supraclavicular areas of the neck.

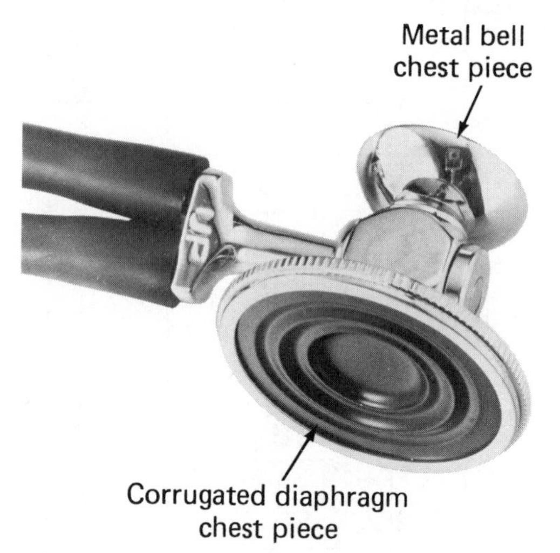

Figure 5.15 Harvey double-headed stethoscope (Model No. 5079–28) with bell and corrugated chest pieces. (Courtesy of Sybron Corporation, Medical Products Division, Rochester, N.Y.)

5.6 THE PHONOCARDIOGRAM

Ausculation of the normal heart usually reveals two sounds and occasionally three. Recording of the normal heart sounds by means of phonocardiography can reveal as many as four sounds.

The *phonocardiogram* is a graphic representation of the precardial heart sounds generated intracardially and transmitted through the tissue to the thoracic surface. Such sounds are acoustic and are caused by changes in pressure resulting from physical displacement of the dynamics of the heart chambers. Like musical notes, heart sounds are distinguishable by their pitch or frequency. Pure tones do not exist.

Since pure tones do not exist, most heart sounds can be lumped into frequency groups called *bands* for more effective analysis and diagnosis. Such an electronic system includes a sensitivity transducer, preamplifier and/or amplifier, band-pass filter, and a recorder. The phonocardiogram is reproduced more faithfully by using the electronic band-pass filter, which accepts certain frequencies and rejects others. Our concern is the region below 1000 Hz. One all-pass channel of the phonograph is often reserved with little amplitude distortion while the signal amplitude of each discrete channel is adjusted separately.

The phonocardiogram provides a technician, nurse, or physician with a visual record of sound frequency and the intensity of the recorded signal. Analysis of phonocardiogram records has enabled us to relate acoustic properties to cardiac disorders. Valvular stenosis and regurgitation, heart failure, and congenital defects are routinely detected in the properly recorded and interpreted phonocardiogram.

In making a recording of the phonocardiogram, the patient should be in the supine position and stripped to the waist. Before beginning any part of the procedure, the patient should be relaxed by explaining the purpose and simplicity of the test. To obtain a good carotid pulse tracing, it is advantageous to position the patient's head over a pillow and hold the carotid transducer in place with a plastic clamp or an inflatable cuff.

Ausculation should be performed before taking a phonocardiogram to determine the optimum locations for microphone placement. Routinely, heart sounds and murmurs are recorded at the cardiac apex, fourth and third intercostal spaces at the left sternal border, and pulmonic and aortic areas. If the patient is sitting up, murmurs due to aortic insufficiency and coarction of the aorta should be recorded for such anomalies.

Precise timing measurements of certain systolic time intervals are an integral part of the analysis of phonocardiographic recorders, as shown in Fig. 5.16.

In a typical three-channel phonocardiogram of the normal heart sound at the apex, one channel records the heart sound, the second channel records the ECG, and the third channel records the carotid pulse. In adults, the first two heart sounds, S_1 and S_2, only are usually heard. In infants and children, the third and fourth sounds also are frequently heard and displayed on the recorder chart.

There can be six to nine channels available for different frequency ranges when making spectral phonocardiograhpy measurements. In Fig. 5.16, the systolic time intervals include the left ventricular ejection period, the pre-ejection period, and the isovolumetric contract time. A six-channel phonographic tracing of the normal first and second sounds recorded at the apex is shown in Fig. 5.17.

The first heart sound (S_1) follows the QRS interval of the ECG and is normally louder than the second sound (S_2) at the apex. As the microphone is moved from the apex toward the base of the heart, the intensity of S_1 decreases and S_2 increases.

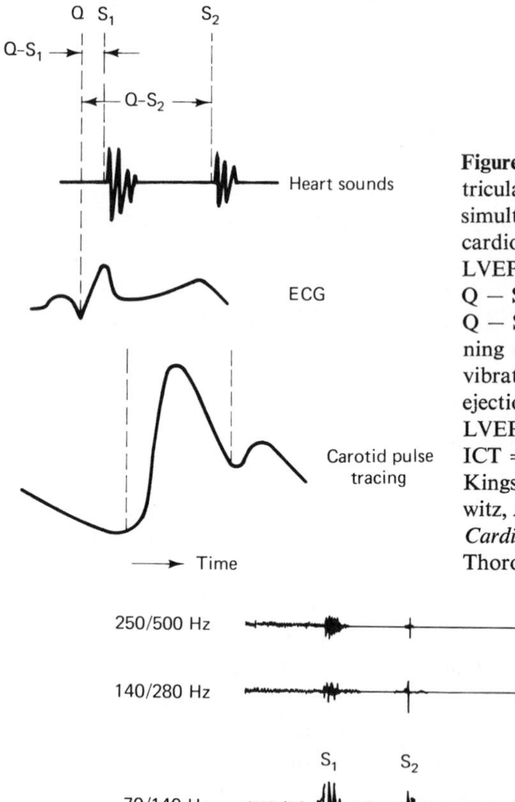

Figure 5.16 Measurements of left ventricular systolic time intervals with simultaneous phonocardiogram, electrocardiogram, and carotid pulse tracing. LVEP = left ventricular ejection period; $Q - S_2$ = electromechanical systole; $Q - S_1$ = the interval from the beginning of the Q-wave to the first major vibration of the first heart sound; preejection period (PEP) = $(Q - S_2) -$ LVEP; isovolumetric contraction time ICT = (PEP) − $(S - S_1)$. (From B. Kingsley, J. W. Linhart, and P. Kantrowitz, *Advances in Noninvasive Diagnostic Cardiology*, Charles B. Slack, Inc., Thorofare, N.J., © 1976, p. 48.)

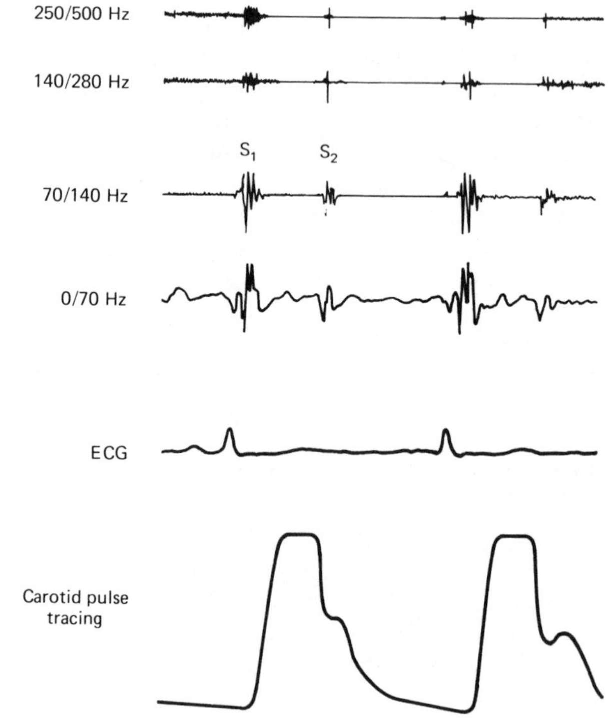

Figure 5.17 Typical six-channel phonocardiogram of a normal heart sound at the apex. Four different frequency channels of heart sounds, one lead of electrocardiogram, and an indirect carotid tracing are recorded simultaneously. (From B. Kingsley, J. W. Linhart, and P. Kantrowitz, *Advances in Noninvasive Diagnostic Cardiology*, Charles B. Slack, Inc., Thorofare, N.J., © 1976, p. 49.)

5.7 REVIEW QUESTIONS

1. Describe an electrocardiograph and list five uses for it in cardiology.
2. Describe the conduction mechanism of the heart.
3. Describe the standard bipolar limb lead, augumented unipolar limb lead, and precordial chest leads configuration of the ECG.
4. Draw a typical ECG. Label all parts and give the duration of the P wave and the Q-T interval for the ECG.
5. Describe problems associated with continuous monitoring of the ECG in the coronary care unit.
6. Describe a method for eliminating 60-Hz interference on the ECG.
7. Describe good ECG electrode practices.
8. Describe the use of heart sounds and phonocardiograph.
9. Describe a method of determining the heart beat from the ECG.
10. Describe the meaning of the Einthoven triangle.
11. Describe how a technician recognizes a normal ECG.

5.8 REFERENCES

1. Guyton, A. C.: *Textbook of Medical Physiology*, 5th ed., W. B. Saunders Company, Philadelphia, 1976.
2. Narula, O. S., ed.: *His Bundle Electrocardiography and Clinical Electrophysiology*, F. A. Davis Company, Philadelphia, 1975.
3. Ganong, W. F.: *Review of Medical Physiology*, 6th ed., Lange Medical Publications, Los Altos, Calif., 1973.
4. Jacobson, B., and Webster, J. G.: *Medicine and Clinical Engineering*, Prentice-Hall, Inc., Englewood Cliffs, N.J., 1977.
5. Berne, R. M., and Levy, M. N.: *Cardiovascular Physiology*, 3rd ed., The C. V. Mosby Co., St. Louis, Mo., 1977.
6. *ECG Techniques. Application Manual*, Application Notes AN 721, Hewlett-Packard Co., Palo Alto, Calif., Sept. 1975.
7. Kingsley, B., Linhart, J. W., and Kantrowitz, P.: *Advances in Noninvasive Diagnostic Cardiology*, Charles B. Slack, Inc., Thorofare, N.J., 1976.
8. Luisada, A. A., and Liw, C. K.: *Intracardiac Phenomena in Right and Left Heart Catheterization*, Grune & Stratton, Inc., New York, 1958.
9. Wooley, C. F.: Intracardiac Phonocardiography, Intracardiac Sound and Pressure in Man, *Circulation*, Vol. 57, No. 6, June 1978.
10. *Coronary Care, Arrhythmias in Acute Myocardial Infarction*, American Heart Association, Dallas, Tex., 1976.

6

BLOOD PRESSURE AND FLOW MEASUREMENT SYSTEMS

6.1 INTRODUCTION TO BLOOD PRESSURE MEASUREMENTS

The arterial blood pressure is the lateral force exerted on a unit area of the blood vessel wall due to the pulsing action of blood flow. This pressure changes levels during each heart beat because the heart is a cyclic pump that forces the blood through the circulatory system. The highest pressure reached in this cycle, the *systolic*, occurs when the heart contracts, propelling blood outward into the aorta and thus maintaining circulation. The lowest pressure reached is the *diastolic*, which occurs when the heart relaxes, expands, and fills with blood.

The numerical difference between the systolic and diastolic pressures is called the *pulse pressure*. The average of the two blood pressure readings is the *mean pressure*. Mean pressure increases nonlinearly with age.

The blood pressure is expressed as a fraction whose nominal value is 120/80 and is referenced to the blood pressure taken at the ascending aorta. The numerator of this fraction is the systolic pressure and the denominator is the diastolic pressure.

For young adults in good health, the normal average blood pressure in different parts of the circulatory system are:

1. Brachial artery: systolic, 110–120 mmHg; diastolic, 65–80 mmHg
2. Capillaries: 20–30 mmHg
3. Veins: 0–20 mmHg

The measured arterial blood pressure of an individual depends on that person's cardiac output, peripheral vascular resistance, blood viscosity, and arterial wall elasticity. Blood pressure is also related to heart rate.

Blood pressure is important because it may indicate changes in circulation due to hypertension, hypotension, strokes, emotional stress, trauma, arterial sclerosis, and shock. In a few words, the body's blood pressure is a primary indicator of physiological distress.

6.2 CLASSIFICATION OF BLOOD PRESSURE MEASUREMENTS

Blood pressure measurement techniques may be classified as shown below:

1. Invasive, direct
   (a) Sensor placed within the arterial system
   (b) Pressure acts directly on sensor
2. Invasive, indirect
   (a) Sensor placed within the arterial system
   (b) Sensor measures wall displacement as a function of pressure
3. Noninvasive, indirect
   (a) As transmitted vibrations
      (1) Stethoscope
      (2) Piezoelectric
      (3) Magnetic
      (4) Others
   (b) As vibrations at origin
      (1) Ultrasonic
      (2) Others
   (c) Other techniques

Both invasive and noninvasive techniques for measuring blood pressure are used.

Direct methods of blood pressure measurement use strain gage sensors, linear variable differential transducers, and piezoelectric or capacitance sensors, usually located outside the patient and driven through a catheter inserted through the skin into a blood vessel.

Indirect or noninvasive methods use occlusive-Korotkoff techniques. Automated devices, ultrasonic measurement of arterial wall movement, and arterial tonometry based on wall motion are other techniques used to measure blood pressure.

6.3 SPHYGMOMANOMETRY

The most common blood pressure measurement techniques using the indirect noninvasive method combine vibration pickup, microphone sensor, or stethoscope with an inflatable cuff connected to a mercurial or aneroidal manometer.

In Fig. 6.1, the brachial artery is encircled with such an inflatable cuff, and a stethoscope (auscultatory sensor) is placed under the cuff at the brachial artery that descends down the arm. This cuff is connected to manometers as shown in Fig. 6.2.

The blood pressure measuring device, or *sphygmomanometer*, is actually a measurement system composed of functionally interrelated components. As shown in Fig. 6.2, the measurement system consists of:

1. An inflatable compression bag, enclosed within an inextensible cuff, for application of pressure to the artery.
2. An instrument to measure and indicate the applied pressure, the manometer.
3. An inflation bulb to create pressure in the system.
4. An adjustable valve through which deflation of the system can be controlled at a desired rate.

Of these four elements, the two most important from the standpoint of critical effect on accuracy are the cuff and the measuring instrument. Numerous investigations have shown that the accuracy of systolic and diastolic blood pressure readings is materially affected by the length of the arterial segment compressed. These studies show that for optimum results the effective width

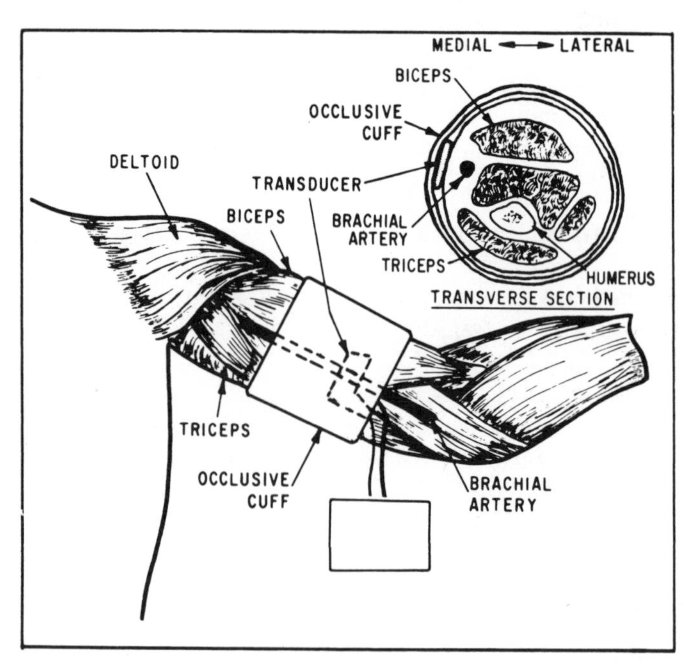

Figure 6.1 Positioning of a sensor (transducer) over the brachial artery of the upper left arm.

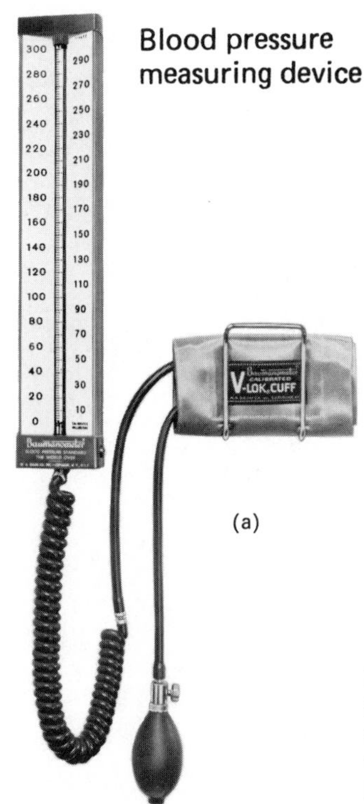

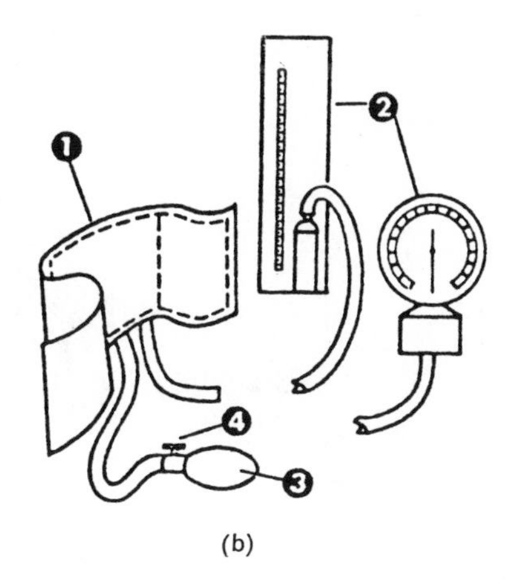

Figure 6.2 (a) Blood pressure measuring device pictorial. (b) Most common blood-pressure measuring apparatus: (1) compression cuff, (2) manometer, (3) inflation bulb with (4) adjustable valve. (Courtesy of W. A. Baum Co., Inc., Copiague, N.Y.)

of the compression bag should be approximately 20% greater than the diameter of the limb to which it is applied. In accordance with this finding, standard cuff sizes have been established and are listed in Table 6.1.

TABLE 6.1 Standard Sphymomanometer Cuff Sizes

Age	Width of Inflatable Bladder
Adult	13 cm (5.1 in.)
4–8 years	9 cm (3.5 in.)
1–4 years	6 cm (2.3 in.)
Newborn	2.5 cm (1 in.)

The cuff must be able to produce an even pressure across its entire width. This means that it must be constructed so that inflation of the bag will not cause bulging or displacement, as either defect would contribute to an erroneous reading.

As shown in Fig. 6.3, the brachial, femoral, or popliteal artery may be used to measure blood pressure. The cuff is placed around the arm or leg and inflated to a pressure usually above 150 mmHg, where no pulses may be heard on the stethoscope. The pressure exerted by the cuff is slowly reduced until

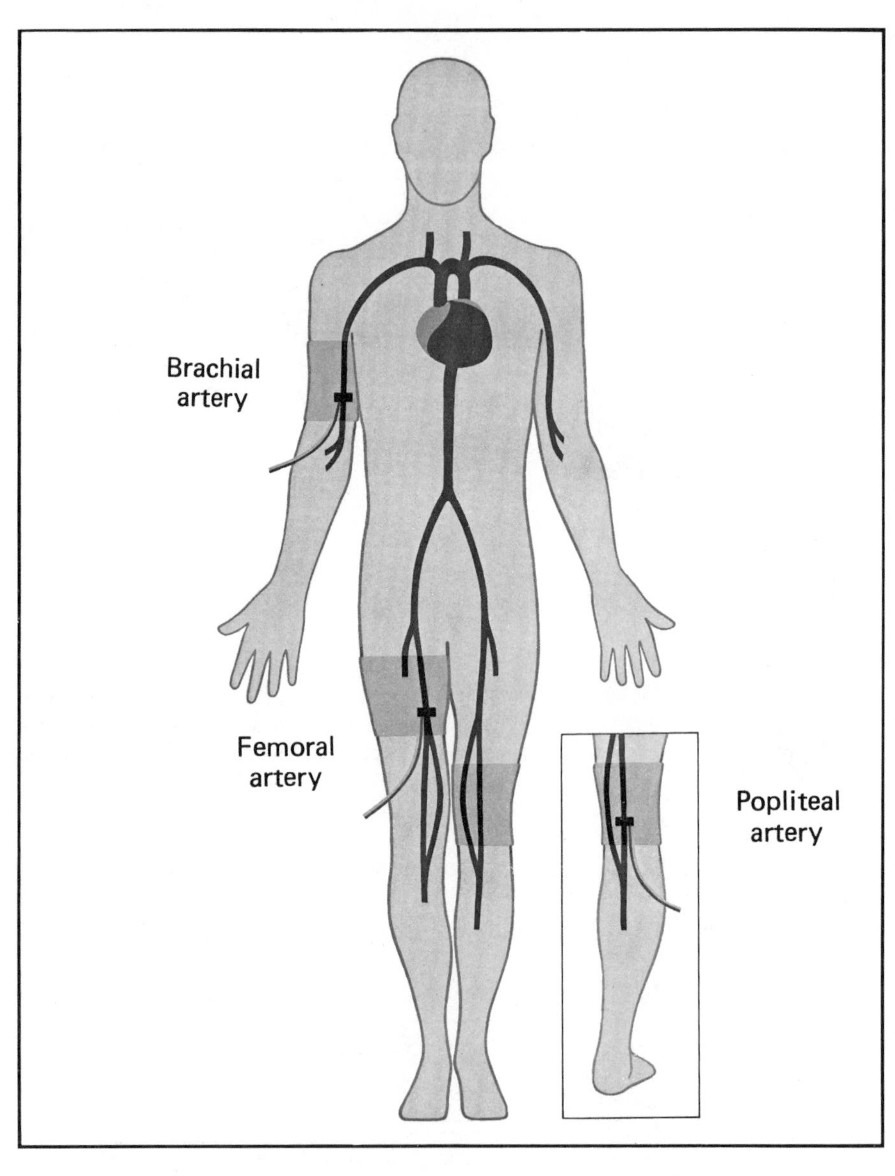

Figure 6.3 Points of measurement. (Courtesy of Roche Medical Electronics Inc., Cranbury, N.J.)

the first pulse noise is heard. This point is the systolic pressure and is indicated in millimeters of mercury (mmHg) on the manometer. The cuff pressure is allowed to drop slowly again until there is no longer any pulse sound. Just at the point where all sound stops, the manometer will indicate distolic pressure. As the blood flows through the system, it makes identifiable sounds. If a microphone were placed under the cuff and the exact procedure for blood pressure measurements were repeated, the resulting "pulse" sounds, called *Korotkoff sounds*, would appear as shown in Fig. 6.4. As the pressure in the cuff reached the systolic pressure, a sharp thud would be heard. As the cuff pressure was reduced further, the sounds would become progressively softer, until finally they would just fade away.

The method of measuring blood pressure using an occlusive cuff is only accurate to approximately ± 5 mmHg, because of the following:

1. Blood pressure readings vary with the location and height of the transducer in relation to the heart blood pressure. Readings not taken at heart level should be compensated to correspond to readings at heart level.
2. If a stethoscope is used, the hearing of the user may affect the reading. Any stethoscope used should be of a standard variety within the institution and be in good working order.

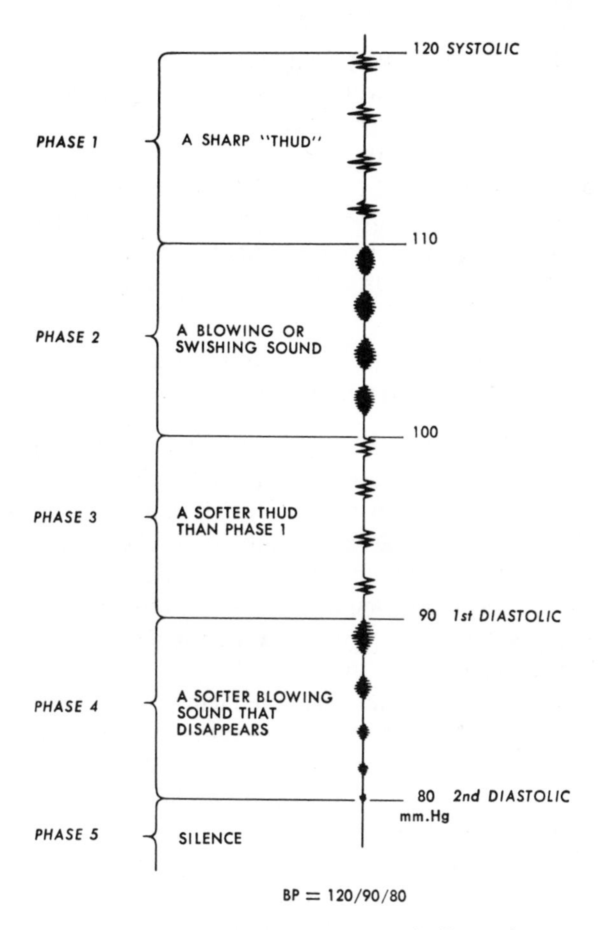

Figure 6.4 Various phases of Korotkoff sounds.

3. Motion artifacts may occur, depending on the type of vibration introduced into the system: making a fist, bending or moving the arm, body movement, and so on. Blood pressure measurements are particularly sensitive to movement if the patient is in shock because the pulse is weak and the amplitude of Korotkoff vibrations is low.
4. Touching the arm (pressor effect) can alter the reading from both normal and emotionally stressed individuals. Hyperventilation, on the other hand, may reduce the pressor effect.
5. Faulty measurement techniques include improper positioning of the extremity, improper deflation of the compression cuff, failure to have the mercury column vertical, ausculatory gap, and improper application of the cuff. The *ausculatory gap* is defined as follows: Occasionally during deflation of the cuff, the seconds appear, disappear, reappear, and then disappear, the first disappearance occurring with a cuff pressure between systolic and diastolic pressure.

If the cardiac rhythm is very irregular, as in the case of dysrhythmias, blood pressure determinations will be inaccurate because stroke volume and blood pressure may vary from one cardiac cycle to the next.

For difficult patients, it is recommended that three or four measurements be taken and an average reported as the final value. Some physicians recommend that their patients monitor their own blood pressure at home. This is particularly true for hypertensive patients, where the patient-recorded blood pressure values may prove useful to the physician in treating and controlling the disease.

6.4 SEMIAUTOMATED AND AUTOMATED BLOOD PRESSURE MEASUREMENT SYSTEMS

Noninvasive semiautomated and automated blood pressure measurement are becoming accepted clinical tools used to monitor systolic and diastolic pressures. These instruments are gaining wider use for patients. This is especially true for individuals suffering from cardiac conditions, strokes, and trauma.

Measurements taken with an automatic device have to be correlated with a standard because automated devices usually sense first diastolic pressure readings (Phase 4) at the onset of muffling, not at the onset of silence (Phase 5), which is detected manually.

The repeatability of pressure measurements by use of a piezoelectric, ultrasonic, or strain gage sensor with semiautomated and automated blood pressure systems has been reported as being greater than those made manually with a stethoscope because the auditory senses of the physician or technician are not as reliable as the sensor used.

Semiautomated and automated blood pressure devices are measuring tools used in federal agencies such as NASA, as well as hypertension and health agencies, and are illustrated in Figs. 6.5 and 6.6.

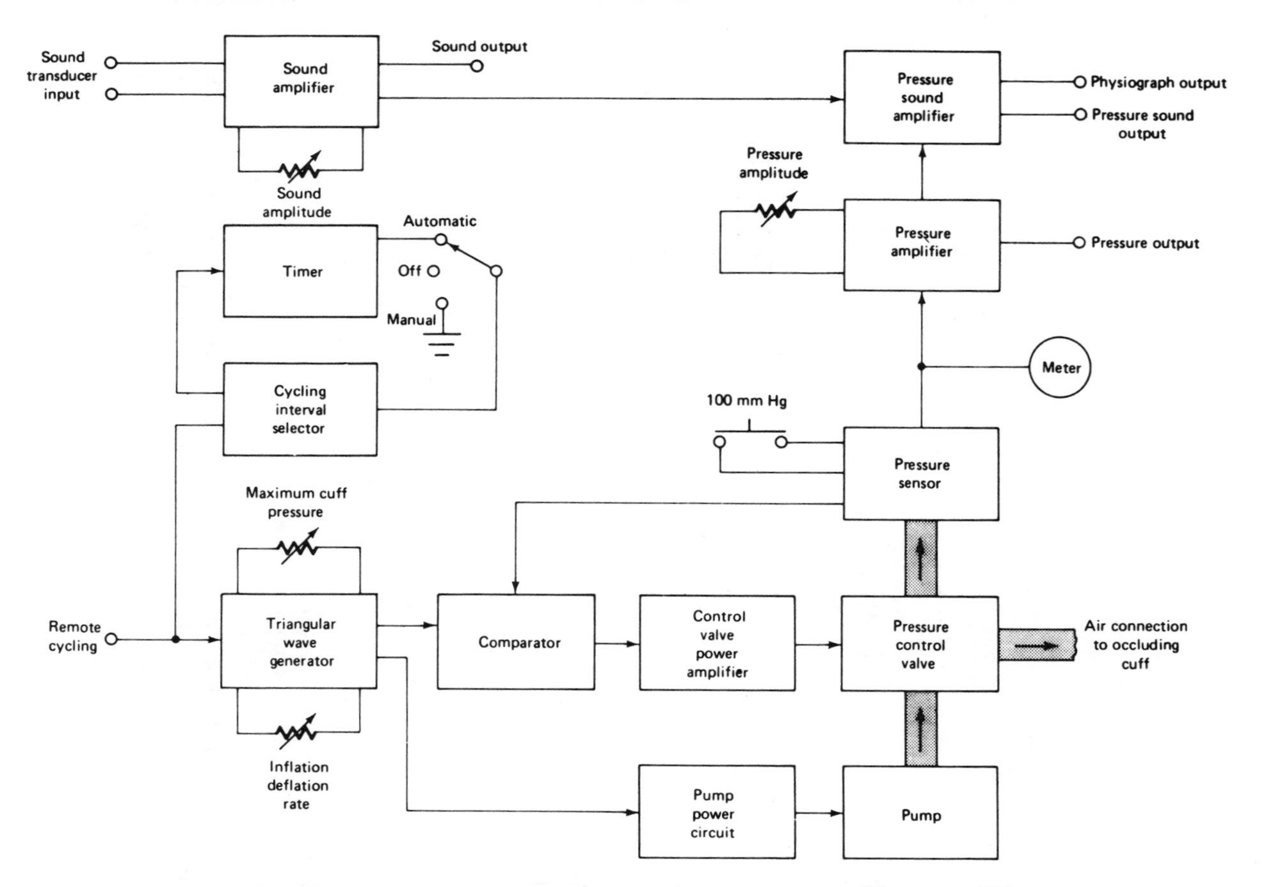

Figure 6.5 Narco programmed PE 300 electrosphygmomanometer. (Courtesy of Narco Bio-Systems, Houston, Tex.)

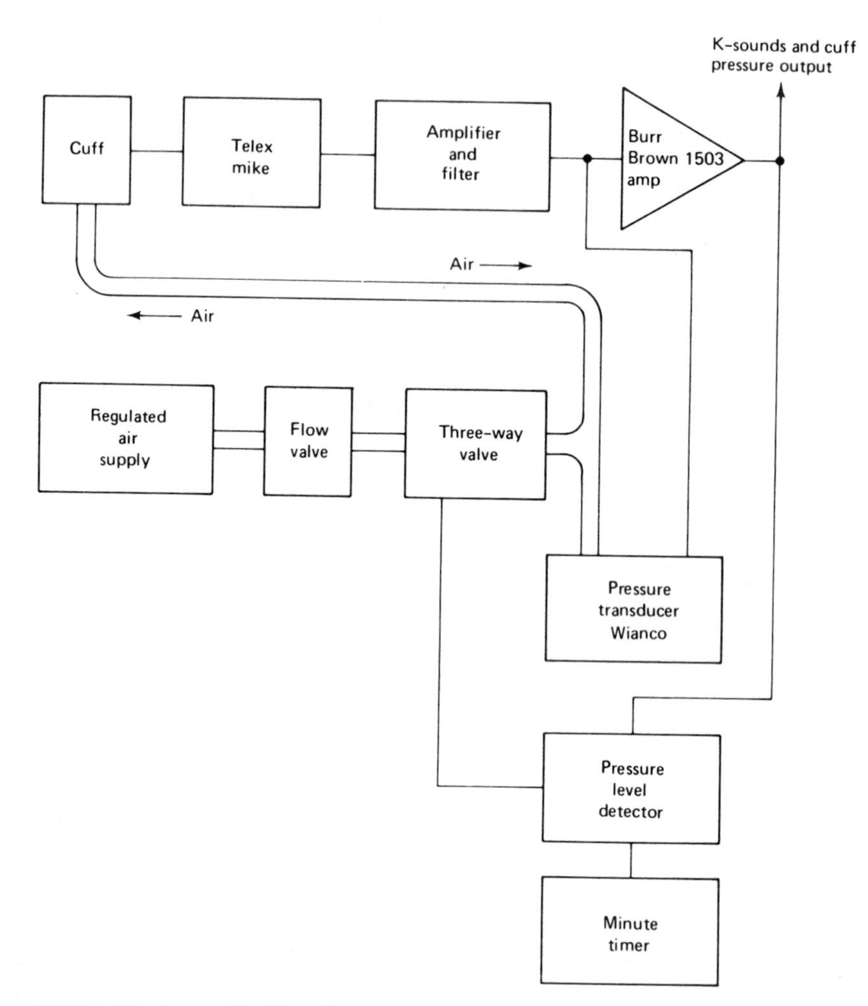

Figure 6.6 Automatic blood pressure system. (Courtesy of USAF School of Aerospace Medicine, Brooks AFB, Tex.)

When using semiautomated and automated pressure devices, the compression cuff at the brachial artery is inflated to about 30 mmHg above the pressure at which the brachial pulse disappears. The cuff is deflated at a rate of 2 to 3 mmHg per heartbeat.

The chart from the graphic recorder (Fig. 6.7) shows ascending and descending lines which form a triangle. On the descending line starting at 200 mmHg, a pulse train is seen. The first pulse is the systolic arterial pressure and the last readable pulse is the diastolic pressure.

All semiautomated and automated blood pressure measuring devices are sensitive to motion artifact because of the cuff sensor/skin coupling scheme

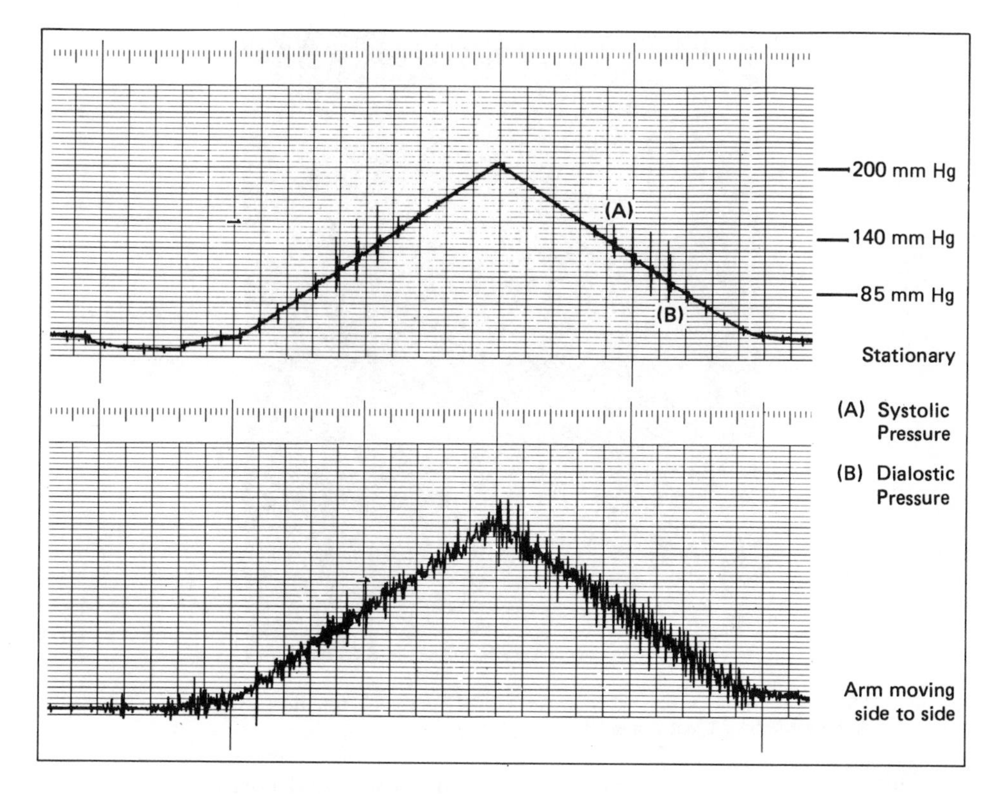

Figure 6.7 Brooks Air Force Sensor (Houston, Texas) connected to a Narco PE 300 with (a) arm stationary and (b) arm moving side to side. Note motion artifacts in the bottom recording.

presently available. Since all automated devices are sensitive to motion artifacts, arm or finger movement will result in burying the desired signal noise.

Automated blood pressure devices based on the ultrasonic Doppler effect are also commercially available. In these, however, motion artifact is a serious problem because the brachial artery moves and creates a secondary Doppler effect.

Figure 6.8 shows a semiautomated blood pressure device with a digital display. A recording of the pressure waveform may be viewed on an associated recorder.

6.5 BLOOD FLOW MEASUREMENT: HEMODYNAMICS AND HEMODYNAMIC SYSTEMS

Hemodynamics is the study of the circulation of blood within the human body. Blood flow means that a quantity of blood passes a given point in the circulatory system over a period of time. William Harvey was responsible for the initial measurement of blood circulation. The most rapid blood flow within the body

Figure 6.8 SPHYGIE noninvasive blood pressure monitor. (Courtesy of Abbott Medical Electronics, Houston, Tex.)

is in the aorta and is 300 milliliters/sec. The flow rate decreases as it enters other blood vessels.

During open heart surgery, a cardiopulmonary bypass system is used. Often called a heart–lung machine, this pump, which performs the function of the heart, and an oxygenator, which performs the functions of the lungs, are used to maintain circulation within the body. The most commonly used heart–lung machine is a roller model type, although more complex pulsatile pumps which mimic the heart rhythm have been developed and used in operations where patients require support longer than 6 hours. Disposable bubbler-type oxygenators are the standard short-term artificial lung components of the system. In long operations, membrane oxygenators, which protect the blood from exposure to raw oxygen, are necessary.

The circulation of blood is also important in renal dialysis. Artificial kidneys are used to cleanse the blood of toxic waste materials for patients suffering from kidney failure.

The kidney dialysis machine utilizes plastic tubing through which blood passes. A cellophane-like semipermeable membrane separates the blood from a rinsing solution called dialysate. The patient's circulatory system is taped at a major vein; the blood is extracted and pumped through the tubing into and out of the dialyzer assembly. Dialysate is pumped into and out of the opposite side of the dialyzer. The passing of the blood and dialysate on opposite sides of the membrane creates a partial pressure imbalance, resulting in transfer of the toxins from the blood into the dialysate.

6.6 MEASUREMENT OF CARDIAC OUTPUT

Cardiac output may be one of the most important parameters of the circulator system. Cardiac output is a measure of the available blood flow for all tissues of the body. Diseases of the heart usually result in a decrease of cardiac output, leading to inadequate support of body tissue. The basic problem physicians report today is the lack of a simple reproducible, noninvasive method for measuring cardiac output at the bedside.

Cardiac output, in general, may be expressed as the product of heart rate and blood volume pumped per stroke of the heart. Assuming a consistent heart rate and stroke volume,

$$\text{Cardiac output}\left(\frac{\text{liters}}{\text{min}}\right) = \text{heart rate}\left(\frac{\text{beats}}{\text{min}}\right) \times \text{stroke volume}\left(\frac{\text{liters}}{\text{beat}}\right) \tag{6-1}$$

Cardiac output may also be expressed as

$$\text{Cardiac output or blood flow} = \frac{\text{mean blood pressure}}{\text{peripheral resistance}} \tag{6-2}$$

In human beings, several techniques have been developed and are currently in use for measurement of cardiac output for patients at rest. They include the direct use of the Fick principle, the dye dilution technique, the thermodilution technique, the pressure pulse contour technique, the pulse gradient method, ballistocardiography, x-ray cardiometry, and the photocalorimetric oxygen detector technique. All of these techniques require cardiac catheterization procedures involving arterial and venous punctures.

In the *direct use of the Fick principle*, the blood flow through the lungs is calculated as follows:

$$Q_{CO} = \frac{Q_{O_2}}{V_{O_1} - V_{O_2}} = \frac{\text{ml } O_2/\text{min}}{\text{ml } O_2/\text{ml blood}} = \frac{\text{ml blood}}{\text{min}} \tag{6-3}$$

where Q_{CO} = cardiac output
Q_{O_2} = oxygen consumption (ml O_2/min)
V_{O_1} = concentration of oxygen in left side of the heart (ml O_2/ml blood)
V_{O_2} = concentration of oxygen in right side of heart (ml O_2/ml blood)

A block diagram to show cardiac output measurements by the direct Fick method is shown in Fig. 6.9.

The Fick principle (Fig. 6.10) involves the measurement of the oxygen picked up by the blood as it passes through the blood per unit time by means of a spirometer which measures oxygen consumption in air breathed in and out through a face mask; and the differences in oxygen concentration in the blood as it passes through the lungs by means of catheters placed into both

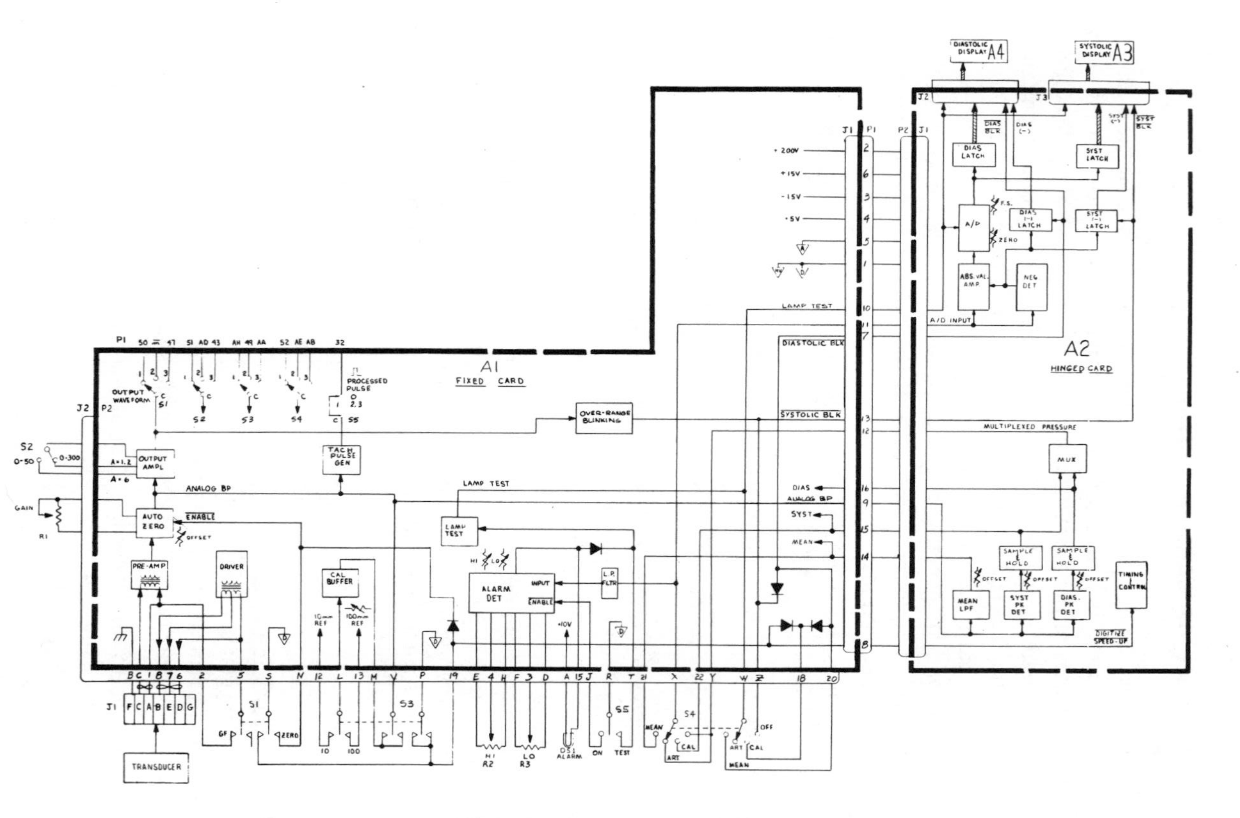

Figure 6.9 Overall module block diagram. (Courtesy of Abbott Medical Electronics Co., Houston, Tex.)

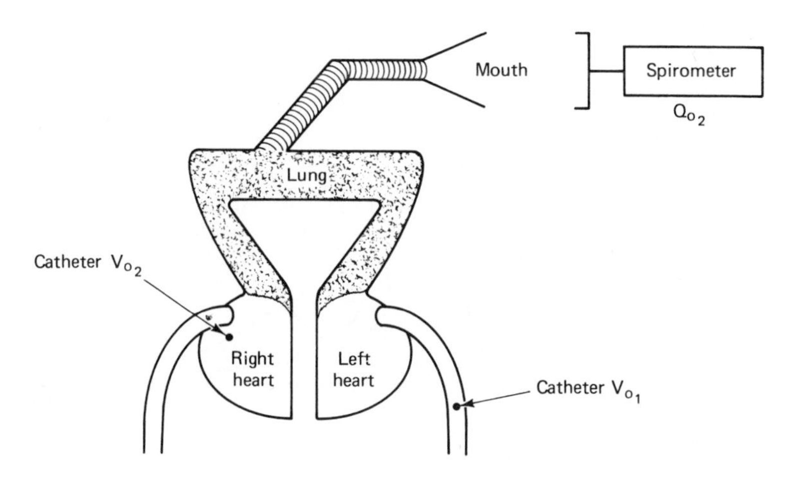

Figure 6.10 Diagram of direct Fick principle.

the right atrium of the heart and in the arteries near the left side of the heart. Readings taken over a period of about 15 minutes are required and the blood samples withdrawn from each catheter must be measured with a blood-gas analyzer to determine oxygen concentration on either side of the heart. In addition, while oxygen concentration in blood samples from any artery are nearly the same, oxygen concentration in different veins may differ greatly. For this reason, the venous catheter is usually placed in what is called the pulmonary artery wedge position. The result is believed to be the best "mixed" venous sample possible. This method is obviously not useful for continuous monitoring of cardiac output. Very significant errors result whenever respiratory or circular conditions change.

A more widely used application of the Fick principle is the *dye dilution* method, shown in Fig. 6.11. This procedure involves the injection of a bolus of dye into the circulatory system and determination of its downstream concentration after mixing is complete. If the dye is injected into the veins feeding into the right heart by means of a catheter, and continuous monitoring on the arterial side of the heart is performed, the resultant curve will be recorded as shown in Fig. 6.11.

Cardiac output is then expressed by the cardiac output formula

$$\frac{\text{liters of blood}}{\text{min}} = \frac{\text{mg dye injected}}{\begin{pmatrix}\text{average concentration of dye}\\ \text{in each ml of blood for the}\\ \text{duration of the curve}\end{pmatrix} \times \begin{pmatrix}\text{duration of}\\ \text{the curve}\\ \text{in seconds}\end{pmatrix}} \qquad (6\text{-}4)$$

$$\text{cardiac output} = \frac{60 \text{ mg} \times \text{dye injected}}{T_1 \times d_2} = \frac{\text{mg}}{\dfrac{\text{mg} \times \text{sec}}{\text{liter}}} = \frac{\text{leters}}{\text{min}}$$

In order to perform the calculations of cardiac output, the amount of dye injected, the time it takes to pass the monitoring point and the mean

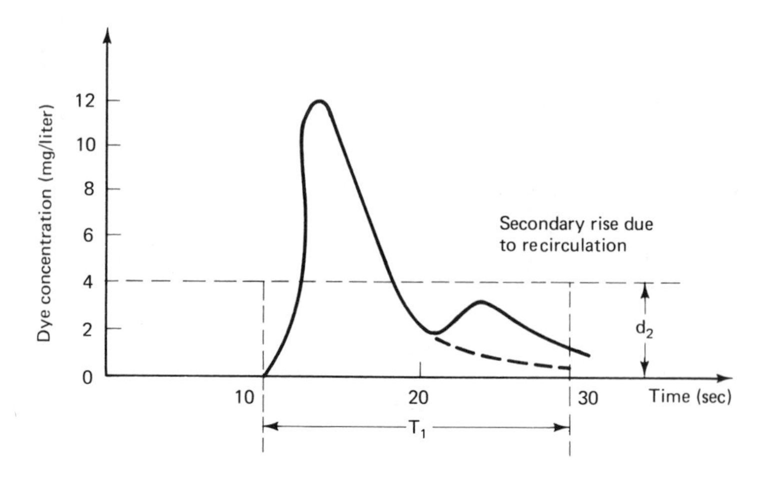

Figure 6.11 Dye concentration in mg/liter as a function of time in seconds. T_1, time in seconds from the first detection of dye to the extrapolated last time; d_2, integration of the area under the extrapolated curve, excluding the secondary rise.

concentration during that time must be known. The curve provides this data.

The exponential decay of the first peak is also shown extrapolated to zero. The time taken to pass the monitoring point is shown as distance T_1 on the curve. The area under this curve must be obtained correct for recirculation in units of (mg $\times$ sec)/liter.

Figure 6.12 shows a cardiac output system using a *thermodilution* technique, consisting of a catheter, catheter cable, computer, and recorder. Cardiac output by the thermal dilution is based on an injection of a known amount of a cold solution into the right atrium or superior vena cava. Cardiac output is inversely proportional to the integral of temperature change. A thermistor in the pulmonary artery measures the subsequent change in blood temperature, and cardiac output is calculated by a computer computation.

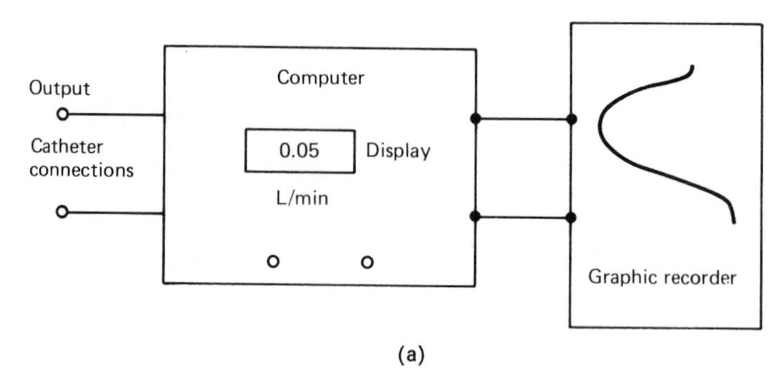

Figure 6.12 Cardiac output system using the thermodilution system. (a) Block diagram of thermodilution system.

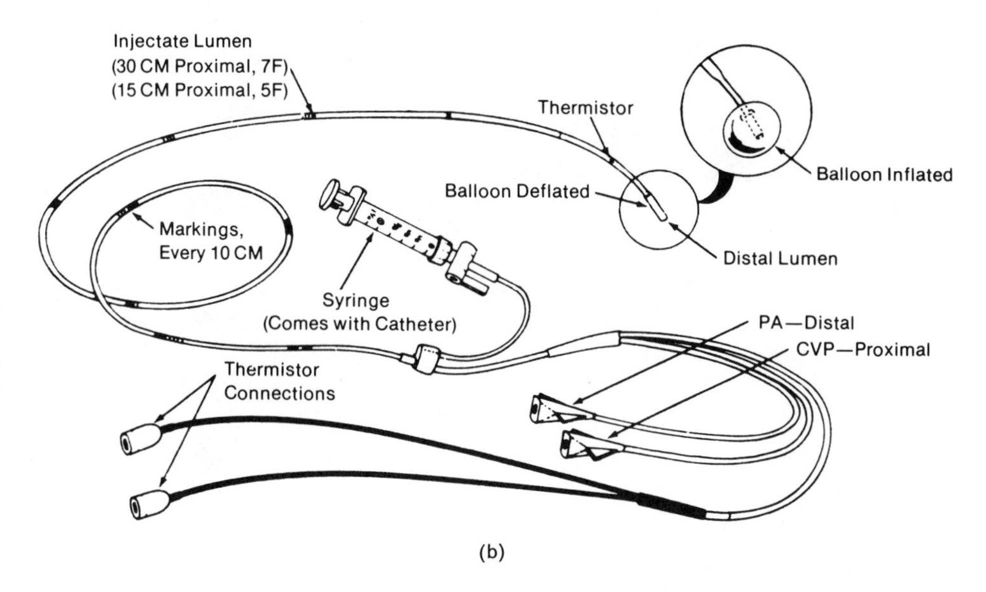

Figure 6.12 (b) Connections for the thermodilution system.

The thermodilution technique has the advantages that it does not require peripheral arterial invasion, blood withdrawal, or lengthy laboratory procedures. It also correlates well with both the dye dilution and direct Fick methods.

Diagnosis of low cardiac output usually depends on recognition of its effects rather than on direct measurement. Many techniques have been developed for the estimation of cardiac output, but most of these involve venous and arterial punctures and are not applicable to bedside monitoring. All cardiac output measurement techniques are indirect estimations based on physical and mathematical assumptions. Therefore, each technique contains inherent errors that must be recognized.

Echocardiography is a simple, noninvasive bedside technique based on reflected ultrasonic waves for measuring and monitoring cardiac output.

As shown in Fig. 6.13, the *impedance plethysmograph* operates as follows: Two band electrodes are placed around the patient's neck, a third band around the thorax at the level of the xiphosternal joint, and a fourth band around the lower abdomen. This four-electrode system is similar to that originally used by Wheatstone to measure low-resistance wires and is commonly used in solid-state physics to measure the resistivity of semiconductor material. The spacing between the two band electrodes around the neck is approximately 1 in. and the spacing between the third and fourth band electrodes is approximately 3 in. The electrodes at the upper neck and abdomen are excited by a 100-kHz sinusoidal current of approximately 6 mA and the resultant voltage (impedance) changes occurring with the cardiac cycle are monitored from the other two electrodes. By Ohm's law, the voltage across the inner bands is

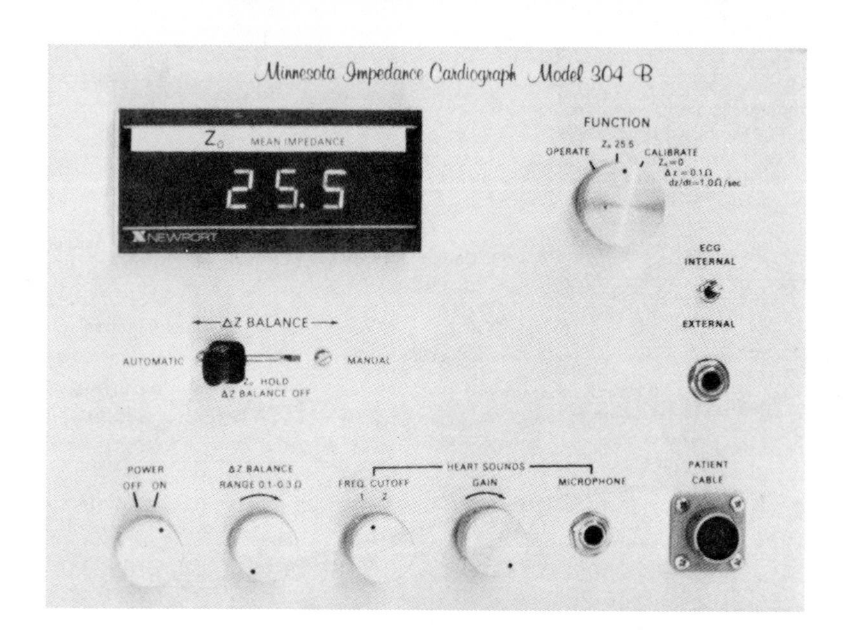

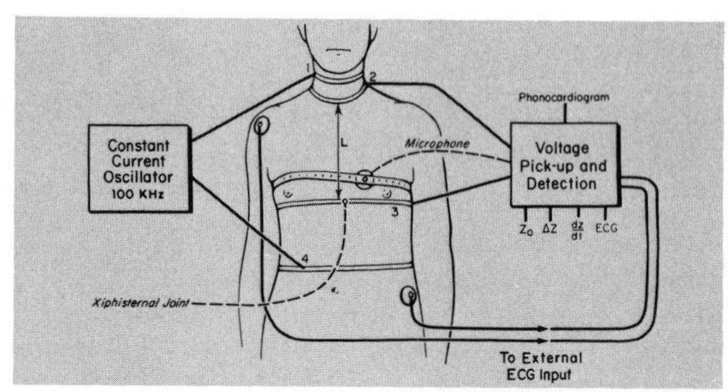

(a)

The basic circuits and electrodes

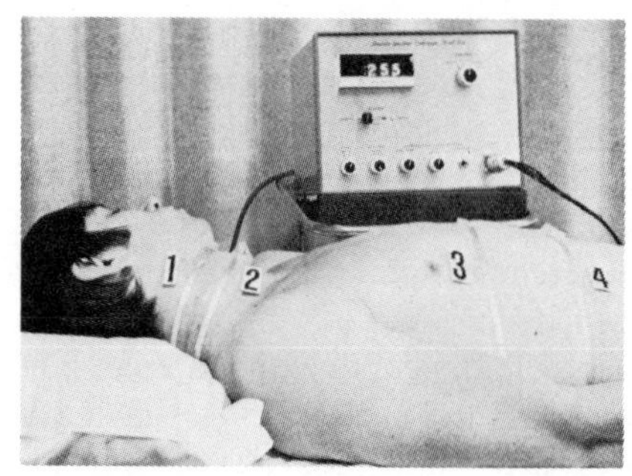

A view of the Impedance disposable electrodes

(b)

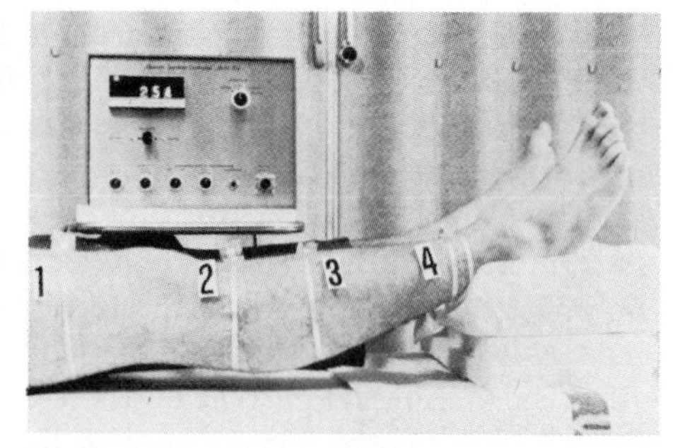

The Impedance electrodes for peripheral circulation

(c)

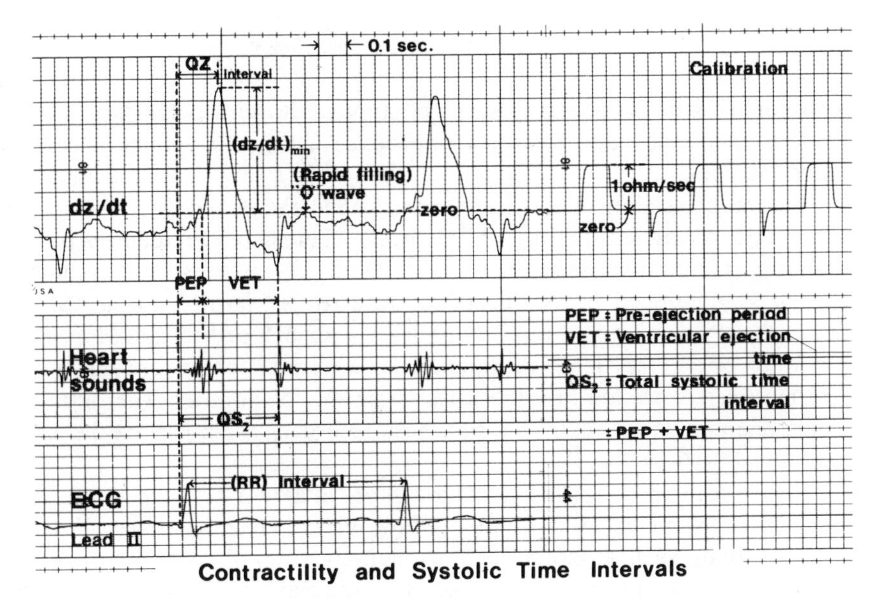

A recording on a normal subject

(d)

Z_0 = Total impedance between leads 2 and 3

$\triangle Z$ = Impedance change during Cardiac Cycle

dz/dt = First time derivative of $\triangle Z$

ECG signal from leads 1 and 4 or external leads (A)

Phonocardiogram with two frequency bands

VET = Ventricular Ejection time

Figure 6.13 Four-electrode impedance plethysmograph measures stroke volume. Provisions for the phonocardiogram and ECG are also provided. The Minnesota Impedance Cardiograph Model 304B provides an optional ECG input by a simple switch position on the front panel. When the ECG switch is in the internal position, an ECG signal is obtained from the impedance electrodes 1 and 4. In some individuals this yields an inadequate ECG signal. By placing the ECG switch on the external position, the special ECG input leads are connected to standard-type ECG patch electrodes as illustrated above. Other ECG electrode locations can be selected if desired. The external ECG input must always be used with the standard impedance patient leads connected to the impedance electrodes 1, 2, 3, 4. Impedance lead 3 functions as the center terminal for the ECG amplifier to provide common-mode rejection. (Courtesy of W. G. Kubicek, President of Surcom, Inc., Minneapolis, Minn.)

proportional to the impedance when the applied current is constant. Stroke volume is then calculated from the impedance change to volume in a conducting medium.

The latest version of this instrument, called the *impedance cardiograph,* is automatic, operates with only one recording channel, and minimizes patient discomfort. Continuous monitoring is possible. Once the stroke volume is found, it is multiplied by the heartbeat rate to get cardiac output. Stroke volume should be summed up over at least 4 to 6 heartbeats. In Fig. 6.14, the outputs of the impedance cardiograph on a normal subject are illustrated. These parameters are fed into the cardiogram microcomputer. The output of the microcomputer is fed to a strip chart recorder, as shown in the block diagram in Fig. 6-14.

Figure 6.15 shows principles and a block diagram of a sine wave electromagnetic flowmeter used in blood flow measurement systems. When a conductive fluid, such as blood or saline, cuts the lines of force of a magnetic field, an electromotive force is generated in the fluid which is perpendicular to both the magnetic lines of force and the direction of motion of the fluid. This electromotive force (flow signal) is directly proportional to the intensity of the magnetic field, the distance between the sensing electrodes, and the integral of the fluid velocity.

The transducer consists basically of an electromagnet to generate the magnetic field and two electrodes to sense the flow signal. They are encapsulated in epoxy in the form of a cuff to fit around the blood vessel. The lumen fixes the cross-sectional area of the vessel, changing the transducer to a flow-rate measuring instrument, although basically a velocity transducer. The electrodes make contact with the vessel wall; contact is directly with the blood in cannulating and extracorporeal transducers. Calibration of the transducer is independent of temperature, pressure, density, or viscosity of blood, within physiological ranges.

The flowmeter supplies the current to energize the transducer. The derived flow signal is then amplified, discriminated from artifacts, and conditioned for display, recording, or computation.

6.7 CARDIAC CATHETERIZATION

In 1905, Fritz Belichroeder passed a catheter into his own veins. Forsmann in 1919 passed a catheter up his own forearm into the right atrium under fluoroscopic control in front of a mirror held by a nurse. Since selective angiography was developed by Chavez in 1946 using dye injection, *cardiac catheterization* has become an important surgical procedure to determine cardiac abnormalities.

Many hospitals and medical centers have a catheterization laboratory for

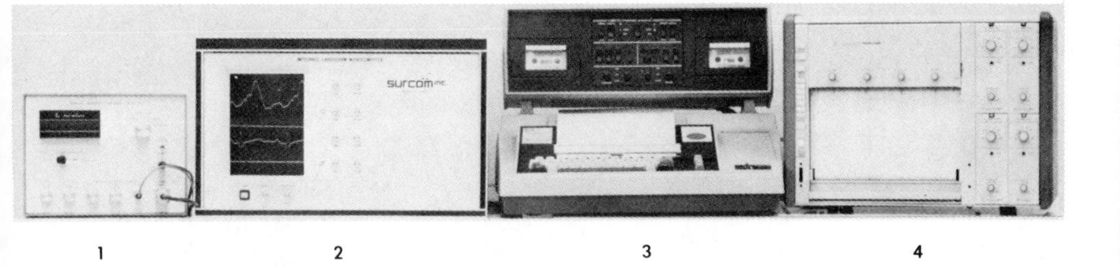

1. **The Minnesota Impedance Cardiograph™**
2. **The new Impedance Cardiogram Microcomputer**
3. A **Texas Instruments Silent 700 Printer** with dual cassette tape deck and acoustic coupler provide for rapid patient data acquisition, storage and retrieval.
4. A high quality strip chart recorder provides a permanent record of dz/dt waveform, ECG and phonocardiogram for pre and post therapy comparison.

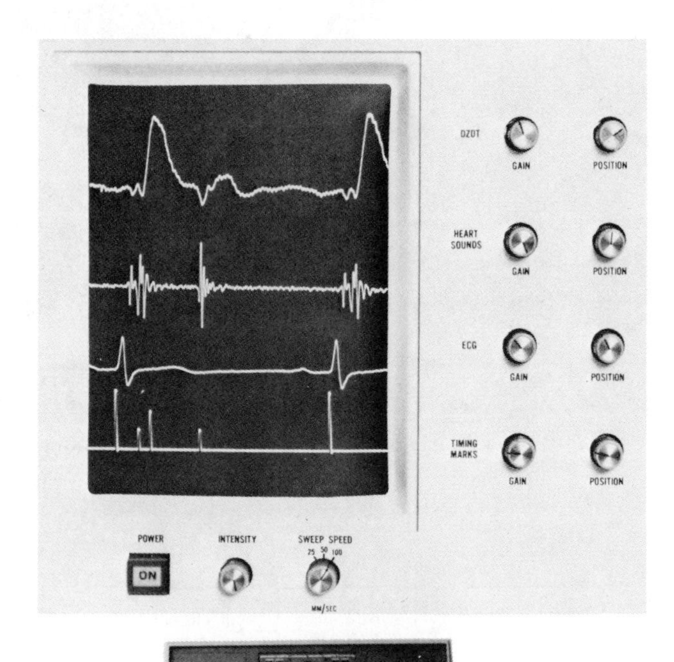

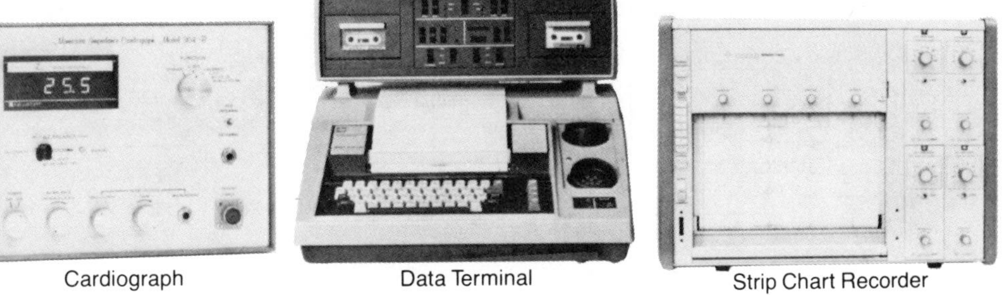

Figure 6.14 Impedance cardiogram microcomputer. (Courtesy of W. G. Kubicek, President of Surcom, Inc., Minneapolis, Minn.)

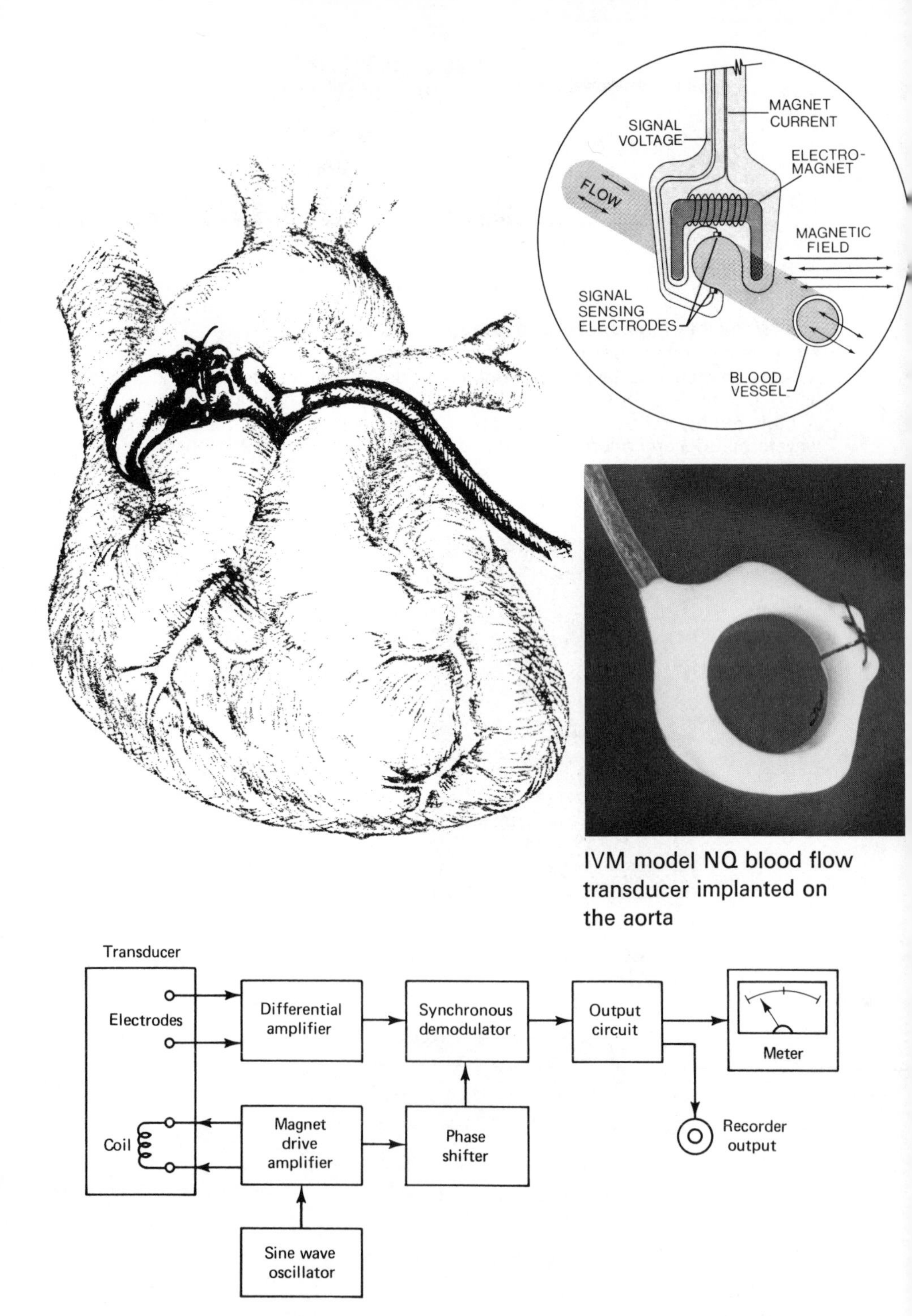

Figure 6.15 Principles and block diagram of a sine wave electromagnetic flowmeter. (Courtesy of IVM, Healdsburg, Calif.)

diagnosis of problems in patients suffering from heart disease. In these laboratories any chamber of the heart, as well as the major vessels near the heart, can be entered to measure blood pressure and/or flow on a detailed beat-by-beat basis. The basic clinical technique utilized in studying the hemodynamics of the heart is cardiac catheterization. Catheters, which are long, thin mechanical tubes equipped to sense pressure and/or flow, are introduced into the peripheral arteries or veins either by surgical exposure, by penetration of the vessel, or by puncture using special needles and/or inserters. The catheter is then pushed along inside the vessel until it reaches the region of interest.

The newest extensions of cardiac catheterization are cardiac angioplasty and percutaneous intra-balloon pumping. In cardiac angioplasty, a balloon catheter is inserted in the femoral artery and pushed up to the clotted fatty deposits in the heart. The balloon at the tip of the catheter is enlarged with oil and the fatty deposits pushed back. Percutaneous intra-balloon pumping devices are inserted through the femoral artery and pushed up to the heart. The purpose of this device is to share the work load placed on the heart.

Figure 6.16 indicates the common methods for catheterizing the left heart and the aorta. Either the main artery of the arm (the brachial artery) or a major artery in the groin (the femoral artery) is entered. The pressure and/or flow

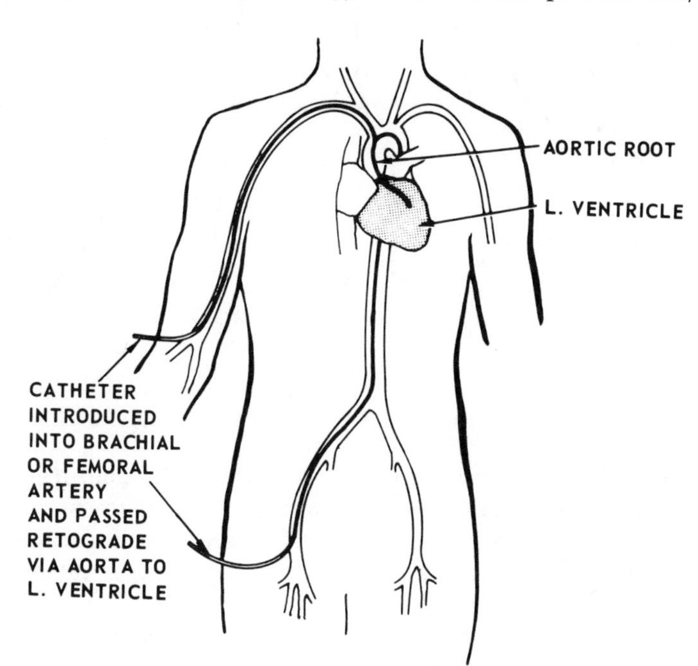

Figure 6.16 Catheterization for left heart and aorta. (Redrawn from Fig. 7, p. 162, courtesy of the *IEEE Transactions on Industrial Electronics and Control Instrumentation*, Vol. IECI-17, April 1970, © 1970, IEEE.)

catheter is passed through the artery into the aorta and hence to the aortic root, through the aortic valve, and into the left ventricle. Similar catheterization techniques for studying the right heart are indicated in Fig. 6.17.

Cardiac catheter sensors in this work must be small and mechanically safe with respect to the interior of vessels and heart chamber walls.

Table 6.2 lists the standard catheter diameters that are clinically used. The size is frequently given in terms of its French number, an increase of one

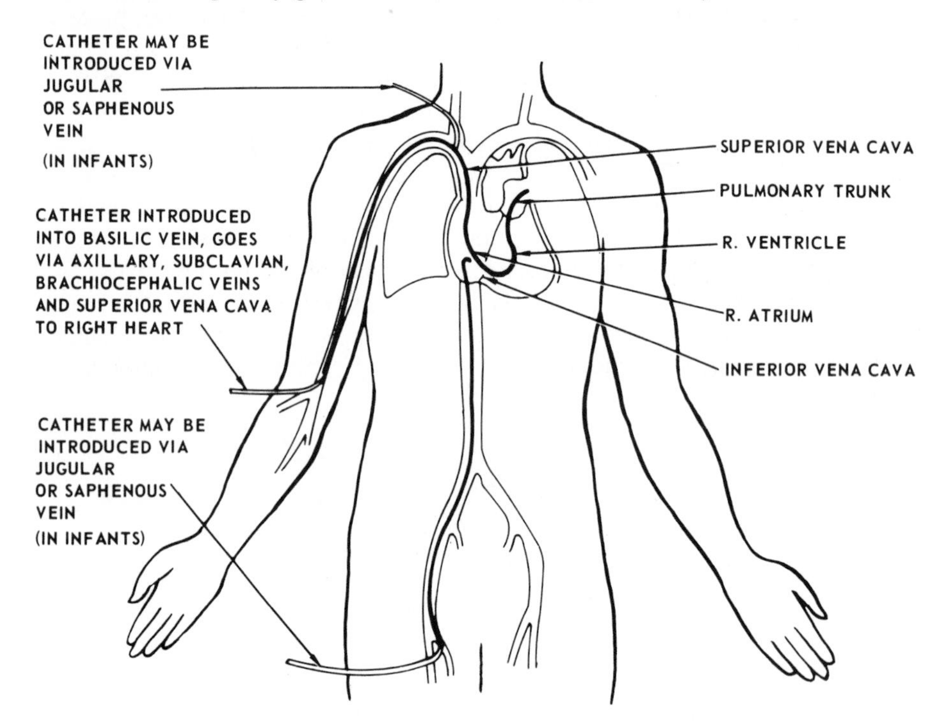

Figure 6.17 Catheterization for right heart and pulmonary artery. (Redrawn from Fig. 8, p. 163, courtesy of the *IEEE Transactions on Industrial Electronics and Control Instrumentation*, Vol. IECI-17, April 1970, © 1970, IEEE.)

number indicating an increase in outside diameter of 1/3 mm. It is important to standardize catheter diameters to these sizes. A major reason is that when entering an artery, which is a high-pressure vessel, special inserters are used to minimize bleeding and damage to the artery. An arterial catheter must fit properly in the inserter.

Cardiac catheterization of infants and small children is necessary in some cases of congenital heart disease. The sizes required for pediatric use are very small, and the problem can become severe, particularly if a flow sensor is desired. Catheters must be structurally strong and flexible and absolutely free of tendencies to kink or break.

TABLE 6.2 Catheter Sizes in Cardiology

French	Outside Diameter		Comments
	Millimeters	Inches	
3	1.00	0.039	Preferable in infants
4	1.33	0.052	
5	1.67	0.065	Largest for infants
6	2.00	0.078	Preferable in adult arteries
7	2.33	0.091	
8	2.67	0.104	Largest arterial inserter
9	3.00	0.118	Primarily for venous catheterization
10	3.33	0.131	

Artifact during left or right heart catheterization is usually due to:

1. Faulty connections or recordings. Any minute air bubble alters the pressure tracing or problems with the ECG before preparing for catheterization.
2. Blood clots.
3. Movement of the catheter.

The proper transducer–catheter system (e.g., Millar pressure transducers) gives satisfactory results.

Figure 6.18 shows a pictorial of blood flow through the heart. One can see the valves and can easily pinpoint the heart areas where a disorder can exist.

Symbol	Meaning	Normal pressure readings systolic/diastolic
RA	Right atrium	$\overline{3}$
LA	Left atrium	8
RV	Right ventricle	22/4
LV	Left ventricle	120/80
—	Aorta	120/80
Pulmonary artery	Pulmonery artery	25/8

(a)

Figure 6.18 (a) Normal blood pressure readings in the heart. (b) Blood flow through the heart. (Redrawn from Fig. 6, p. 161, courtesy of the *IEEE Transactions on Industrial Electronics and Control Instrumentation*, Vol. IECI-17, April 1970, © 1970, IEEE.)

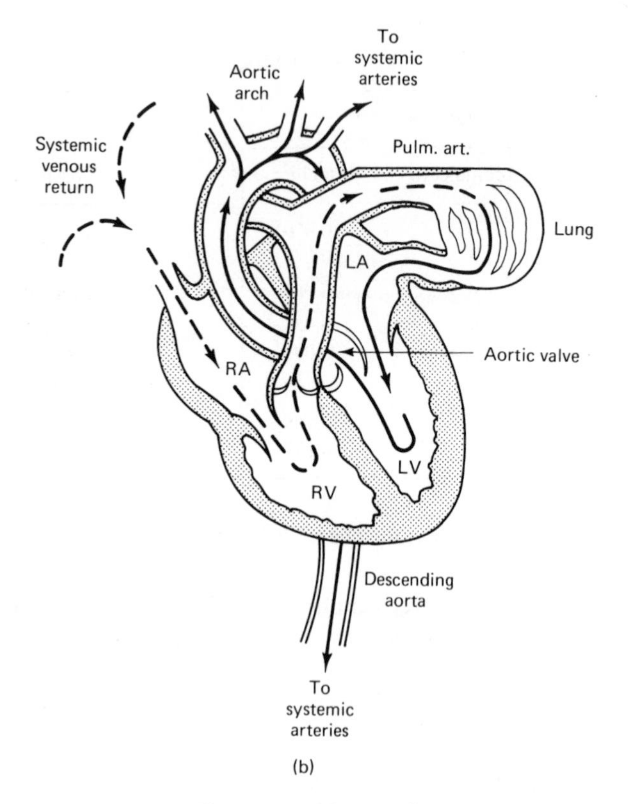

Figure 6.18 (*Continued*)

6.8 REVIEW QUESTIONS

1. Discuss direct and indirect methods to measure blood pressure.
2. Discuss the direct use of the Fick principle of measuring cardiac output.
3. Discuss the dye dilution and thermal dilution techniques to measure cardiac output.
4. Discuss echocardiography to measure cardiac output.
5. Discuss cardiac catheterization.

6.9 REFERENCES

1. Kantrowitz, P.: Noninvasive Blood Pressure—State of the Art, *Medical Electronics and Data*, September–October 1976.
2. Bruner, J. M. R.: *Handbook of Blood Pressure*, PSG Publishing Co., Inc., Littleton, Mass., 1978.
3. Kantrowitz, P., and Kingsley, B.: A Survey of the Clinical Measurement of Cardiac Output, *Medical Research Engineering*, Vol. 10, No. 6, December 1971.

4. Rushmer, R. F.: *Cardiovascular Dynamics*, 2nd ed., W. B. Saunders Company, Philadelphia, 1975.

5. Guier, W. H.: Cardiovascular Sensors in the Clinical Environment, *IEEE Transactions on Industrial Electronics and Control Instrumentation*, Vol. IECI-17, No. 2, April 1970.

7

ELECTROENCEPHALOGRAMIC MEASUREMENTS

7.1 INTRODUCTION

The functions of *electroencephalographs* (EEGs) are characterized by detection and amplification of bioelectric potentials associated with the neuronal activity of the brain by electrodes placed on the surfaces of the scalp. These signals are traced, usually by ink-pen stylus driven by D'Arsonval movements, on ruled recording paper.

For the past few years, electroencephalographs have been applied extensively in the diagnosis of epilepsy and other brain-related disorders. This is accomplished through sleep studies, waking studies, or a combination.

The EEG, like the electrocardiograph (ECG) and electromyograph (EMG), senses bioelectric potentials, but unlike the other instruments, the frequency of the output signal is the important characteristic, not the signal's wave shape. The EEG is nonperiodic in that the output signal continuously shifts in amplitude, phase, and frequency. For this reason, an EEG recording must be taken over a considerable period of tims. The EEG output frequency bands are classified as given in Table 7.1.

The first EEG measurement has been attributed to Hans Berger of the University of Jena in Austria in 1929. In 1934, with the advent of the differential amplifier, the present EEG machine was produced. Although the circuit components have changed, the basic theory and design principles remain the same.

The human brain makes us act as individuals and we perform similar acts in individualistic manners. As a result, the activity of the brain varies from individual to individual. No two people generate exactly the same EEG record, which makes standardization difficult but not impossible.

TABLE 7.1 Frequency Bands of Encephalographic Signals

Signal	Frequency Band
Delta	0.5–3.5 Hz
Theta	3.5–7.5 Hz
Alpha	8–12 Hz
Beta	13 Hz and above

The EEG is used extensively in the following clinical areas:

1. *Neurology.* Neurologists rely heavily on EEGs. As a clinical tool, used in conjunction with other tests, such as EMGs, echo cardiograms, and neurological examinations, it is used to define patient's brain pathology.
2. *Neurosurgery.* Neurosurgeons use EEGs to help localize pathological abnormalities such as tumors that they will remove from the brain surgically.
3. *Anesthesiology.* Anesthesiologists use EEGs to determine levels of anesthesia administered to patients. This is especially true for a patient undergoing cardiac surgery or for patients who are difficult to monitor.
4. *Psychiarty.* EEGs are used to try to determine the presence or absence of an organic brain disease in order to diagnose a mental disorder with more certainty.
5. *Pediatrics.* EEGs, together with other tests, such as averaged evoked potential, are used to determine hearing and visual problems in the newborn.

7.2 THE CLINICAL EEG EXAMINATION

There are two very important parameters that must be considered when making a clinical EEG examination: the patient's age and state of consciousness. Both affect the EEG pattern. The age variable is directly related to frequency and inversely related to amplitude. As age increases, so does frequency; as age increases, amplitude decreases. A child's EEG will show high-amplitude slow waves as a normal EEG, and an adult will show low voltage and faster frequencies. State of consciousness is equally important. During sleep, an adult's EEG will show high-voltage slow waves, which would be a very abnormal finding in the awake state.

In a clinical EEG examination, the patient is greeted and seated in the examination room with the EEG technician. The patient is put as much at ease as possible and the test is briefly explained. The technician quickly scans the EEG referral from the physician; the referral gives a brief history of the patient as to medical complaints: seizures, headache, trauma, or others. Having read the referral and knowing a bit about the patient, the EEG technician again questions the patient to obtain a specific medical history. For instance, in the case of seizures, the technician asks questions regarding the types of seizures, their frequency of occurrence, how they start, how they progress, whether or not tongue biting occurs, clinic or tonic movements, and so on. In the case of headaches or trauma, the technician tries to find localization, type of pain, frequency of occurrence, and related items.

Electroencephalographs, presently in wide use, utilize essentially identical measurement techniques. The signal is detected by electrodes (surface or subdermal) placed in a standard pattern (i.e., the 10–20 electrode placement system) on the scalp. The leads from the electrode are coupled to a differential amplifier, where they are amplified, and the resulting signal is traced out on chart paper.

There are usually 10 pairs of electrodes, any two of which can be switched to the input of the amplifier to obtain a corresponding trace. The use of a single channel would be time consuming and laborious. Thus, EEG apparatus is almost always multichannel, with from 6 to 16 independent channels available. A six-channel instrument is shown in Fig. 7.1. Simultaneous recordings from 6 to ten pairs of leads is therefore possible. The most common EEG machine is the eight-channel clinical unit, as opposed to the six-channel portable unit.

The electroencephalograph consists of a series of single-channel differential amplifiers, with variable gain and frequency band-pass circuits, which are connected to the patient via a switching junction box. The output of each of the variable gain–bandwidth amplifiers is connected to one of the individual ink-pen recorder movements. The pen movements are independent of each other but appear on the same chart paper, which is driven by a single motor-driven chart mover.

EEGs have proved to be extremely reliable devices, due to the high quality of their construction. They do, however, have an extreme artifact problem. Their most serious artifacts are similar to those found with ECG and EMG, but there is however an additional artifact not found with the other physiological instruments. This is the cardiac artifact, which essentially the ECG pattern as detected at the EEG electrodes. Unfortunately, when present at a pair of electrodes, it is nearly impossible to reduce this artifact.

Line-frequency artifacts are greatly reduced in EEG apparatus if the lined cord is armored with a double-woven shield.

The next state of an EEG test is the application of the electrodes, as shown in Fig. 7.2. Between 18 and 24 electrodes are generally applied in the clinical examination. These electrodes are applied in a standard area, according to

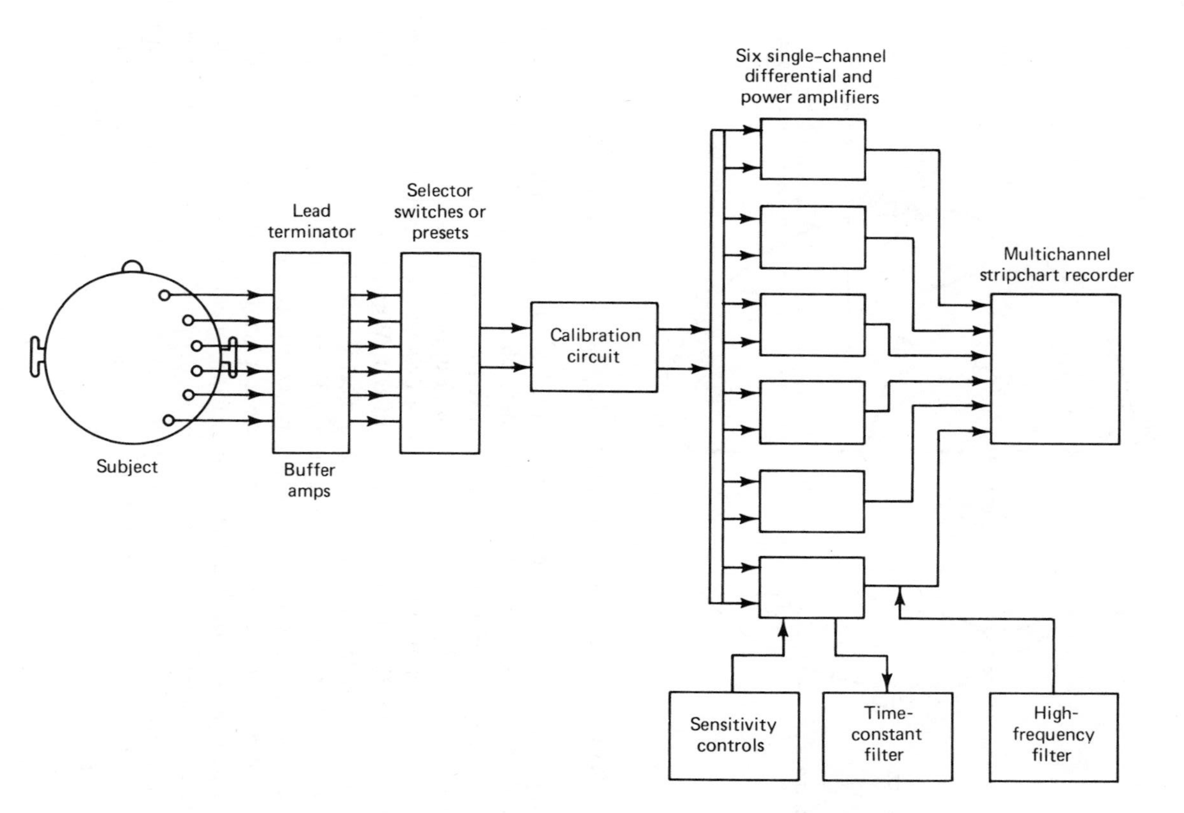

Figure 7.1 Simplified block diagram of a six-channel EEG.

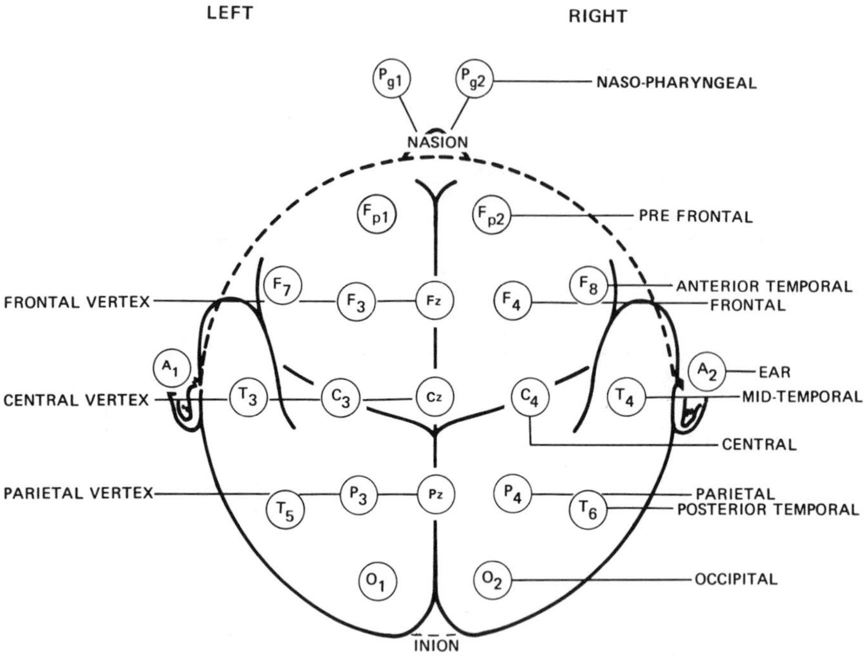
LEFT
RIGHT
NASO-PHARYNGEAL
NASION
PRE FRONTAL
ANTERIOR TEMPORAL
FRONTAL
FRONTAL VERTEX
EAR
MID-TEMPORAL
CENTRAL VERTEX
CENTRAL
PARIETAL VERTEX
PARIETAL
POSTERIOR TEMPORAL
OCCIPITAL
INION
Pg1
Pg2
Fp1
Fp2
F7
F3
Fz
F4
F8
A1
A2
T3
C3
Cz
C4
T4
T5
P3
Pz
P4
T6
O1
O2

Figure 7.2 International 10–20 system for EEG electrode placement.

anatomical definition. This definition is called a 10–20 *electrode system*, and the placement is based on certain anatomical characteristics of the head: for instance, the nasion and the inion and the external auditory meatus.

Scalp areas to which the electrodes will be applied are usually marked with a red pencil. The technician then uses alcohol or acetone or redux and cleanses the area to which the electrodes will be affixed. When this is completed, the technician starts actual application of the electrodes. There are several techniques used in electrode application; among them are the paste and collodion techniques. Small, subdermal needle electrodes are also used in some laboratories. In the collodion technique, two main applications are used. The first is the gauze-pad method. In this case, the electrode is held firmly on the head with one hand and a small amount of collodion is dropped on a small gauze pad. The collodion-soaked gauze pad is then placed over the electrode and dried with compressed air. A small amount of electrolyte jelly is then inserted through a hole in the gauze pad and into the electrode. At the same time as the electrolyte jelly is inserted, the scalp is abraded with a blunt syringe. The other application of collodion electrodes is similar to this, except that the gauze pad is omitted and the collodion is applied directly to the electrode–head interface.

In the paste technique, a small amount of adhesive electrode paste is placed in the electrode cup. The electrode is then placed on the head, and the technician puts a small amount of electrode paste over the electrode. The paste is then covered with a cotton ball or gauze pad. Variations this technique also exist; each technician works out his or her own method of application.

When all electrodes are affixed to the head, the patient relaxes in a chair or is asked to lie on a bed in the room. The electrodes are connected to the lead terminator box and checked by an impedance meter. The technician then calibrates the EEG instrument. The room lights are dimmed and the patient is told to relax with his or her eyes closed. The technician can use the master selector switch or hand select the individual electrode run. This electrode run, or montage, is the selection of electrodes from which the technician wants to record. These electrode montages are usually preestablished and are a standard part of each recording. In a normal examination, there are usually 8 to 16 preset electrode patterns from which recordings are made. The technician may go from one to the other on a random basis or use a set pattern. All areas of the scalp are sampled.

Figure 7.3 illustrates the 10–20 electrode system and the identification of the recording channels using bipolar recording. Bipolar recording is the recording between two active scalp electrodes with each channel of the EEG machine. Each channel records only the difference in potential between two electrodes led to it. Bipolar linkages should include simple chains running anteroposteriorly or transversely.

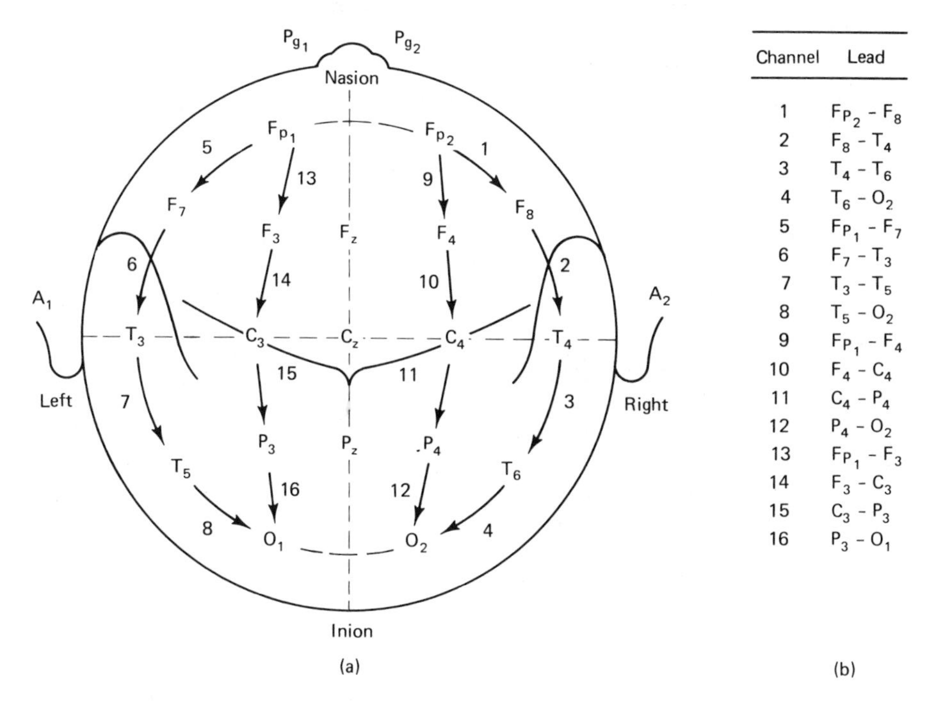

Channel	Lead
1	$Fp_2 - F_8$
2	$F_8 - T_4$
3	$T_4 - T_6$
4	$T_6 - O_2$
5	$Fp_1 - F_7$
6	$F_7 - T_3$
7	$T_3 - T_5$
8	$T_5 - O_2$
9	$Fp_1 - F_4$
10	$F_4 - C_4$
11	$C_4 - P_4$
12	$P_4 - O_2$
13	$Fp_1 - F_3$
14	$F_3 - C_3$
15	$C_3 - P_3$
16	$P_3 - O_1$

Figure 7.3 (a) 10–20 international electrode system, showing a bipolar montage for a 16-channel recorder. (b) Identification of recording channels.

The collodion electrode technique is used to obtain alpha, beta, delta, and theta waveforms, shown in Figs. 7.4 to 7.6. In these curves, five single horizontal boxes represents 1 second and one vertical box represents 50 μV. Channels 1 to 4 represent the front of the head to the back of the head on the right side, channels 5 to 8 represent the front of the head to the back of the head on the left side, channels 9 to 12 represent the front of the head to the back of the head on the right side, and channels 13 to 16 represent the front of the head to the back of the head on the left side. Channels 1 through 8 represent the even numbers on the right-side positions of the nasion/inion and channels 9 through 16 represent the even-number positions of the nasion/inion (Figs. 7.2 and 7.3).

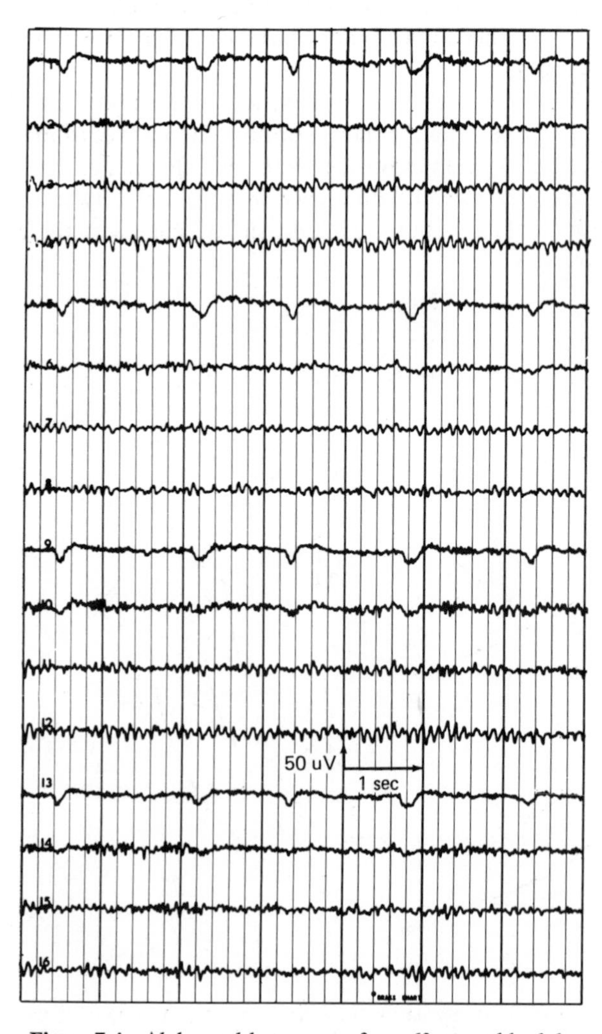

Figure 7.4 Alpha and beta waves for a 63-year-old adult.

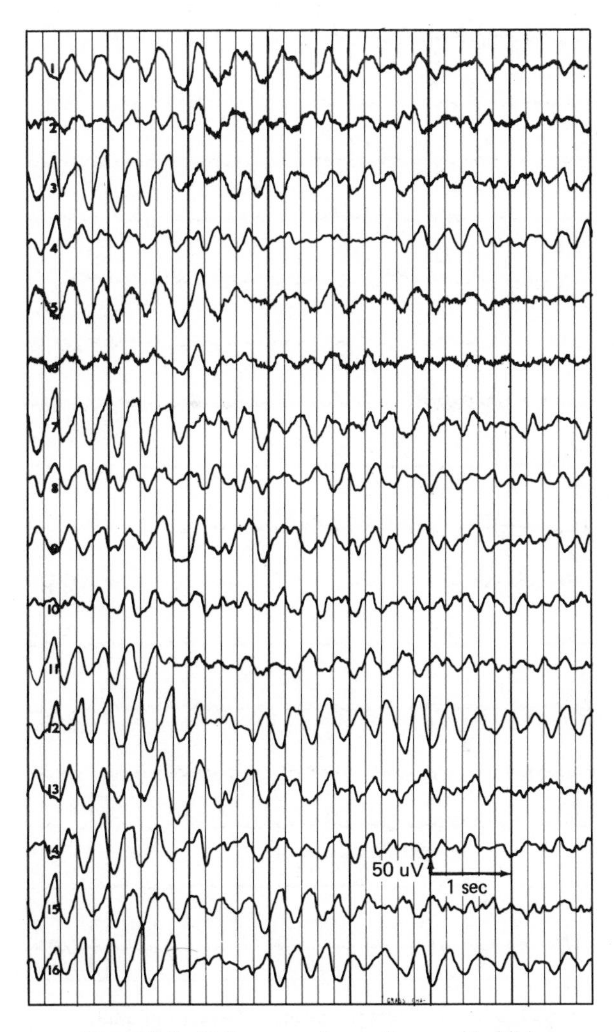

Figure 7.5 Delta waves for an 8-year-old child.

In Fig. 7.4, alphawaves have the highest signal-strength waveform in channels 3, 4, 7, 8, 11, 12, 15, and 16. Looking at channel 16, we can calculate about 9.5 wiggles between two red lines or about 10 wiggles per second or 10 Hz. Beta waves are seen in channels 1, 5, 9, and 13 of Fig. 7.4.

An alpha wave possesses the following features:

1. It is a sinusoidal wave.
2. It should be bilateral, synchronous, and relatively symmertical, but may be of slightly higher amplitude over the right hemisphere.
3. It is present at rest with the eyes closed.

When testing alpha and beta waves, the patient is alert and relaxed.

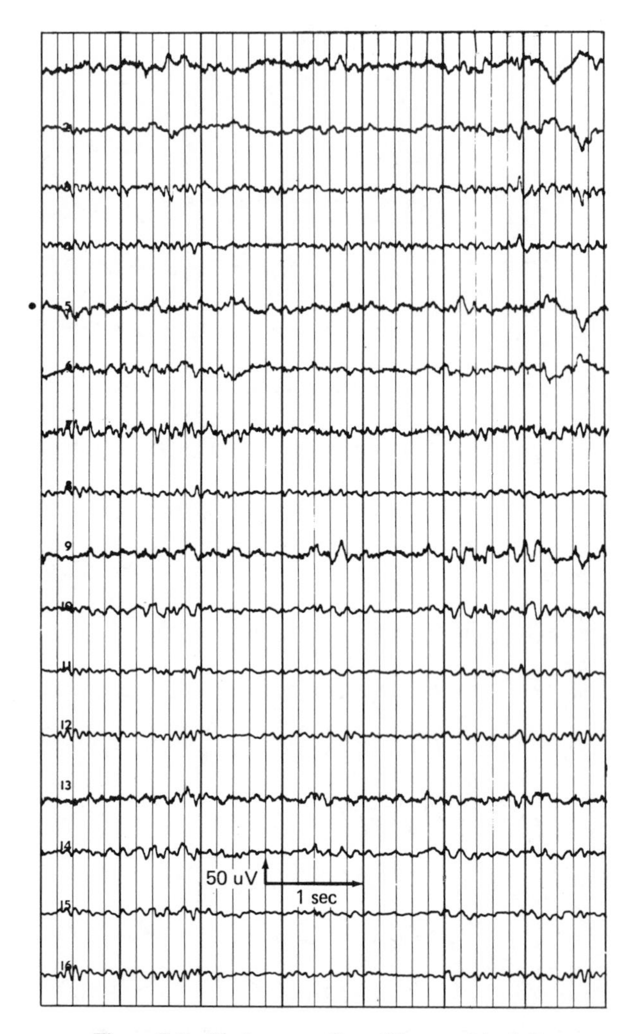

Figure 7.6 Theta waves for a 25-year-old adult.

In Fig. 7.5, delta waves are seen with a frequency range of 4 to 7 Hz. In Fig. 7.6, theta waves are seen with a frequency range of 1 to 3 Hz. Both delta and theta waves are tested on sleeping patients, during hyperventilation in the young, and in patients with dysfunction disorders.

In most laboratories, it is considered advantageous for the technician to allow the patient to fall asleep. The same montages will usually be used to record during the early and light stages of sleep. Sleep in this case is considered an activating technique because certain EEG patterns will be elicited or will show during the early states of sleep which will not appear in the awake record. Other activation techniques that are used in clinical recordings are hyperventilation and photic stimulation. Hyperventilation usually consists of 3 minutes of very deep breathing by the subject. This deep breathing essentially lowers the CO_2 level, thus constricting the blood vessels of the brain and decreasing the amount of oxygen available. This, in turn, may create large slow waves or slow patterns in the EEG. Photic stimulation is also used, usually at given frequencies to elicit a driving response. A photoconvulsive response to photic stimulation is shown in Fig. 7.7.

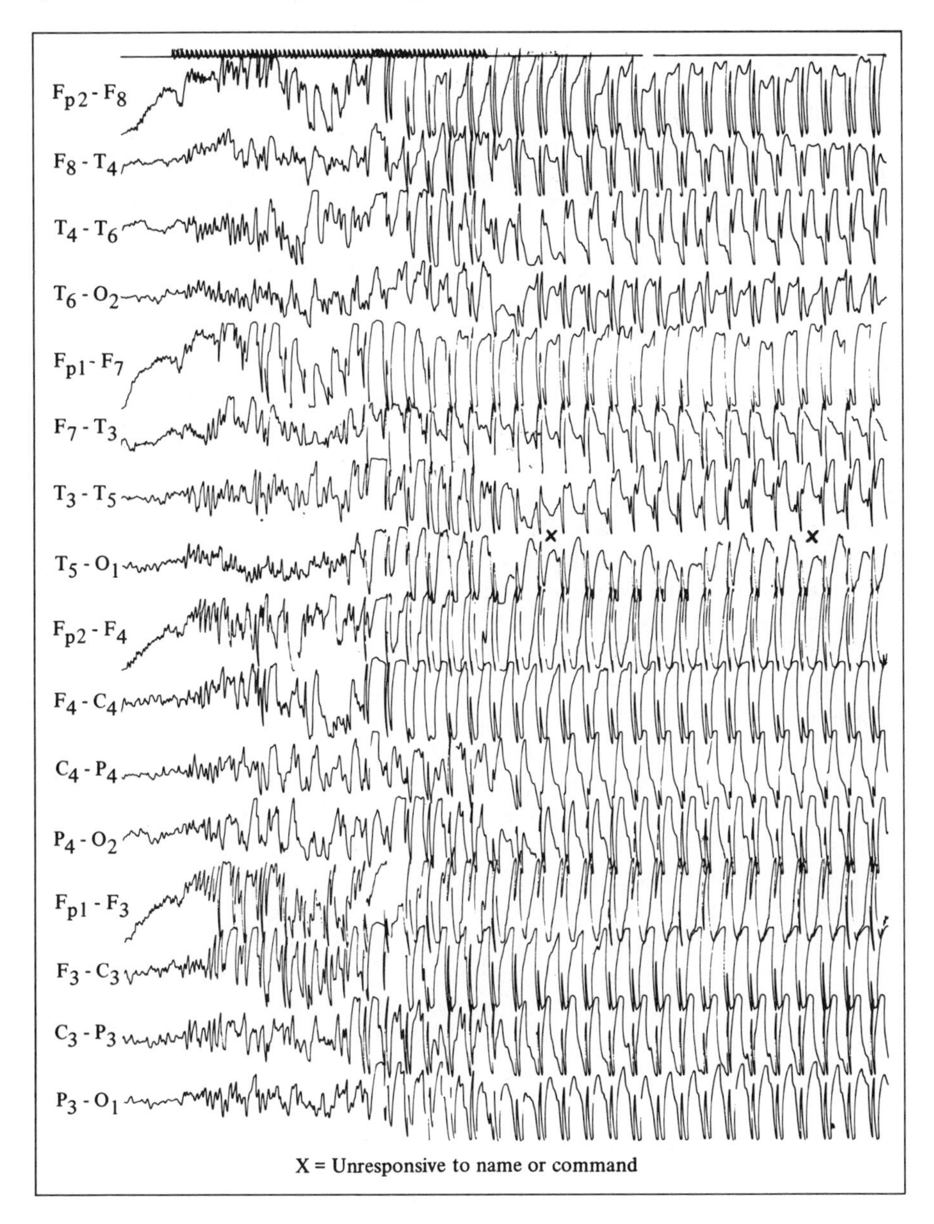

Figure 7.7 Photoconvulsive response to photic stimulation.

Both hyperventilation and photic stimulation are considered good activating techniques for certain types of epilepsy and must be used with discretion. Clinical EEG examinations are also verified by ultrasonic brain scans discussed in Chapter 1.

7.3 ELECTRONIC PROBLEMS AND HAZARDS ASSOCIATED WITH THE EEG

Very high input impedance is usually desirable in that it minimizes the instrument's effect on the EEG signal as detected at the scalp. If an instrument has a low input impedance, it will load the signal, causing distortion and making the recordings inaccurate. The input impedance of the instrument should be approximately 25 MΩ.

High input impedance is necessary so that when a signal is present on the electrodes, it will be unaltered and ready for amplification. However, the electrodes act as an antenna in that they become conductors for radio, TV, and most surely, the 60-Hz signal radiated from power lines. The 60-Hz signal is most important because it falls in the frequency spectrum of the EEG signal to be recorded.

The common-mode rejection of differential amplifiers used in the EEG help reduce the 60-Hz signal. The common-mode rejection is usually stated as a ratio, such as 1000 : 1, which means that a common-mode signal appearing at the input will have its amplitude decreased by a factor of 1000 at the output of the amplifier. If the common-mode signal radiated from the power lines happened to measure 10 mV at the electrodes, only 10μV would appear on the chart. The higher the common-mode ratio, the better the instrument can record in hostile or noisy environments. Typically, EEG instruments have a common-mode rejection of 5000 : 1, which means that the interference signal amplitude is decreased 5000 times by the time it gets to the pens.

There is another less desirable way in which EEG instruments have been designed to handle the 60-Hz interference problem. This is through the use of a 60-Hz or notch filter. The 60-Hz filter attenuates the 50-Hz component selectively. Unfortunately, it will also attenuate frequencies just below and above 60 Hz, so these devices should be used with caution.

Confusion may also exist because of gain and sensitivity, which, although related, are not identical. *Gain*, sometimes called *amplification*, is the ratio of the output signal amplitude to the input signal amplitude of an amplifier. In EEG, the input is the signal level taken from the electrodes and the output is the voltage present at the pen recoder.

Example

If the EEG signal is 100 μV and the output of the EEG amplifiers is 10 V,

$$\text{gain} = \frac{\text{output}}{\text{input}} = \frac{10\ \text{V}}{10\ \mu\text{V}} = 100{,}000$$

Because gain is actually a ratio, it has no units; it is stated simply as a number.

Sensitivity is the signal level required at the input of a system to derive a specified output from the system. In the case of EEG, it is the input voltage required to create a stylus deflection on the chart.

$$\text{sensitivity} = \frac{\text{input voltage}}{\text{output deflection}} = \frac{50\ \mu\text{V}}{7\ \text{mm}} = \frac{7.1\ \mu\text{V}}{\text{mm}} \quad \text{or} \quad 7.1\ \mu\text{V/mm}$$

Gain affects sensitivity directly. Although the input voltage has not changed, by increasing the output voltage we have increased the gain and thus made the instrument more sensitive; 7 μV/mm is a more sensitive setting than 14 μV/mm. Note that it is a necessity for the electroencephalographer to know the sensitivity at which the record is run, which is why most technicians run pages of calibration signals at the beginning and end of the EEG. But, when the gain is changed during the recording, the electroencephalographer does not know the sensitivity unless he or she makes a note of it. Simply marking a change in sensitivity by writing "up gain" or "down gain" on the record can be very misleading, as the gain is sometimes changed several times in each direction. Also, if the EEG'er reads EEGs recorded from instruments of different manufacture, the increase or decrease in sensitivity is probably not the same. Also, it is *imperative* that the sensitivity be marked on the record in EEG recordings in cases of suspected cerebral death.

To obtain the gain and sensitivity accurately, the original sensitivity of the instrument must be known as well as how to calculate the sensitivity after every change in gain. The gain increase is usually different for instruments of different manufacture.

Consider an instrument calibrated for a sensitivity of 50 μV/7 mm and having gain steps of 2. This means that the output voltage of the amplifiers is doubled, as is the sensitivity, with every step increase in gain:

Calibrated sensitivity:	50 μV/7 mm = 7.1 μV/mm
From original settings, increase gain 1 step:	sensitivity = 3.6 μV/mm
From original settings, decrease gain 1 step:	sensitivity = 14.2 μV/mm

Note that 14.2 μV/mm is a less sensitive setting than 7.1 μV/mm. Conversely, 3.6 μV/mm is more sensitive than 14.2 μV/mm. The fewer microvolts "contained" in each millimeter of pen deflection, the more sensitive is the instrument to a voltage change across the electrodes.

Portable EEG machines are inclined to be hazardous. Although not in constant motion throughout the hospical as electrocardiographs are, they are moved from the EEG lab to surgery and the emergency room. When they are moved, great care must be exercised to prevent physical damage to the machine, electrode-cable sets and line cords.

Further complicating the fact that the EEG machine is used in diverse areas is the fact that it can be used on a patient who is already connected to a wide range of other devices. Any one of these, such as physiological monitors, respirators, electrosurgical units, hypohyperthermias, or volume-controlled ventilators, when used in conjunction with the EEG, can create situations of extreme hazard to both patient and operator.

Special noise problems arise when the signal to be recorded contains so much noise that the output signal is distorted and does not display the proper information. This is frequently true for evoked potential responses. Evoked EEG responses can be elicited by visual or acoustical stimuli. This can be achieved by a flash of light or when an audible click is measured. To determine the response to the stimulus from ongoing EEG activity, the EEG signals are time-locked to the stimulus pulses, so that the evoked response is reinforced with each stimulus presentation, while any activity is not synchronized. In order to display the signal mixed with noise, averaging techniques are used.

7.4 SLEEP RECORDINGS AND PATTERNS IN EEG MEASUREMENTS

Sleep recordings are considered by many to be an essential part of any EEG test. The literature generally indicates that nearly 50% of all spike discharges are seen only during sleep and that an additional 30% of spikes are best demonstrated in sleep.

Sleep recordings are useful in assessing maturation in children, and in some adults they provide supportive evidence of abnormality. For example, it has been found that the asymmetry of sleep spindles is at times the only detectable abnomality in cases of suspected and proved subdural hematoma.

It has been found that depriving a patient of a significant portion of the night's rapid eye movement (REM) sleep may sometimes be effective in activating discharges not produced during routine procedures.

The patient is awakened from netural night sleep 3 hours earlier than usual and is kept awake until early afternoon, when the EEG is recorded and spontaneous sleep is obtained. The transitional periods of drowsiness and light sleep are the most useful stages for activation. It is therefore essential that the instrument be kept running while waiting for sleep. Paper speed is often reduced to 1.5 cm/sec during this period, until the patient becomes drowsy. A relaxed and quite atmosphere is essential.

Defining the level of sleep and wakefulness in the newborn period and obtaining artifact-free recordings are often extremely difficult. Infants have frequent movements of the limbs and trunk, grimacing and sucking movements of the face, and fluttering of the eyelids. Respiration is sometimes irregular. The use of eye movement monitors (EOG), chin muscle monitors (EMG), respiratory monitors, and cardiac monitors should be incorporated as part of the routine encephalogram to aid in interpretating the patterns recorded.

Sleep pattern characteristics include the following:

1. *Sleep spindles (sigma)*. Rudimentary central sleep spindles of 13 to 14 Hz are often present from birth. Initially and for several months they may occur quite asynchronously on the two sides. They change with age, increasing in amplitude and synchrony. During the first month of life, spindles increase in amplitude and duration up to 3 seconds. By 3 months of age they appear fairly synchronously and by 11 to 12 months of age should be completely synchronous. After the age of 6 months, frontal dominant sleep spindles may be present as well as central sleep spindles. However, frontal spindles often appear at a frequency of 11 to 12 Hz and may remain asynchronous into late childhood.
2. *Central or vertex transients*. Central transients become evident as low-voltage diphasic waves in central derivations at approximately 6 weeks of age. Initially they are poorly defined, but become quite prominent by about 3 months of age. By 3 years of age, these central transients are often high in amplitude and sharp in form. Amplitude decreases after the age of 3 years. In children, these central complexes may have a notched apperance and at times occur in runs. In adults the central transients may appear as a single sharp potential followed by a short run of spindles.
3. *V waves: monophasic occipital theta of Hess, or POSTS* (positive occipital sharp transients of sleep). V waves first appear after the age of 10 months, persist through childhood, and are maximal in incidence in the early 20s. They are diphasic or biphasic, with a tendency to sharpness in appearance. They frequently occur in runs at a frequency of 4 to 5 Hz and are present from occipital derivation during light to medium stages of sleep. They look like lambda waves and may be prominent enough to be mistaken for occipital spikes.
4. *K complex*. This phenomenon is an evoked response to a nonspecific stimulus during light sleep. It is present from the sixth week of age throughout life. In infants, stimulation will evoke a response resembling a biparietal hump or central transient, but with maturation will assume a more definite form composed of a biphasic potential and sleep spindle. Sound is the best stimulus for the production of the K complex. Repeated stimuli of the same quality will fail to evoke a response after the second or third stimulus (habituation). A change of stimulus, however, will elicit a response even after habituation.
5. *Beta activity during sleep*. Fast rhythms are normally seen in drowsiness and light sleep only after the age of 6 months. In infancy fast activity in the waking or sleep record is present only in the immediate newborn period. After 6 months of age low-voltage beta activity up to 30 Hz is common from temporal and occipital regions. From 1 to 2 years of age

fast activity increases in amplitude and may be seen more diffusely. After this there is a decrease in amplitude and amount and by 6 years, fast activity is uncommon in sleep. Fast activity is present during drowsiness, increases in light sleep, disappears in deep sleep, and increases on arousal.

Two newer EEG techniques used in clinical sleep studies include the following:

1. *Polysomnography* is a complex multichannel continuous recording of several physiological functions during an extended period of sleep, used for diagnostic purposes. A polysomnogram reveals a patient's sleep pattern and normal amount of sleep, thus permitting comparison with those of other patients and with other patterns of insomnia.
2. *Human chronobiology* provides EEG signals for nonpsychiatric diagnosis and treatment of sleep disorders. It provides a sleep–awake cycle over a continuous period of time.

7.5 REVIEW QUESTIONS

1. What are the advantages and disadvantages of needle electrodes?
2. State the common-mode rejection ratio for EEG instruments.
3. Describe in detail how to measure delta rhythm.
4. What is a K complex? Where is it seen, and when?
5. In what year did Hans Burger publish his findings?
6. In what situations would you record a simultaneous EEG?
7. Do EEG systems use high or low impedance?
8. What hazards, if any, are there in performing an EEG on a patient with **(a)** an internal pacemaker; **(b)** an internal cardiac catheter?
9. List five areas in which EEG systems are used, and explain how the EEG is used for each.
10. List the frequency bands of EEG signals.

7.6 REFERENCES

1. Craib, A. R., and Perry, M.: *Beckman EEG Handbook*, 2nd ed., Beckman Instruments, Inc., Fullerton, Calif., 1975.
2. Wooldridge, D. E.: *The Machinery of the Brain*, McGraw-Hill Book Company, New York, 1963.
3. Strong, P.: *Biophysical Measurements*, 1st ed., 3rd printing, Tektronix, Inc., Beaverton, Oreg., 1973.
4. Jacobson, B., and Webster, J. G.: *Medicine and Clinical Engineering*, Prentice-Hall, Inc., Englewood Cliffs, N.J., 1977.

5. Thompson, R. G., and Patterson, M. M.: *Bioelectric Recording Techniques*: Part B, *Electroencephalography and Human Brain Potentials*, Academic Press, Inc., New York, 1974.

6. Glasser, O.: *Medical Physics*, Year Book Medical Publishers, Inc., Chicago, 1961, pp. 361–370.

7. *The CIBA Collection of Medical Illustrations*, Vol. 1: *Nervous System*, Summit, N.J.; and in the Canadian journal *Spike*, "Spike and Wave," Special Issue No. 6, June 1972.

8. Smith, R. G., Houge, J. C., and Webster, J. G.: Portable Device for Detection of Petit Mal Epilepsy, *IEEE Transactions on Biomedical Engineering*, Vol. BME-26, No. 8, August 1979.

8

ELECTROMYOGRAMIC MEASUREMENTS

8.1 INTRODUCTION

The first attempt to obtain biopotential tracings in cases involving peripheral nerve paralysis, by R. Proebster in 1928, has been credited with initiating clinical electromyography (EMG). EMG has been a valuable clinical tool for muscular disorders since 1960.

Muscles fall into three general classifications: skeletal, cardiac, and smooth. The EMG is the bioelectronic measurement of limbs, thorax, heart, intestines, and involuntary muscles.

Myography records the mechanical effects of muscular contractions caused by depolarization of muscle fibers, *electromyography* records the electrical effects of muscular contractions, and *electroneurography* records peripheral nerve action potentials.

A simplified diagram of an electromyographic device is shown in Fig. 8.1. Semiconductor amplifiers are used to magnify the small voltage input picked up by the electrodes to a level adequate for the operation for a readout device, such as an oscilloscpoe or loudspeaker. The stimulating electrode signals are fed to a stimulator, which is synchronized with a CRT and a camera.

The oscilloscope trace seen when the electromyograph is turned on but not connected to a patient, or when a patient is connected but completely relaxed and not producing discernible bioelectric potentials, is called the *baseline*. Normally, the baseline will be a relatively flat, sharp, undeflected line. *Baseline artifacts* are those deflections in the baseline caused by extraneous

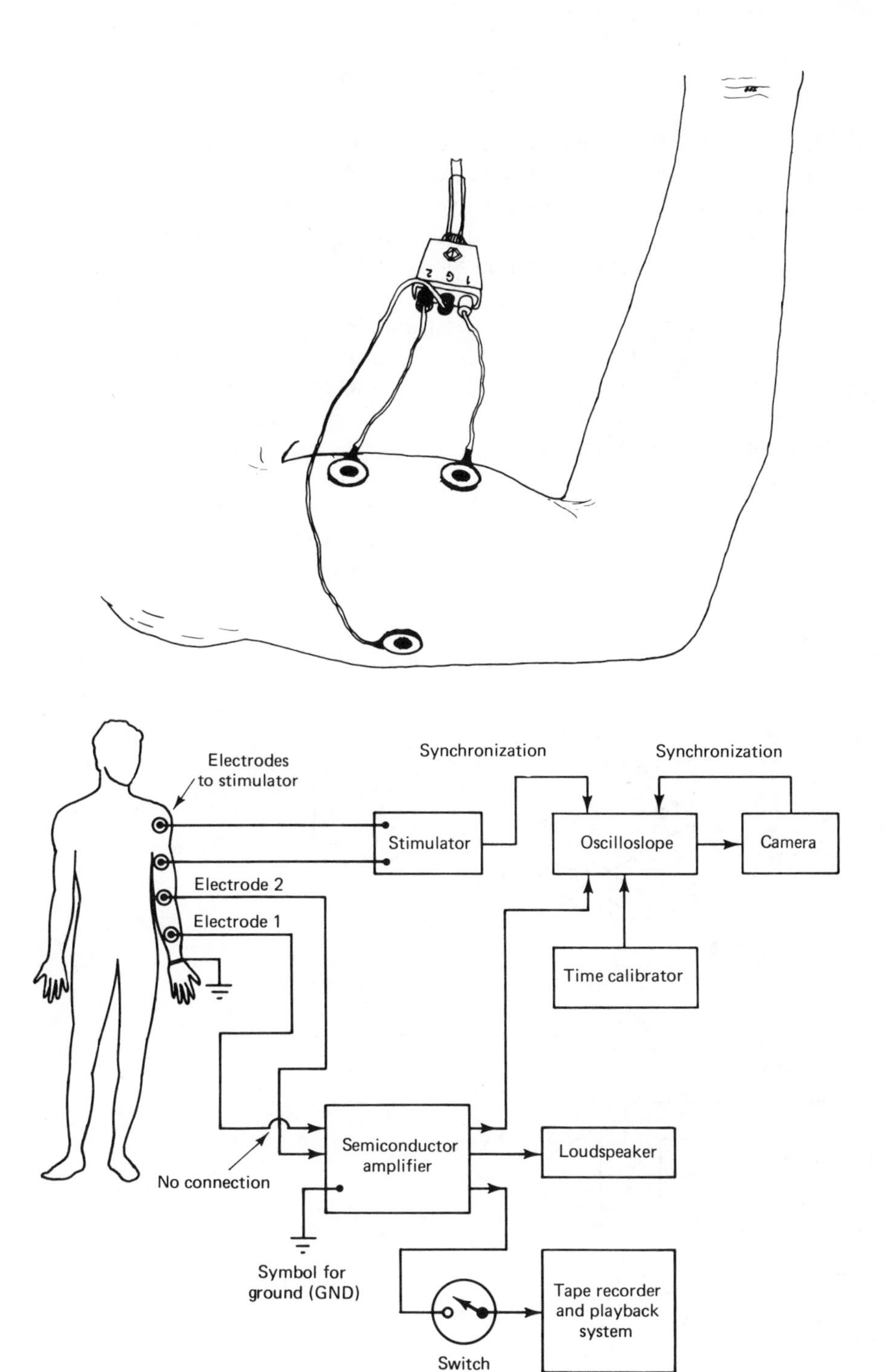

Figure 8.1 Simplified block diagram of an electromyographic device. The top pictorial side shows electrode placement for recording EMG activity from the arm biceps.

interference signals from such sources as ac writing, fluorescent lights, diathermy, and radio and television transmissions, or caused by muscle action potentials originating some distance from the active needle electrode.

A *calibration wave* is a signal whose amplitude and behavior in time are known and which can be introduced at the input of the electromyograph as a standard for calibration.

By inserting a needle in the skeletal muscle mass, an EMG biopotential output can be observed, as shown in Fig. 8.2. Surface electrodes similar to those used for ECG analysis provide an EMG with a record of the electrical activity from many muscle fibers, which are then averaged. In research, needle electrodes are frequently employed because they allow smaller regions to be observed and can be used to monitor a better defined group of muscle fibers.

An ungounded bioplar EMG lead system uses a differential amplifier input. The electrodes are located on a line perpendicular to the long axis of the limb. This reduces the possibility of interference from the cardiac cycle.

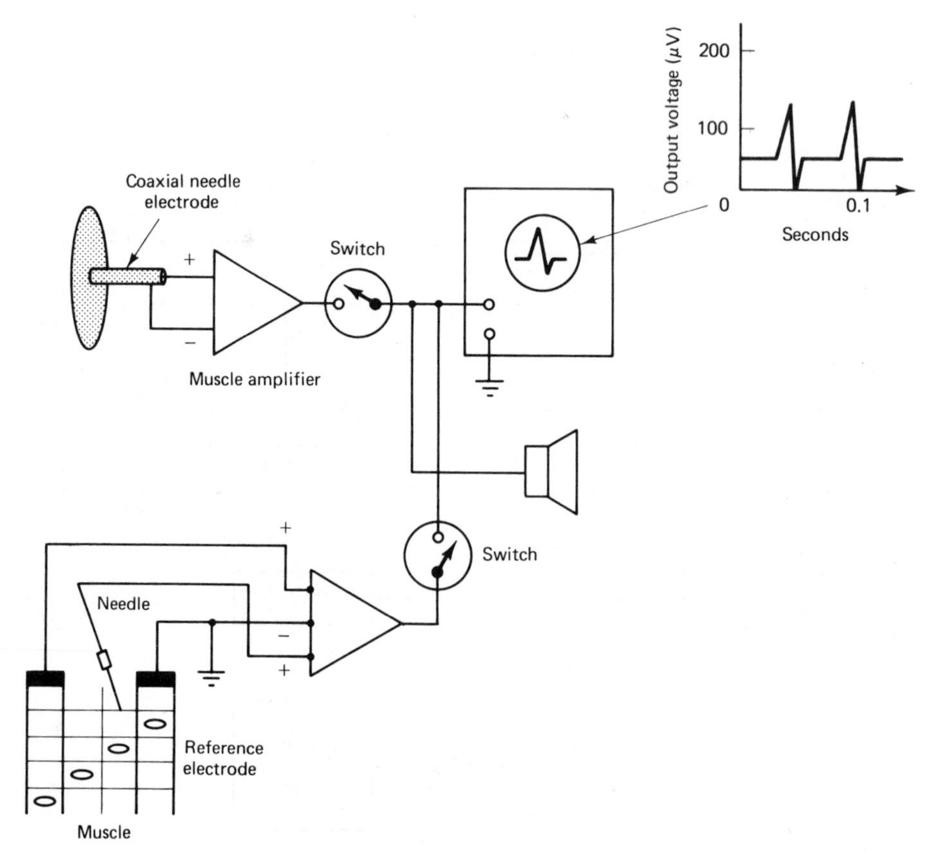

Figure 8.2 Electromyogramic output from a skeletal muscle mass.

The frequency of the EMG signals is 10 Hz to 2 kHz, with an amplitude of 20 μV to 10 mV, depending on the location of electrodes and the preparation of the skin.

The EMG signal amplifier requires a high input impedance and gain equal to that of the EEG amplifier. For good EMG studies, the frequency response of most direct-writing recorders is not sufficient; it is necessary to use an oscilloscpoe display for the 1000-Hz spikes.

8.2 PHYSIOLOGICAL RESPONSE TO MODERATE CURRENTS

Moderate currents will excite nerves and muscles. The harmful effects of this type of excitation result from uncoordinated activity of the muscles, which may cause the cessation of respiration, the jerking of arms or legs, or the clasping of objects with subsequent inability to let go. If the current goes through the brain, it can produce convulsions or damage the cardiovascular and respiratory control centers, with subsequent dysfunction of the respiratory and circulatory systems.

In Fig. 8.3 we see how a muscle cell develops a 100-mV potential. All living mamallian cells exhibit a potential difference between the inside and the outside of the cell: the outside is positive with respect to the inside. This potential is approximately one-tenth of a volt, or 100 mV.

If a microelectrode is placed within the substance of a living cell as shown in Fig. 8.3, an electrical potential difference, called the *resting potential*, will be found to exist between the interior and extracellular environments. Measurements of from 50 to 100 mV are commonly encountered, with the interior polarized negatively. This voltage is thought to be caused by differences in the concentration of certain ions in the intracellular and extracellular fluids. Especially implicated are sodium, potassium, and chloride ions, with sodium and chloride more highly concentrated in extracellular fluid and potassium in intracellular fluid. This difference in concentration appears to be maintained by selective permeability of the cell membrane to these ions, and active transport of sodium and probably potassium by the cell membrane.

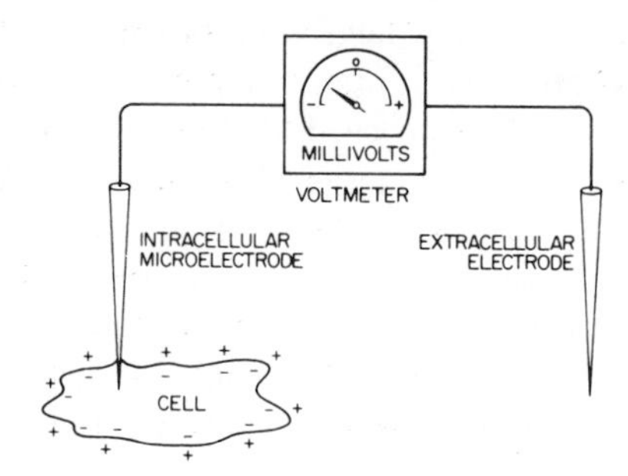

Figure 8.3 Muscle cell develops a 100-mV potential.

If such a cell is subjected to adequate stimulation (or excitation) by any of several types of change in environment, the cell membrane may undergo a sudden change in permeability behavior. This allows a change in net ion migration through the membrane and partial redistribution of ion concentrations. During the period of this change, the original potential across the membrane will be found to have been lost—in fact, it generally has reversed polarity of from 10 to 30 mV. The cell membrane promptly undergoes a recovery process in which it regains its original permeability characteristics and meneuvers the ion concentrations to their original states. As might be expected, this recovery process requires more time than the original alteration.

As this event is occurring in a portion of the cell surface, the voltage changes at the site cause migration of ions (because they carry an electrical charge) and a change in the resting potential of adjacent areas of membrane. This disturbance is normally sufficient to cause the alteration previously described to occur at this new site. In the same way, the perimeter of the involved area widens until the entire cell surface is so affected.

The significance of this disturbance and its manner of spread is most apparent in greatly elongated cells such as muscle fibers and neurons, where appreciable lengths of time are required for involvement of the entire cell surface. In these cells the disturbance is exhibited by the sarcolemma and axolemma, respectively, and is spoken of as a *propagated impulse*.

The impulse is currently thought to be the minimal unit of communication in the peripheral nervous system. It can be initiated at any point on the cell surface and can proceed in either direction with equal facility. Once started in a given direction, however, it cannot back up. This is probably due to the fact that during depolarization and the initial period of recovery, the membrane is *absolutely refractory;* that is, no stimulus of any intensity can bring about an additional impulse. Also, an additional period, called the *relatively refractory* period, follows in which stimuli of much greater than normal intensities are required to initiate another impulse. In this way, areas of membrane in the wake of a propagated impulse are protected from reexcitation. These effects do not prevent impulses from being conducted simutaneously in both directions from the stimulus site somewhere along the course of a nerve or muscle fiber.

For an electrical stimulus to be effective, its polarity must be such as to decrease the resting membrane potential, and a minimal intensity is required. If these conditions are met at a given site in a single nerve fiber (axon), a propagated impulse will proceed along its length (usually in both directions).

For any given nerve fiber and orientation of an accompanying pair of stimulating electrodes, this minimal value of stimulus voltage (or current) is spoken of as *threshold.* The term *subthreshold* is applied to stimuli of insufficient intensity to produce a propagated impulse. Increases in intensity above threshold level will produce an impulse indistinguishable from any other impulse in that fiber. In most applications it is the whole nerve that is being stimulated.

In typical peripheral nerves, fibers are encountered having conduction velocities ranging from less than 1 to more than 100 m/sec. The fibers are distributed unevenly throughout this range into several overlapping groups. Normal fibers supplying motor units in human skeletal muscles tend to have velocities in the range 50 to 90 m/sec.

Nerve cells are specialized for the conduction of information from one part of the body to another. Muscle cells are specialized for contraction, or force development. Both nerve and muscle cells have a membrane potential of approximately 0.1 V, comparable to the potential seen in the generalized cell presented in Fig. 8.4. A Figure 8.4 shows a typical nerve cell, its connection with a muscle, and the gap between the nerve and the muscle, called the *myoneural junction.*

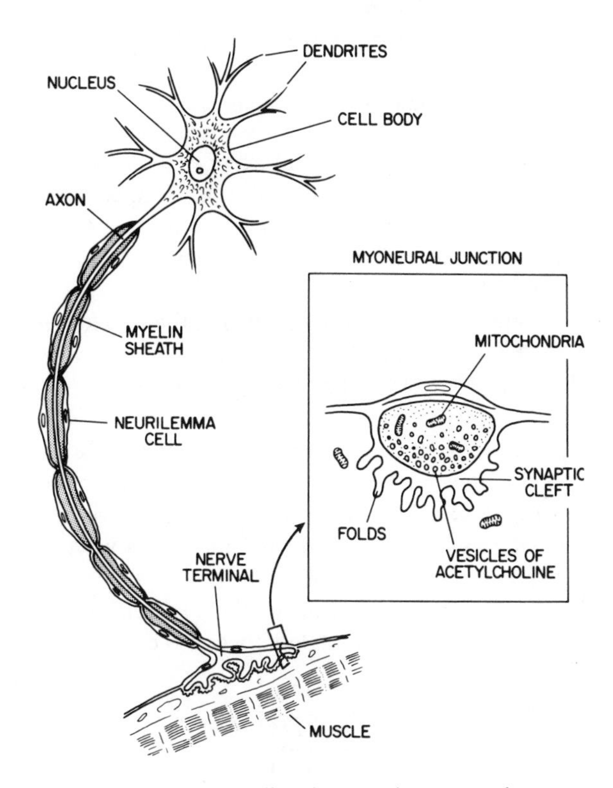

Figure 8.4 Nerve cell and connection to muscle.

In Fig. 8.5 are shown the polarities of a cell during rest, depolarization, and repolarization. Nerves are excited when the resting membrane potential is reduced to a critical level. This critical voltage level is approximately −60 mV. When the inside of the membrane changes from its resting level of −100 mV to −60 mV the membrane becomes unstable and depolarized.

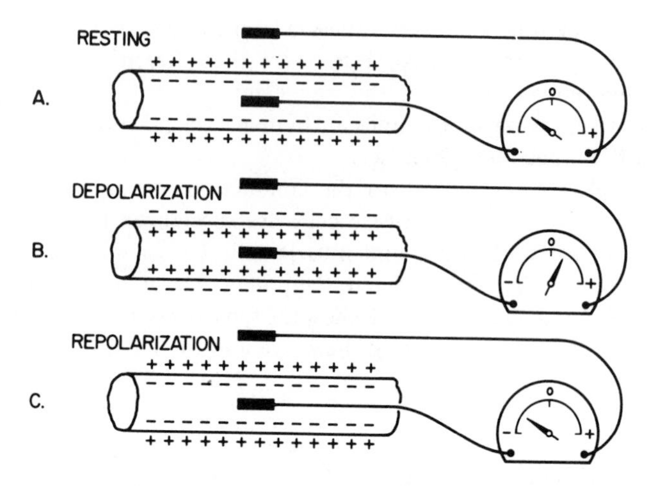

Figure 8.5 Polarities during resting, depolarization, and repolarization.

As shown in Fig. 8.5, the potential across the membrane is altered; the inside is now negative and the outside positive. Immediately thereafter, the membrane returns to its initial resting level. This sequence of events from the resting levels to depolarization and back to repolarization, called an *action potential*, is shown in Fig. 8.6.

The local changes in the membrane during the development of an action potential are propagated along the whole nerve or muscle membrane. This wave of depolarization travels along the entire nerve or muscle body until it reaches the end of the nerve or muscle. The sequence of events is started when a cathodal current is applied to the nerve or muscle membrane. At the end of the nerve, a chemical (acetylcholine) is released (see the myoneural junction region shown in Fig. 8.4). This chemical combines with the muscle membrane in the immediate vicinity of the nerve membrane, and a wave of depolarization proceeds down the entire muscle membrane.

Immediately following depolarization of the muscle membrane, the muscle contracts. The muscle can be stimulated to contract by activating the nerve

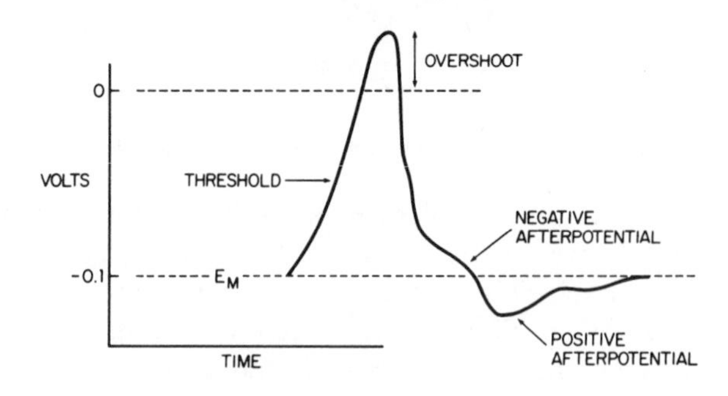

Figure 8.6 Action potential.

connected to the muscle or by stimulating the muscle directly with electrical current. The properties that define the excitation of nerve and muscle are described by the strength–duration curve, where the strength of current and the duration of time necessary to produce excitation are plotted against each other, as shown in Fig. 8.7. It is important to note that a current acting for a very short period of time cannot produce excitation. The left-hand portion of the graph shows that at small durations of stimulus, the strength current needed to excite rises to infinite levels. For this reason, the frequency of stimulation is important in terms of the physiological effectiveness of a current in producing excitation. An optimal frequency for producing excitation of nerve or muscle is between 60 and 100 Hz. Currents that are applied to the body at a very high frequency and short duration cannot be excited.

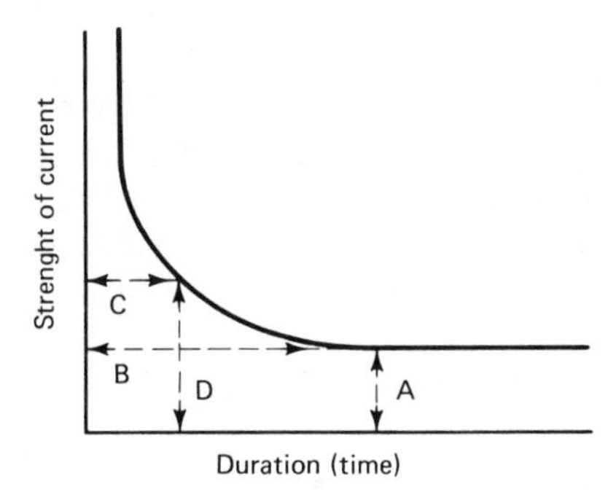

Figure 8.7 Current–time relation in muscle excitation.

8.3 ELECTROMYOGRAPHIC BIOELECTRONIC DEVICES

The electrode system commonly used in clinical electromyography (4) employs a shielded patient cable by which an active and a reference electrode are connected to the two inputs of a differential preamplifier. This cable also has a ground electrode connected to the ground input terminal or the chassis of the EMG. Most electromyographers today connect the active electrode so that the initial, positive deflection of a normal motor unit wave will appear as an upward deflection on the oscilloscope screen. In the three-electrode differentiating system, the action potentials that are observed electromyographically are the voltage differences developed between the reference electrode and the active electrode.

The active needle electrode, used to probe or explore the muscle to pick up the action potentials to be observed, is also called the probing, exploring, pickup, or recording electrode, terms that have a roughly equivalent meaning.

A reference electrode would be unnecessary in the EMG electrode system if there were no extraneous signals to interfere with the pickup and display of action potentials. In practice, a reference electrode is usually paired with the active electrode to help minimize interfering radio, television, diathermy, and 60-Hz radiations, and undesired potentials from bioelectric activity not under study in the patient.

Active electrodes may be either needles or surface (percutaneous) devices. Active surface electrodes, small metallic disks or circular cups applied with electrode cream or jelly on the skin over the belly of the muscle to be studied, are suitable only for detection and quantitation of gross muscle activity. Individual details of motor unit or fiber activity are not detectable with active surface electrodes. For that reason active needle electrodes are used by most clinical electromyographers. The principal types of active needle electrodes are as follows:

1. Monopolar (unipolar) needle electrodes consist of either beveled or atraumatic straight needles coated with insulating material except for a minute area at the tip.
2. Single coaxial (concentric) electrodes consist of a fine insulated wire threaded through and embedded in a hollow needle (22 to 27 gage). The inner conductor is beveled with the needle and becomes the active electrode, the needle itself becoming the reference or the ground electrode. The coaxial needle electrode enables the electromyographer to limit the sampling to a much smaller volume of muscle tissue than the pickup area of a monopolar electrode.

Multiple coaxial needle electrodes consist of two or more fine insulated wires, comprising active electrodes, threaded through and embedded in a hollow needle. An advantage of coaxial electrodes is the durability of their steel exterior surface. The construction of multiple coaxial needle electrode is shown in Fig. 8.8. Figure 8.9 illustrates a surface electrode placement used in EMG as well as EEG measurements.

Since muscle action potentials vary from a low of about 10 μV to a high of several millivolts, averaging 100 to 500 μV, they must be greatly amplified with minimum distortion before they can be displayed.

Differential amplification systems permit attenuation of signals other than actual voltage differences between active and reference electrodes. Since many interfering signals are of this common-mode type, the differential amplifier affords considerable improvement in the ratio of the response to the desired signal (the muscle potential) compared with the response to the interfering

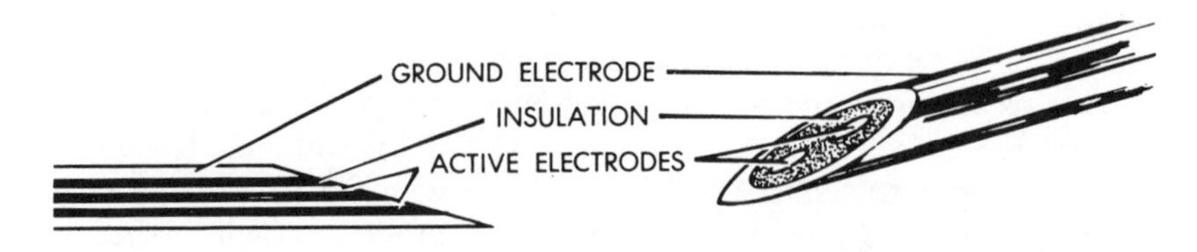

Figure 8.8 Construction of a multiple coaxial needle electrode.

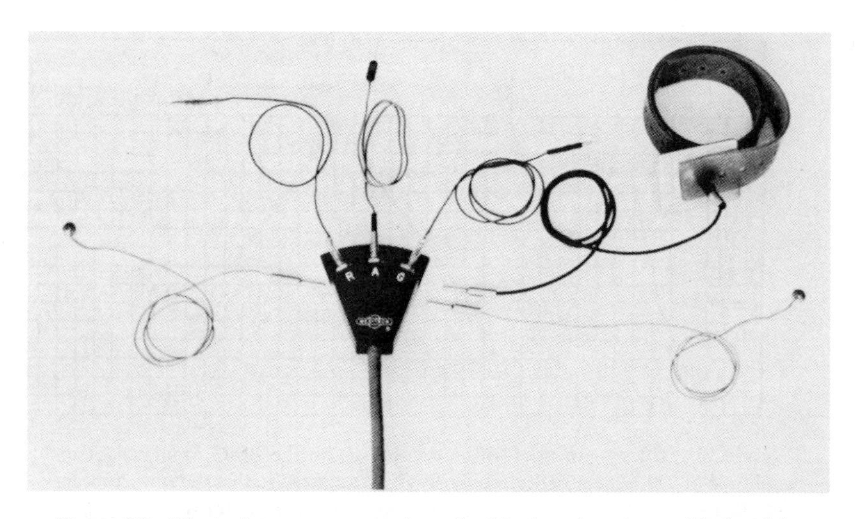

Figure 8.9 Electrodes are connected to the Marinacci patient cable head by phone tips. The active electrode shown in place is an insulated monopolar needle electrode. The $\frac{1}{2}$-in. noninsulated EEG needles, or the rectangular plate, or the 6-mm perforated disks, may be used for ground and reference electrodes. The surface electrodes may be held in place by a perforated rubber strap, or by adhesive, or cellophane tape.

extraneous common-mode signals. In the better quality amplifiers, the desired signal is amplified as much as 100,000 times the common-mode signal. Such bioelectronic systems are used for many input systems.

Calibration is accomplished by introducing into the oscilloscope circuit a known signal suitable for comparison with the bioelectric activity to be measured. Most calibration signals consist of a periodic changing voltage (usually a sine or a square wave) of known amplitude. By means of a sensitivity control, the calibration wave is adjusted to a vertical deflection distance convenient as a standard of comparison. If, for example, a 200-μV signal is adjusted to occupy 1 in. vertically on the oscilloscope screen, the deflection sensitivity is said to be set at 200 μV/in. Or if a 1000-μV signal is caused to occupy a vertical distance of 10 cm, the deflection is said to be 100 mμV/cm.

Most commercially manufactured clinical electromyographs have built into them one or more precalibrated horizontal or time axes, referred to as *sweep times*. However, any sweep source can be calibrated by applying a periodic changing voltage in which the intervals between events or cycles are known. This can be the same signal as that used for vertical calibration. As an example, if a 60-Hz sine wave is employed, the interval between cycles will be 16.7 msec, and if two succeeding cycles are caused to be separated by 1.67 in., the sweep will be known to be traveling at 10 msec/in., as shown in Fig. 8.10.

Many bioelectronic measuring systems are versatile. Eight physiological modules can also be used. One channel monitors EEG, two channels are used for respiration, two channels are used for ECG studies, one channel can moni-

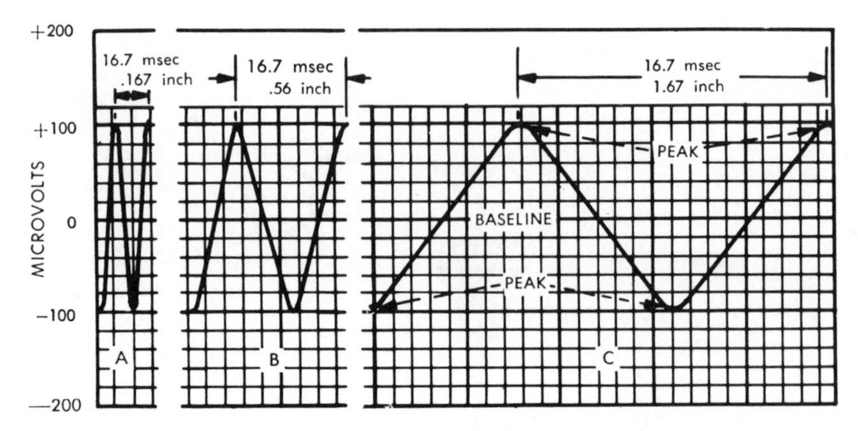

Figure 8.10 60-Hz wave used for calibration. With the EMG input set for an amplitude of 200 μV, the deflection sensitivity is adjusted so that the wave appears 1 in. high peak to peak. (a) At a sweep of 100 msec/in., a single cycle of the wave covers 0.167 in. (b) At 30 msec/in., a single cycle covers 0.556 in. (c) At 10 msec/in., it confers 1.67 in. Regardless of sweep-speed setting, a full cycle takes 16.7 msec.

tor the EMG, and two channels can monitor eye movements. Measures of eye movement, called *electro-oculograms* (EOG), are used in studying sleep patterns, especially REM sleep, in which rapid eye movement lasts for 5 to 30 minutes.

8.4 ELECTROMYOGRAPHIC EXAMINATION USING BIOELECTRONIC MEASUREMENTS

EMG electrodes must be kept in perfect condition for successful use. Electrode jelly must not be allowed to cake on surface electrodes. The points of needle electrodes must be kept sharp and clean and the insulation intact. Coaxial needle electrodes can be sharpened with a whetstone just like hypodermic needles. However, greater care is required to prevent bits of metal from becoming embedded in the insulation between the inner conducting wire and the cannula of the needle, which could cause the electrode to short. Monopolar needle electrodes are inexpensive and are expendable. They should be checked frequently and discarded if the insulation is found to be ineffective or if the insulation has receded from the needle tip as much as 1 mm. Even tiny leaks in the insulation cause the needle to become "noisy," producing baseline disturbances.

Mere visual inspection will not reveal minute defects in needle electrode insulation. Accurate checking can be done only by use of a sensitive ohmmeter. One ohmmeter lead is connected to the needle electrode to be tested, and a small cotton swab is fixed to the end of the second ohmmeter lead. Soak the swab in saline solution remove the excess moisture, and rub the swab along and on all sides of the insulated shank of the needle electrode being tested. If the insulation is broken, the ohmmeter will show a flow of current. This test will also detect leaks between the inner conductor and the outer shaft of coaxial needle electrodes.

The active needle electrode should be sterilized by autoclaving (if the electrode can withstand it) or by immersing in Zephiran chloride 1 : 1000 aqueous solution for about 20 minutes. If subcutaneous needles instead of surface plates are to be used for reference and ground electrodes, they should be similarly sterilized. Before inserting needle electrodes, the skin should be thoroughly cleansed with Zephiran chloride 1 : 1000 or alcohol.

Attach the patient cable to the EMG input socket. Locate the muscle to be examined. Place the reference electrode near the site selected for exploration and over the same muscle whenever practicable, to minimize possible detection of motor unit activity in adjacent muscles. The ground electrode should be conveniently placed, usually some inches from the reference electrode. If surface electrodes are used for reference and ground, lift up the plates and massage the skin underneath very thoroughly with a generous amount of electrode jelly. Replace the electrodes and make sure that they are strapped on tightly enough to remain firmly in place throughout the examination. With all electrodes connected to the patient cable attached to the EMG, the electromyographer is ready to insert the active needle electrode and begin the examination. Surface electrodes such as those used in EEG examination are also frequently used for the EMG examination.

Almost all of the skeletal muscles can be examined electromyographically if the examiner has an adequate knowledge of anatomy and kinesiology, a skillful examining technique, careful training in electromyographic interpretation, and well-designed equipment in good operating condition. Knowledge of kinesiology is important, because only by proper testing of the muscles with the exploring needle electrode in place can the examiner determine which muscle he or she is examining. Not only should each individual muscle be adequately examined, but a minimum number of muscles should routinely be examined. The extensiveness of the examination will be determined by the examiner's awareness of the medical problem and knowledge of the peripheral neuroanatomy of the area.

Most high-frequency extraneous interference signals are easily recognizable as artifacts on the oscilloscope screen and are dealt with by properly grounding the patient and instrument, and by eliminating. filtering, or screening out the extraneous signal source. Artifacts from electrocardiographic signals are low in frequency and seldom cause any confusion, requiring merely to be recognized. If these should be of troublesome magnitude, however, they may be minimized by more compact electrode placement or by the use of coaxial needle electrodes.

Baseline disturbances due to poor electrical contacts, improperly applied skin electrodes, or a defective active needle electrode sometimes resemble electromyographic activity, although the auditory components of these disturbances are usually less confusing than the visual components.

Figure 8.11 shows a complex motor unit potential for a patient with musclar dystophy. The complex motor unit potentials are produced when a myopathic process leads to diffuse random destruction of muscle fibers. Myopathic motor units tend to have a smaller than normal number of muscle fibers. The degree of asynchronism tends to be unaffected, however, and a shift toward polyphasic forms occurs.

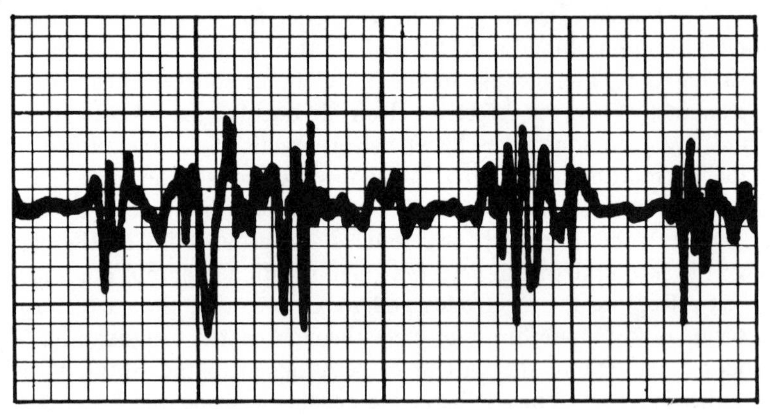

Figure 8.11 Complex motor unit potentials—myopathic. Recorded from a patient with muscular dystrophy. (Courtesy of Robert B. Pearson, M.D.)

8.5 BIOFEEDBACK ELECTROMYOGRAPHY

Biofeedback as used in EMG constitutes a sophisticated solid-state electronic device that accurately detects minute changes in localized motor nerve/muscle fiber activity and feeds them back in modes excellently suited for biofeedback electromyographic training.

The modular EMG instrument is ideal for research and clinical applications of electromyographic feedback. Possible applications include the control of general strees reactions through deep muscle relaxation, systematic desensitization to phobias and other emotional problems, muscle rehabilitation, muscle tension reduction in localized areas, the detection of subvocalization, and general monitoring and recording of muscle tone for diagnosis. Specifications for commercial a biofeedback EMG[1] follow:

1. *Amplifier:*
   (a) Common-mode rejection: over 100 dB
   (b) Additional filtering: over 50 dB (at 60 Hz)

[1]Biofeedback Instrument Co., Model P-775.

(c) Input impedance: over 1 MΩ
(d) System noise level: 0.08 μV rms or 0.21 μV peak to peak
(e) Sensitivity: 0.1 μV peak to peak
(f) Input range: 0.2 μV peak to peak to 500 μV peak to peak
(g) Input current: 1 μA
(h) Filters: 3-dB points: 65 and 360 Hz: 12 dB/octave

2. *Outputs:* Strip chart recorder outputs: rectified EMG signal (full-scale meter reading = 1.5 V output), filtered and raw EMG signals (0 to 0.3 V peak to peak). External speaker jack (headphone jack).
3. *Meter scale ranges:* 0 to 5 μV, 0 to 10 μV, 0 to 20 μV, 0 to 50 μV, 0 to 100 μV, 0 to 200 μV, 0 to 500 μV (seven scale multiplier factors).
4. *Electrodes:* Silver–silver chloride electrodes (ideal for placement anywhere on the body).

Using muscle electromyographic feedback, patients with digestive-system disturbance, depression, eczema, and neurodermatities have been treated. Muscle feedback has also been used successfully for systematic densensitization in the treament of dysponetic and phobic responses.

Electromyographic feedback has also been used to induce a specific relaxation response in the treatment of migraine tension headaches, ulcerative proctitis, and spastic colon.

Muscle feedback is also a potent tool in rehabilitating patients with neuromuscular problems, especially mild paralysis of spasticity. EMG feedback is used in cases of spasticity to let patients "hear" muscle spasms and thereby learn to decrease them through decreasing the feedback tone.

8.6 BIOELECTRONIC DEVICE TO RELIEVE PAIN

Headache victims and sufferers of chronic muscle tension use bioconductive therapy to measure points of low bioconductivity in the nerves of problem areas of patients.

This bioelectronic device works as a bioelectric receptor—finding areas of low energy and applying microcurrents until a proper conductive level in the area is achieved. Proper conductivity appears to relieve pain.

Bioconductive therapy is painless (no needles penetrate the skin) and drugless.

8.7 ELECTROTHERAPY

The concept of electrotherapy in the treatment of pain dates back more than 2000 years. The early Greeks and Romans used energy from electrical fish to treat headache and gout.

Centuries later, electrical stimulation was rediscovered by physicians, who built bulky devices to generate electricity and transmit it to patients. Electrotherapy became so popular that it succumbed to the cult of personality. Benjamin Franklin's medical experiments included electrifying patients suffering from paralysis and epilepsy.

Since 1965, when Melzack and Wall proposed the gate control theory of pain, electrical stimulation has emerged as a new therapeutic agent. According to the theory, electrical stimulation can close a "gate" in the spinal cord, blocking transmission of pain signals to the brain.

Today, many types of pain can be relieved by electrotherapy. Electrodes can be applied around the spinal cord, placed over specific nerves on the skin, or surgically implanted within the body. Applications cover all types of acute and chronic pain, ranging from common athletic knee injuries to phantom limb (persistent pain in the amputated limb).

8.8 A WIDEBAND EMG TELEMETRY SYSTEM[2]

One useful technique for determining the orthopedic corrections required by people suffering from cerebral palsy is to monitor their gaits electronically. In this method, electrodes are placed on six opposing muscle groups on each leg, and pressure-triggered step-indicator switches are mounted on both shoe soles. These electromyographic (EMG) and pressure sensors are monitored as the person walks; the information obtained, in an EMG bandwidth of 20 to 2000 Hz, is used to correlate the activity of the leg muscles with each phase of walking.

Telemetering the data has the advantage of eliminating cables between the subject and the recording equipment. Ordinarily, several data channels would be transmitted by using a multiplexed data transmitter and a single radio-frequency (RF) link. However, because of the EMG bandwidth requirement of 2000 Hz and the need for 12 channels, a multiplexed system would be difficult to implement in a low-powered device having a reasonable transmitting range.

A new, wideband EMG telemetry system, developed for this application, consists of miniature, individual, crystal-controlled RF transmitters located at each electrode site. The transmitters are assigned operating frequencies within the 174- to 216-MHz band. They are small enough to be placed at the sensor sites, yet have a linear frequency response from 20 to 2000 Hz and an operating range of 15 m. The receivers are crystal-controlled, so that there is no ambiguity in identifying the channels. The amplifiers have an input impedance of 10 MΩ and are usd with two sensing electrodes per channel.

The use of individual transmitters for each data channel has advantages. By eliminating the need for cables between the electrode sites and a master

[2]Section 8.8 is based on work by S. A. Rositano and R. M. Westbrook of Ames Research Center, Moffett Field, Calif. From *NASA Tech Briefs*, p. 394, Fall 1978.

transmitter, maximum mobility is allowed, preparation time is reduced, and subject comfort is improved. Data channels can be added or subtracted easily.

8.9 COMPUTERS AND THE ELECTROMYOGRAM

Many more sophisticated techniques of data analysis have recently been applied to the EMG because of the availability of computers. Many of these techniques depend on the time history of the EMG record. Autocorrelation function analysis is a time-dependent parameter but has not shown significant information relative to a muscle's behavior.

To extract information concerning the specific abnormality within the neuromuscular system that is of interest to the investigator, a combination of many of the techniques described above may be necessary. Frequently, the various processes described for analyzing EMG signals can be implemented by hardware, software, or a combination. Currently available computers are used.

8.10 REVIEW QUESTIONS

1. Define electromyography and myography.
2. List five electrodes used in electromyographic work.
3. Discuss the meaning of baseline and calibration curve.
4. Discuss the meaning of action potential.
5. Discuss how an electromyographic examination is made.
6. Discuss the purpose of a stimulator.
7. Discuss the purpose of any biofeedback EMG electronic device.
8. Briefly discuss electrotherapy.
9. Discuss the uses of an EMG telemetry system.
10. Discuss the uses of computers in extracting useful EMG information.
11. Draw a block diagram of a EMG system and discuss the function of each component.

8.11 REFERENCES

1. Basemajian, J. V.: *Muscles Alive: Their Function Revealed by Electromyography*, 3rd ed., The Williams & Wilkins Company, Baltimore, Md., 1974.
2. Basemajian, J. V., Clifford, H. D., McLeod, W. D., and Nunnally, H. N.: *Computers in Electromyography*, Butterworth & Company (Publishers) Ltd., London, 1975.
3. Cenkovich, F. S., and Gersten, J. S.: Fourier Analysis of the Normal Human Electromyogram, *American Journal of Physical Medicine*, Vol. 42, pp. 192–204, 1963.

4. Pearson, R. B.: *Handbook of Clinical Electromyography*, Meditron Co., El Monte, Calif., 1961. Robert B. Pearson, M.D., is from the Neurographic Testing Medical Clinic, Inc., Downey, Calif.
5. Jacobsen, B., and Webster, J. G.: *Medicine and Clinical Engineering*, Prentice-Hall, Inc., Englewood Cliffs, N.J., 1979.
6. Cohen, B. A.: Review of Acquisition and Analysis of the Electromyogram, *Journal of Clinical Engineering*, pp. 142–148, April–June 1977.
7. Strong, P.: *Biophysical Measurements*, 1st ed., 3rd printing, Tektronix, Inc., Beaverton, Oreg., 1973.

9

RESPIRATORY MEASUREMENTS

9.1 INTRODUCTION

Respiration is essentially the gas exchange in any environment. In human beings the respiratory cycle contains very complex functions. The human body takes in oxygen and gives off carbon dioxide by means of breathing, called *external respiration*. This process actually consists of *inspiration* (breathing air into the lungs) and *expiration* (expelling air from the lungs). The inspired air passes through the nasal chambers, larynx, trachea, right and left bronchi of the lungs, and finally the lung air sacs, called the *alveoli*. The red blood cells, by the diffusion process, takes in oxygen through the alveoli and discharge carbon dioxide back through the alveoli. The expired carbon dioxide returns via the same passageways to the outside environment.

The inspiration force is created primarily by the *diaphragm*, a dome-shaped muscle separating the chest from the abdominal wall and the chest wall. The diaphragm contracts, which increases the size of the thorax (the chest cavity) and reduces the air pressure in the lungs.

The circulating blood gives off oxygen to the tissues and takes carbon dioxide from them. The exchange of gases between the blood and body cells is defined as *internal respiration*.

The two human lungs are not identical in size. The heart and pericardium bulge more into the left side of the thorax. The part of the dome of diaphragm muscle covering the right aspect of the liver is higher than the left side. This results in the right lung being shorter, broader, and slightly larger than the left lung. Both lungs are divided into lobes by fissures lined with pulmonary pleura penetrating into the actual lung substance. There are thus lower lobes, a middle lobe, and an upper lobe. The root of the lung is in the center of the medical surface; here the bronchi, arteries, and veins enter and leave the lung.

The respiratory system is shown in Fig. 9.1. The rate of respiration is controlled by the medulla in the brain; an increase of carbon dioxide in the blood stimulates the medulla. The respiratory system (Fig. 9.1) consists of the following:

1. Nose.
2. Pharynx-epiglottis, which closes the air passages when food passes through.
3. Trachea or windpipe, including the larynx at the upper end, which vibrates and produces sound.
4. Bronchi: two, one in each lung.
5. Lungs: bronchi subdivided into smaller bronchioles, which end in air sacs whose walls are lined with capillaries. Carbon dioxide and water vapor diffuse the other way. The blood carries the oxygen to all cells of the body and picks up carbon dioxide and water vapor from the cells.

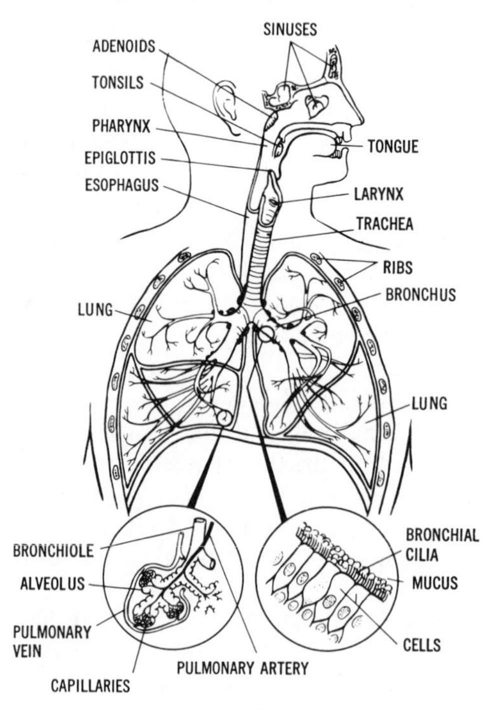

Figure 9.1 Respiratory system. (Courtesy of the American Lung Association, New York, N.Y.)

9.2 MECHANICS OF BREATHING

The adult inhales about 500 cubic centimeters of air with each breath and takes about an average of 15 breaths/min. During moderate exercise a person can inhale as much as 1 liter or more of air with each breath and can take up to 25 breaths/min. At maximum exercise, the lung vital capacity can reach 4 to 5 liters.

In the physiological system shown in Fig. 9.2, the piston corresponds to the diaphragm and chest muscles, the cylinder to the chest wall, and the balloons to the lungs. When the piston moves down, P_2 becomes negative, which causes P_1 to become negative with respect to P_3 (but positive with respect to P_2). If the airway is open, air will flow inward from P_3 to P_1 until $P_1 = P_3$. When the piston moves upward, P_2 becomes positive and acts on P_1, which is then positive with respect to P_3. Air flows from P_2 to P_3 until $P_2 = P_3$.

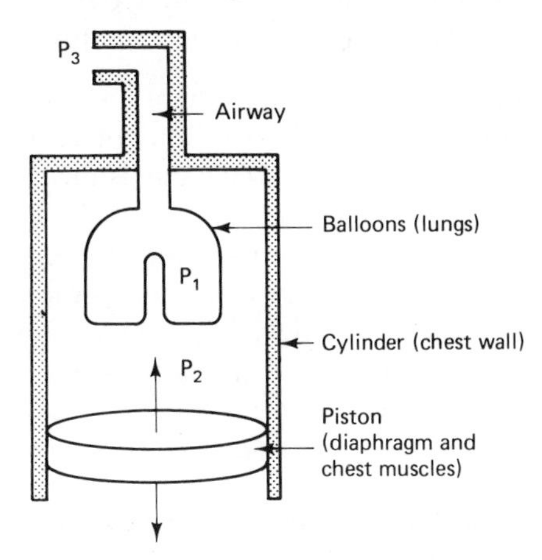

Figure 9.2 Mechanical equivalent drawing of respiratory ventilation.

9.3 LUNG VOLUME AND FLOW RATES

Spirometry is essentially a method for recording the volumes and flow rates of gases entering and leaving the lungs. The recording or spirogram, which is made as a function of time, is used in clinical evaluation of pulmonary functions. The spirogram is to pulmonary diagnosis as the ECG is to cardiovascular diagnosis.

As lung volumes and flow rates are obtained, they must be expressed properly. Gases existing in the lung are at body temperature and ambient barometric pressure; water vapor is present at its saturation pressure. Gas volumes measured under these conditions are expressed as milliliters of body temperature pressure standard (BTPS) and flow rates are measured as ml/min BTPS. When an individual breathes into a spirometer, the gases filling the lungs enter a different environment, specifically one having a lower temperature (usually). The barometric pressure will be unchanged, but the saturated water vapor pressure (due to the water environment inside the spirometer) will be of a different magnitude, due to the altered temperature. These conditions are symbolized as ATPS (ambient). To convert ATPS (spirometer) volumes into BTPS (body) volumes, the following formula is used:

$$V_{\mathrm{BTPS}} = V_{\mathrm{ATPS}}\left(\frac{273 + 37}{273 + t_A}\right)\frac{P_B - P_{\mathrm{H_2O}}}{P_B - 47} \tag{9-1}$$

where t_A = ambient (spirometer) temperature (°C)
P_B = barometric pressure (torr)
$P_{\mathrm{H_2O}}$ = saturated water vapor pressure at t_A
47 = saturated water vapor pressure at body temperature (torr)
37 = body temperature (°C)
273 = absolute scale temperature for the freezing point of water

It is also rather important in some instances to express gas volumes according to standard conditions of temperature and pressure, dry (STPD). This is particularly true when one is measuring volumes of oxygen and carbon dioxide inspired or expired. It is useful to note the molecular effects of these gases in the body. By expressing these volumes at STPD, one can calculate molecular numbers very easily using the relationship that 1 mole of an ideal gas occupies a volume of 22.4 liters. To convert ATPS volumes to STPD volumes, the following formula is used:

$$V_{\mathrm{STPD}} = V_{\mathrm{ATPS}}\left(\frac{273}{273 + t_A}\right)\frac{P_B - P_{\mathrm{H_2O}}}{760} \tag{9-2}$$

Static lung volumes are measures of the quantity of air that can be inspired or expired at various depths of breathing. Vital capacity is often measured as one of the best indicators in this regard, since it indicates the greatest volume of gas that can be voluntarily displaced in and out of the lungs. This volume depends on the physical dimensions of the chest cage, the strength of the respiratory muscles, and the distensibility of the lungs and chest cage.

Flow rates involve measurement of lung volumes as a function of time. Tests measuring these rates are valuable in characterizing situations of increased airway resistance to gas flow or deformative changes in lung tissues; typical disease states include asthma, chronic bronchitis, tuberculosis, and emphysema. These defects fall under the category of obstructive pulmonary disease, where pressures required to achieve flow are increased much above normal.

Nurses and technicians may wish to make spirometer measurements in the pulmonary function lab. The spirometer is used to measure the patient's breathing volume and capacity. In each instance the nurse or technician must be certain that the nose clips and mouthpiece are secure, to prevent leaks of air and thus produce inaccurate readings.

Figure 9.3 is a lung volume and capacity diagram. Measurements in the respiratory intensive care unit may include the following:

1. *Tidal volume* (TV) is the volume of air inspired or expired during ventilation. Normal tidal volume values range from 400 to 600 ml.
2. *Minute ventilation* (MV) is the sum of tidal volume in 1 minute. Normal minute ventilation range from 6 to 8 liters/min.
3. *Inspiratory capacity* (IC) is the maximum volume of air that can be inspired after a normal expiration.
4. *Inspiratory reserve volume* (IRV, or IC − TV) is the maximum volume of air that can be inspired after a normal inspiration.
5. *Expiratory reserve volume* (ERV, or EC − TV) is the maximum volume of air that can be expired after a normal expiration.
6. *Functional expiratory volume* (FEV) is the volume of air present in the lungs following a normal expiration.
7. *Residual volume* (RV) is the volume of air in the lungs when no more air can be exhaled; the lungs do not collapse to an airless state with maximal expiration and cannot be emptied of gas by even the most forceful expiration.
8. *Vital capacity* (VC) is the maximum volume of air that can be expired at a maximal inspiration. Normal vital capacity ranges from 3 to 5 liters.

To predict the vital capacity, the patient's vital capacity should be at least 80% of that predicted for that patient.

Forced efforts are not normally measured in the intensive care unit. The other pulmonary excursions are parameters that measure lung functions and ventilation in hospital examinations, to discover respiratory disorders:

1. *Total lung capacity* (TLC) is measured by having the patient inhale as deeply as possible, then forcefully exhale into the spirometer. Normal vital

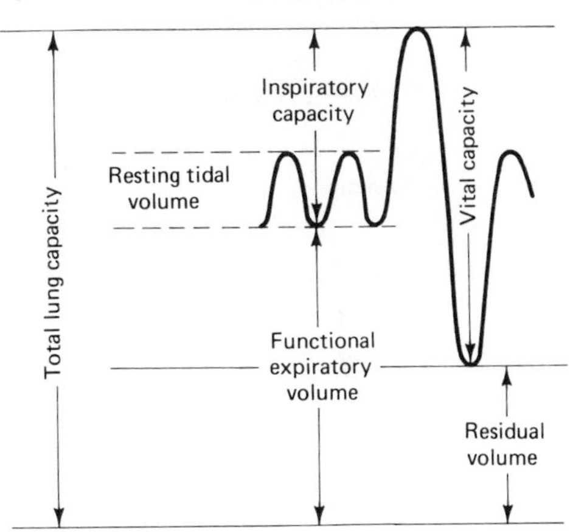

Figure 9.3 Lung volume and capacity diagram.

capacity is 4000 to 5000 liters/min. This figure is markedly decreased when respiratory disorders are present.

2. *Forced expiratory volume* (FEV) is measured on a timed sequence (1–2–3 sec) by having the patient expire forcefully air previously maximally inspired. The normal FEV is 83% of the forced vital capacity in 1 sec, 94% in 2 sec, and 97% in 3 sec. Patients with respiratory disorders require as much as three and four times this amount of time to exhale the total FVC volume. Note that to determine FEV, the patient inhales as deeply as possible and exhales as rapidly as possible.
3. *Maximum voluntary ventilation* (MVV) measures the maximum amount of air a patient can breathe for 15 sec. The minute volume is the maximum amount of air that one breathes in and exhales in 15 sec. The normal MVV is approximately 125 to 150 liters/min for men and 100 liters/min for women. The MVV is decreased in respiratory disorders. Volumes and capacities are approximately 20 to 25% less in women than in men.
4. *Forced expiratory volume* (*timed*) (FEV_t) is the volume of air exhaled in the specified time during the performance of forced vital capacity. FEV_t is expressed in liters per second. The norm is approximately 1800 liters/min.

Table 9.1 shows average values for lung volumes and flow rates for adults expressed in BTPS (mean).

TABLE 9.1 Lung Volume Data and Flow Rates

Function to Be Measured	Average Volume (ml)
Tidal volume	500
Inspiratory capacity	3000
Inspiratory reserve	3050
Expiratory capacity	1750
Expiratory reserve	1250
Vital capacity	5500
Functional expiratory capacity	2750
Residual volume	1500
Total lung capacity	5500

Function to Be Measured	Mean Flow Rate (BTPS)
Respiratory rate	12 per min
RMV	6000 ml/min
MTV	500 liters
MVV	120 liters/min
VC	4000 ml
$FEV_{1/2\%}$	72 liters
$FEV_{1\%}$	85 liters
$FEV_{2\%}$	90 liters
$FEV_{3\%}$	95 liters

Bioelectronic devices used to detect respiratory activity are referred to as *pneuographs* and the bioelectronic recording produced is a pneuogram, that, is an indicator that the patient is breathing. The breathing or respiratory rate can be obtained by spirogram, pneumogram, or a thermistor probe placed in the respiratory airstream near the nostril.

9.4 PRESENT TRENDS IN BIOELECTRONIC SPIROMETRIC MEASURING DEVICES[1]

The number of spirometric measurements performed has greatly increased in recent years, and it is likely that this trend will continue in the future. The desirability of detecting respiratory disease in its early stages has created a need for small, simple-to-use spirometric measurement equipment.

Early spirometric measurement equipment used a direct mechanical response to the flow of respiratory air. Two methods used to measure accumulated flow or volume are shown in Fig. 9.4.

Even with careful counterbalancing, the inertia of the mechanical containment system requires a pressure increase to accelerate the mechanical system in response to flow-rate changes. Some equipment in common use has sufficient inertia to result in a significant reduction in the peak expiratory flow rate (PEF) or the mean forced expiratory flow between 200 and 1200 ml of the forced vital capacity ($FEF_{200\text{-}1200}$) for healthy subjects, in spite of the effect that the resulting back pressure may have on the subject's effort. There is also a small amount of drag, proportional to flow rate or mechanical velocity, which results in back pressure. For the water-seal spirometer, this drag results from

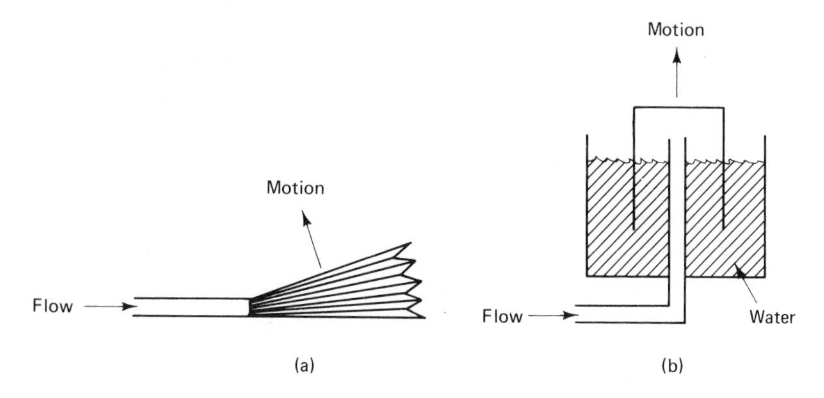

Figure 9.4 Cross section of typical volume-activated spirometers: (a) bellows; (b) water seal.

[1]Section 9.4 appears here courtesy of LSE Corp., Woburn, Mass. Reprinted in part from Rodney F. Edwards, Jr., *Evolution of Improved Capabilities in Spirometric Test Equipment from CVP*, Barrington Publications, Inc., Los Angeles, Calif., March–April 1978.

the relative motion of the tank through the water seal, and for the bellows it is the force required to flex the bellows to a new shape (but excluding the force required to hold a given shape).

Another problem in making dynamic measurements is the proper correction for gas temperature. It is usually assumed that the gas in the container is at ATPS conditions and is corrected to BTPS, since the gas exhaled from the lung (and thus lung volume) is at BTPS. The BTPS gas exhaled at the mouth mixes with ambient gas already in the spirometer. With good mixing and good thermal contact to ambient conditions, the forced vital capacity (FVC) measurement may be close to ATPS conditions. However, the gas in the early part of the exhalation will not immediately reach ATPS conditions, especially since a rubber or plastic tube with low thermal conductivity is used to conduct the gas to the instrument. This lack of immediate cooling results in increasing the indicated volumes, with the greatest percentage increase at the beginning of the test.

Several approaches have been taken in the design of electronic flow sensors, including thermal (both hot-wire and thermistor), pneumotach–pressure transducer combinations, rotating vane or turbine, and ultrasonic. Since the performance of these sensors depends very much on just how they are constructed and on the interface electronic circuitry, complete performance generalizations are not possible. Some common characteristics are identifiable, however, the most important being the need to detect and measure (with good accuracy) flow rates over a range of about 500 to 1, or 12 liters/sec to 25 cc/sec. This is too wide a range for many flow-sensitve phenomena, such as the vortex flow meter or nonlinear pneumotachs, whose pressure increases as the square of gas velocity (thus requiring a 250,000 : 1 operating range in the pressure transducer). Fast response time and low resistance to flow are also required.

Thermal sensors must be very small to achieve sufficiently fast response; thus they tend to be somewhat fragile. Flow is determined by the amount of cooling that results at the heated thermal element. The direction of flow cannot be determined with simple sensors, but flow can be directed by using an upstream–downstream pair of sensors and by correlating the signals. The resistance to flow can be very small.

The best pneumotach–pressure transducer combination uses a pneumotach with a linear flow-pressure relationship, because of the dynamic range problem, and a linear pressure transducer, so that electronic linearization is not required. Common pneumotachs of this type are the mesh screen or a group of parallel small tubes (the laminar-flow element).

Rotating-vane flow sensors must be carefully made so as to rotate at low flow rates while keeping the pressure drop at high flow rates reasonably small. However, a freely rotating turbine does not stop immediately when flow stops and may not come up to speed fast enough to follow accurately a forced exhalation from a health subject. Optical means are used to sense rotation, and the flow rate is represented by a pulse frequency. This may be advantageous when used with digital processing, but requires additional interfacing to be compatible with analog recorders.

The ultrasonic transducer places no obstruction in the flow path, but is more complex to implement. Since the transducer is crucial to the overall performance of the instrument, it is almost certain that further choices, variations, and improvements will evolve in the years to come.

A typical block diagram is shown in Fig. 9.5 for the analog processing required to generate spirometric values from a flow transducer output. A key element in this processing is the integrator, which must have very low drift to keep errors in forced vital capacity within acceptable limits.

The measured values must be stored until they can be displayed and read out. For an analog system this is usually done with capacitors, and the stored value will change slowly with time. To minimize this error, it is thus important to read out measured data promptly after the test is run.

The peak flow value is also stored on a capacitor, which is allowed to charge to the peak transducer voltage achieved, but is not permitted to discharge from this value. The electrical time constant selected can be much smaller than that achievable with mechanical systems.

A digital processing system is a natural choice when the sensor output is in digital form, as with the rotating-vane types. For a typical digital system, as shown in Fig. 9.5, the only calibrations required are for voltage-to-frequency conversion, clock frequency, peak expiratory flow rate (PEF), and digital-to-analog converter if an analog recorder output is provided. Also, measured values can be held indefinitely in digital form so that no additional error will accumulate between running the test and the readout of measurements.

9.5 AUTOMATED PULMONARY FUNCTION MEASUREMENTS[2]

Three basic types of measurements are made in the pulmonary clinic: ventilation, distribution, and diffusion. *Ventilation* deals with measurement of the body as an air pump, determining its ability to move volumes of air and the speed with which it moves the air. *Distribution* measurements provide an indication of where gas flows in the lungs and whether or not disease has closed some sections to airflow. *Diffusion* measurements test the lung's ability to exchange gas with the circulatory system.

The most widely performed measurement is ventilation. Historically, this has been performed using devices called *spirometers*, which measure volume displacement and the amount of gas moved in a specific time. Usually, this requires the patient to take a deep breath and then exhale as rapidly and completely as possible. Called the *forced vital capacity*, this gives an indication of how much air can be moved by the lungs and how freely this air flows.

[2]Courtesy of Maurice R. Blais of LSE Corp., Woburn, Mass. and Hewlett-Packard Co. Reprinted in part from Maurice R. Blais and John L. Fanton, Automated Pulmonary Function Measurements, *Hewlett-Packard Journal*, Vol. 30, No. 9, pp. 20–24, September 1979.

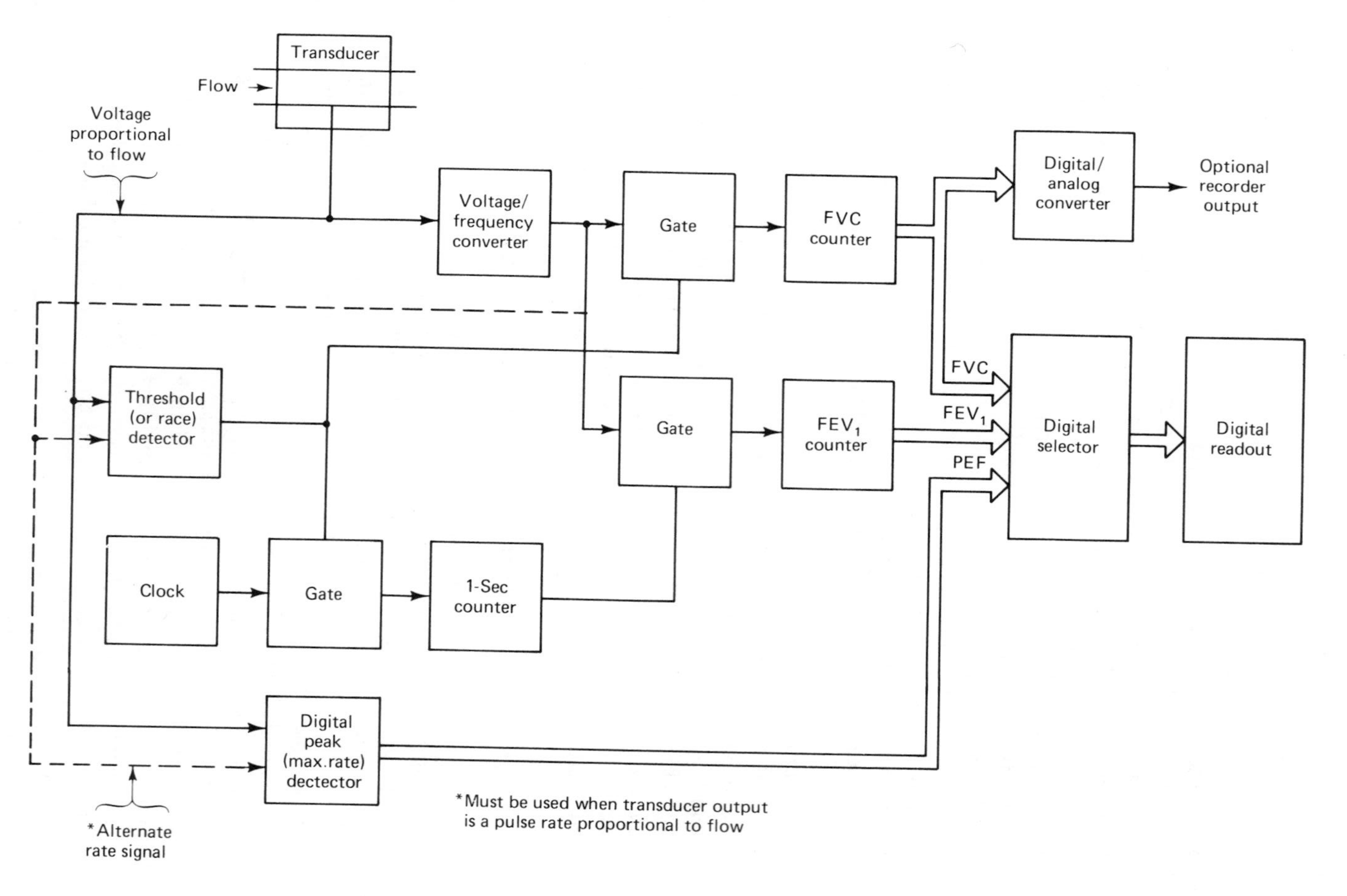

Figure 9.5 Digital processing spirometer.

Distribution measurements quantify degrees of lung obstructions and determine the residual volume, which is the amount of air that cannot be removed from the lungs by patient effort. The residual volume must be measured indirectly, such as with the nitrogen washout procedure.

Diffusion measurements identify the rate at which gas is exchanged with the bloodstream. This is difficult to do with oxygen because it requires a sample of pulmonary capillary blood, so it is usually done by measuring the diminishment of a small quantity of carbon monoxide mixed with the inhaled air.

All these measurements vary widely according to the patient's age and physical size. Therefore, evaluating a measurement requires comparison with a normal value established for people with similar physical characteristics in the particular age group. Normal values are established statistically by measuring a large population and correlating the measured values to such parameters as age, sex, weight, and geographical location (the lung characteristics of people living in Scandinavian countries, for example, differ from those of residents of the United States). Consequently, pulmonary function testing is a world of numbers and graphs.

An automated pulmonary measurement system achieves a considerable reduction in the time required to analyze the data obtained in pulmonary testing. At the same time, it improves the quality of the data by minimizing testing errors and by obtaining higher accuracy in the measurements. Depending on the options chosen, it completely automates the measurements of ventilation, distribution, and diffusion and, after a brief training period, can be operated by pulmonary technicians who have not had previous experience with computers.

A four-channel analog-to-digital converter (ADC) supplies the measurement data to the computer. Inputs to the ADC are from the various measurement devices, which include a pneumotach, which provides a signal proportional to airflow for the various measurements; a nitrogen analyzer, for distribution measurements; and carbon monoxide and helium analyzers, for diffusion measurements. Any combination of these devices, together with some breathing hardware, may be supplied with the system or added later to provide the capabilities desired. All are installed in a convenient desk-height cabinet.

In Fig. 9.6, reports are generated automatically, listing tabular values for comparison to predicted normal values and plotting curves to give graphical interpretation. A computer can print out results on its built-in 16-column strip printer, but an impact printer can be added to the system to give page-width reports and graphs. Software supplied with the system on a cassette consists of over 70 programs and includes a customizing program that enables the user to meet his or her personal requirements by changing values included in the report, the headings on each report, the forms the graphs are in, and so on.

One significant aspect of a software-controlled pulmonary measurement lab is that it allows new programs to be added or existing ones to be modified so that new procedures can be implemented. Pulmonary medicine is a dynamic

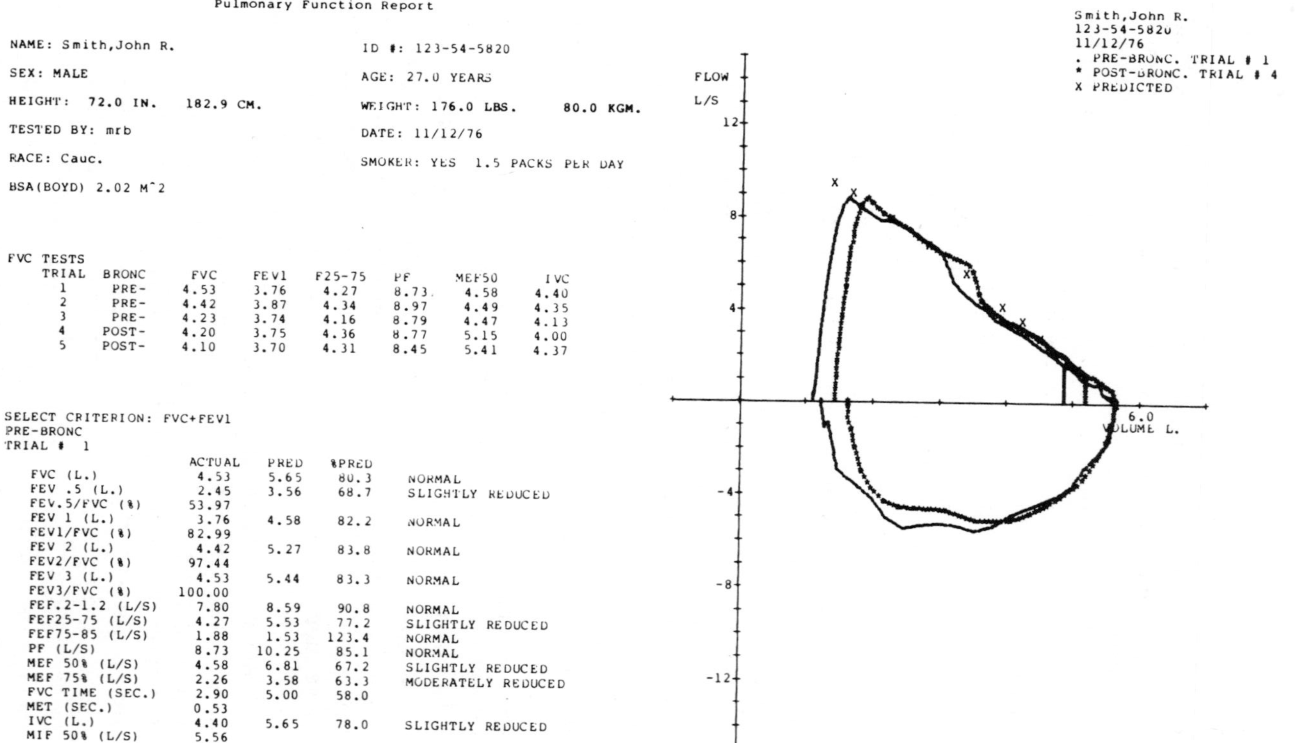

Hewlett-Packard 47804A
Pulmonary Calculator System
Pulmonary Function Report

NAME: Smith,John R. ID #: 123-54-5820

SEX: MALE AGE: 27.0 YEARS

HEIGHT: 72.0 IN. 182.9 CM. WEIGHT: 176.0 LBS. 80.0 KGM.

TESTED BY: mrb DATE: 11/12/76

RACE: Cauc. SMOKER: YES 1.5 PACKS PER DAY

BSA(BOYD) 2.02 M^2

FVC TESTS

TRIAL	BRONC	FVC	FEV1	F25-75	PF	MEF50	IVC
1	PRE-	4.53	3.76	4.27	8.73	4.58	4.40
2	PRE-	4.42	3.87	4.34	8.97	4.49	4.35
3	PRE-	4.23	3.74	4.16	8.79	4.47	4.13
4	POST-	4.20	3.75	4.36	8.77	5.15	4.00
5	POST-	4.10	3.70	4.31	8.45	5.41	4.37

SELECT CRITERION: FVC+FEV1
PRE-BRONC
TRIAL # 1

	ACTUAL	PRED	%PRED	
FVC (L.)	4.53	5.65	80.3	NORMAL
FEV .5 (L.)	2.45	3.56	68.7	SLIGHTLY REDUCED
FEV.5/FVC (%)	53.97			
FEV 1 (L.)	3.76	4.58	82.2	NORMAL
FEV1/FVC (%)	82.99			
FEV 2 (L.)	4.42	5.27	83.8	NORMAL
FEV2/FVC (%)	97.44			
FEV 3 (L.)	4.53	5.44	83.3	NORMAL
FEV3/FVC (%)	100.00			
FEF.2-1.2 (L/S)	7.80	8.59	90.8	NORMAL
FEF25-75 (L/S)	4.27	5.53	77.2	SLIGHTLY REDUCED
FEF75-85 (L/S)	1.88	1.53	123.4	NORMAL
PF (L/S)	8.73	10.25	85.1	NORMAL
MEF 50% (L/S)	4.58	6.81	67.2	SLIGHTLY REDUCED
MEF 75% (L/S)	2.26	3.58	63.3	MODERATELY REDUCED
FVC TIME (SEC.)	2.90	5.00	58.0	
MET (SEC.)	0.53			
IVC (L.)	4.40	5.65	78.0	SLIGHTLY REDUCED
MIF 50% (L/S)	5.56			

Figure 9.6 Automatic generation of reports. (Courtesy of Maurice R. Blais of LSE Corp., Woburn, Mass. and Hewlett-Packard Co. Reprinted in part from Maurice R. Blais and John L. Fanton, Automated Pulmonary Function Measurements, *Hewlett-Packard Journal*, Vol. 30, No. 9, Sept. 1979.)

field with new diagnostic procedures being developed at an increasing rate. Problems with obsolescence are thus minimized.

A key element in this system, basic to all the measurements, is the pneumotach. Early spirometers consisted of a cylindrical enclosure with a counter-balanced plunger. The patient blew into the enclosure, displacing the plunger, and a recording pen connected to the plunger traced the air volume vs. time record. The physician had to calculate rate of airflow from the slope of the trace. Not only was this procedure cumbersome to implement, but inaccuracies arose because of the cooling of the expired air and the condensation of the moisture.

Figure 9.7 shows the resistance to airflow in the pneumotach, which generates a pressure differential that is proportional to flow rate. The pneumotach designed for an automated pulmonary system provides a much more accurate determination of flow rate (the computer integrates the flow rate to determine volume). It consists of a cylindrical enclosure in which a spirally wound sheet of corrugated metal is inserted, essentially creating a bundle of parallel tubes within the enclosure. As air flows through these tubes, a pressure transducer measures the pressure differential caused by the friction of the air against the tube walls, thus giving an indication of flow rate.

At low flow rates, the flow through the pneumotach is smooth or laminar and the pressure drop is proportional to the air flow. At higher flow rates, the flow becomes turbulent and the pressure drop is no longer linearly related

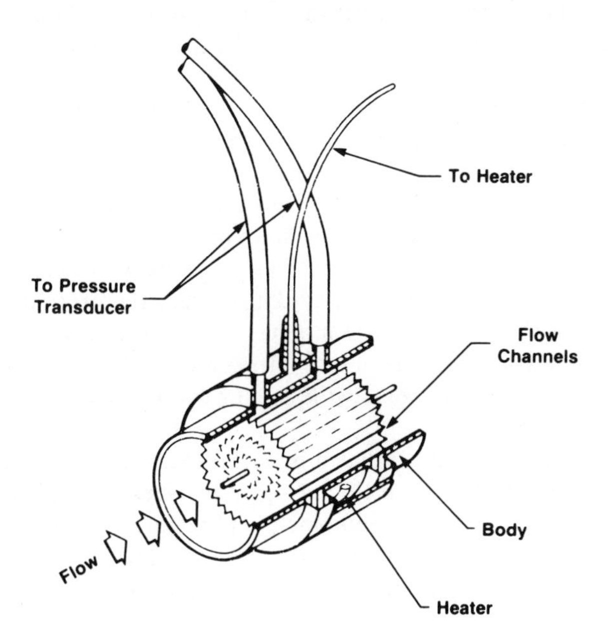

Figure 9.7 Resistance to airflow in the pneumotach.

to flow. However, precision manufacturing techniques achieve uniformity in the corrugations so that a uniform flow profile can be maintained, thereby assuring repeatable flow-to-pressure characteristics. Hence, the flow-to-pressure curve can be linearized electronically.

Most respiratory parameters are reported in BTPS conditions (body temperature, ambient pressure, saturated with water vapor). This is the condition of the air in the lungs and in the mouth. To prevent condensation and maintain the gas under these conditions, the temperature of the pneumotach should be maintained at 37°C. The heater that warms the pneumotach is electrically isolated from the metal case for patient safety, and it is encapsulated so that the entire unit may be immersed in liquids for sterilization. The thermistor that senses the temperature, controlling the heater through a proportional controller, is buried in the metal case.

The accuracy of flow-rate measurements made with the pneumotach is enhanced by correction factors applied by the computer. For example, corrections for gas temperature and viscosity are made on every measurement during a *nitrogen washout test*, as explained in the following paragraphs.

The nitrogen washout test is used in the measurement of functional residual capacity (FRC), the volume of air remaining in the lungs at the end of a normal resting expiration. The FRC can be divided into two other volumes, the *expiratory reserve volume* (ERV), which is the volume of air a person can exhale beyond the resting level, and the *residual volume* (RV), the air that a person cannot force out of the lungs no matter how hard he or she tries. Because of the residual volume, the functional residual capacity cannot be measured by a simple patient maneuver, as are some other parameters, but is measured indirectly using the multiple-breath nitrogen washout technique.

Before a nitrogen washout is started, the gas in a patient's lungs is about 73% nitrogen, which is slightly less than atmospheric because the lungs are saturated with water vapor. During the test, the patient inhales pure oxygen, so that each expiration contains a smaller amount of nitrogen than the preceding. The volume of nitrogen exhaled in each of the breaths is accumulated by the computer until the end of the test, typically when the expired nitrogen concentration is less than 1%, about 3 minutes after the start. Dividing this accumulated nitrogen volume by the initial nitrogen concentration yields the initial lung volume. If the patient was at the resting level when the test was begun, this volume is the FRC.

The amount of nitrogen exhaled in each breath is obtained by measuring the nitrogen concentration with the nitrogen analyzer, multiplying this by the flow rate as measured by the pneumotach, and integrating the product.

To obtain accurate results, the flow (pneumotach) and nitrogen signals must be corrected for a number of factors. The pneumotach responds to the viscosity of the gas passing through it, and viscosity is a function of gas composition and temperature.

The rate of diffusion of oxygen from a respiratory system in the blood depends upon the PO_2 to which the blood is exposed. The greater the PO_2, the greater the dissolved oxygen, and the greater the percent saturation of hemoglobin with oxygen. Since the alveolar-capillary membrane is interposed between the alveolus and capillary, its thickness, surface area, and permeability to oxygen also affect the diffusion rate.

The presence of hemoglobin in blood greatly facilitates the transport of oxygen. At STPD, 1.34 cc of oxygen can combine chemically with each gram. The normal hemoglobin concentration in a healthy resting man is about 16 g/100 cc of blood. Thus, at STPD the O_2 carrying capacity of the blood is $16 \times 1.34 = 21.44$ cc O_2/100 cc blood. At a normal arterial blood PO_2 of 95 to 97 torr, the hemoglobin is approximately 97% saturated with oxygen. Thus, the total bound oxygen at STPD is $16 \times 1.34 \times 0.97 - 19.79$ cc/100 cc. The total dissolved oxygen in arterial blood at PO_2 of 95 to 97 torr at STPD is 0.30 cc/100 cc. Thus, the bound oxygen constitutes $19.79/20.09 = 98.5\%$ of the total oxygen normally carried in the blood.

It is useful to obtain the ratio of oxygen content of hemoglobin to the oxygen-carrying capacity of hemoglobin; this value, when multiplied by 100, is the percent saturation of hemoglobin with oxygen. Both blood oxygen content and percent saturation are important determinations in the respiratory system. Blood oxygen content yields a measure of the effectiveness of the circulating blood in providing oxygen to the tissues. The percent saturation is indicative of the oxygenating ability of the lung.

The *oximeter* (Fig. 9.8) is a bioelectronic measuring device that allows

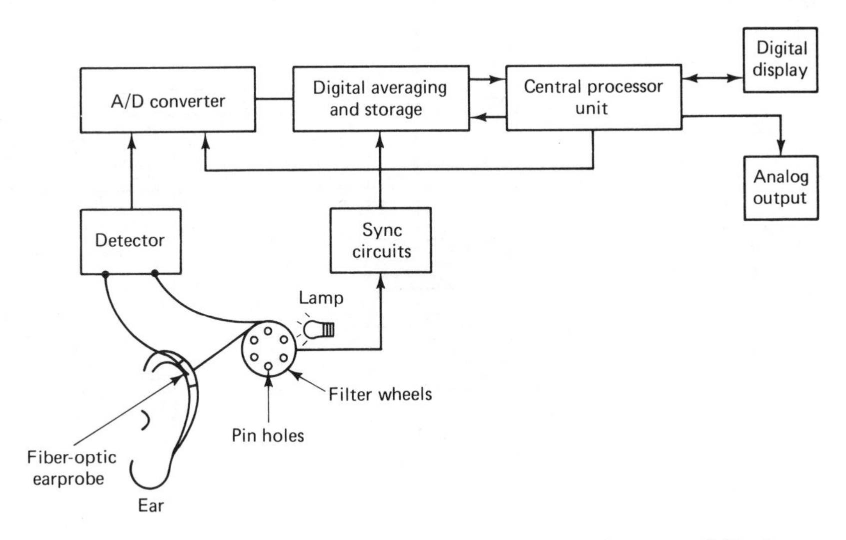

Figure 9.8 Simplified block diagram of an oximeter. (Courtesy of Hewlett-Packard Co. Palo Alto, Calif.)

direct measurement of the percent saturation of hemoglobin with oxygen. The noninvasion oximeter measures the percent saturation of hemoglobin in arterial blood continuously when an ear probe is properly placed on the ear.

9.6 THE CARDIORESPIROGRAPH

The relationship between two parameters—heart rate and uterine contractions—has been used by obstreticians for many years to monitor the well-being of the fetus during birth. Neonatalogists have also found that beat-to-beat heart rate is a function of the condition of the central nervous system. The second vital parameter, respiratory activity, is also controlled from the central nervous system. The *cardiorespirograph* continuously records beat-to-beat heart rate and the respiratory activity waveform, as shown in Fig. 9.9. The resulting trace is called a cardiorespirogram. The cardiorespirogram is a simultaneous recording of beat-to-beat neonatal heart rate and respiration waveform. Heart rate and

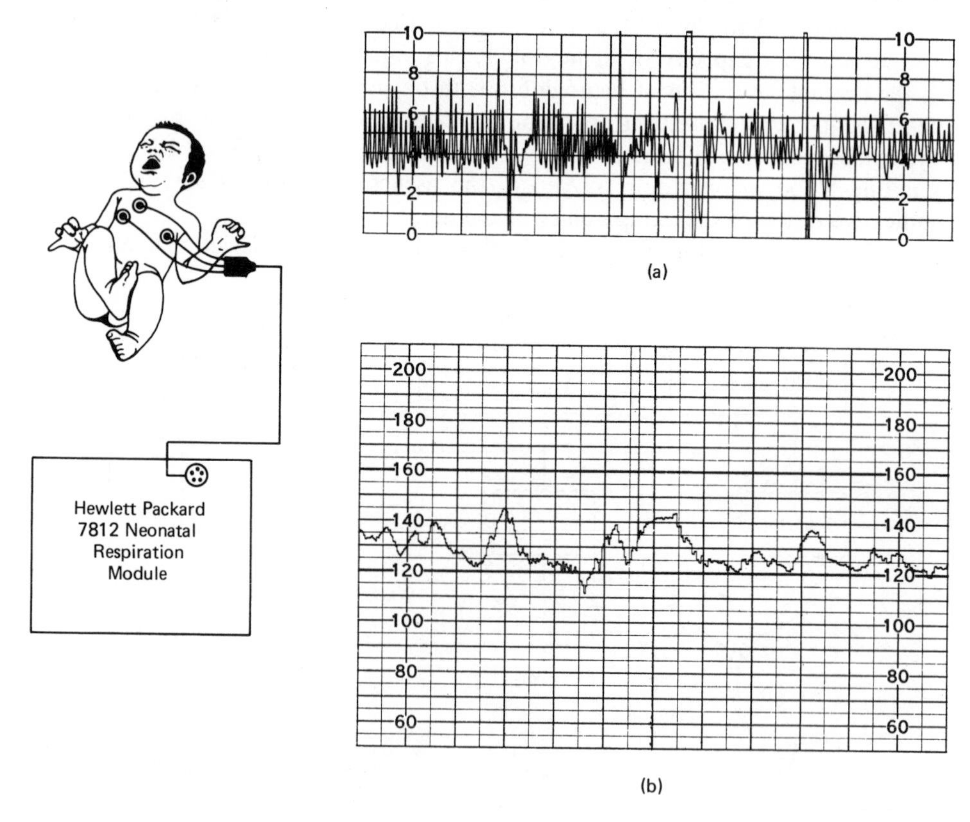

Figure 9.9 Cardiorespirogram: (a) respiratory activity; (b) heart beat rate. (Courtesy of Hewlett-Packard Co., Palo Alto, Calif.)

respiration are controlled by involuntary nervous impulses from the central nervous system. Variations in heart rate from beat to beat are present under normal conditions and contain information that is not seen in an average heart rate value. Respiration waveform, recorded using the impedance method, is a semiquantitative analog of tidal volume and is therefore a good indicator of pulmonary function.

The cardiorespirogram is a record of the correlation of both heart rate and respiration. It has been shown that it is sufficiently sensitive to register and differentiate the severity of life-threatening abnormalities before obvious changes in other parameters occur.

The Hewlett-Packard 7821A neonatal respiration module is shown in Fig. 9.10.

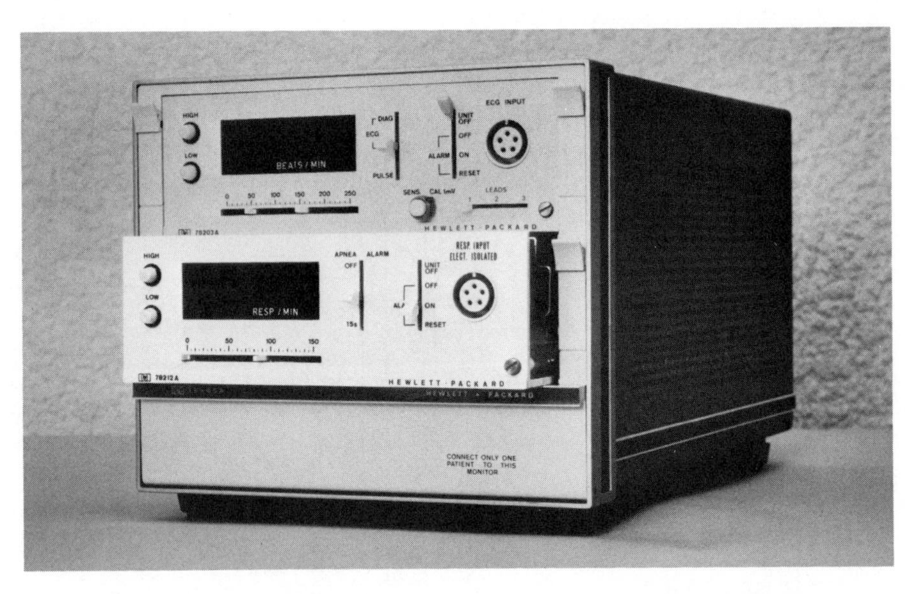

Figure 9.10 Hewlett-Packard 78212A Neonatal Respiration Module. (Courtesy of Hewlett-Packard Co., Palo Alto, Calif.)

9.7 HYBRID RESPIRATION-RATE SIGNAL CONDITIONER

The hybrid impedance pneumograph and respiration-rate signal conditioner measure changes in the impedance of the chest during the breathing cycle. Figure 9.11 shows a hybrid respiration-rate signal conditioner which generates an analog respiration signal as output together with a synchronous square wave that can be monitored by a breath-rate processor.

The circuit in Fig. 9.11 requires two active chest electrodes that are driven with balanced 50-kHz current. A reference (common) electrode is also provided.

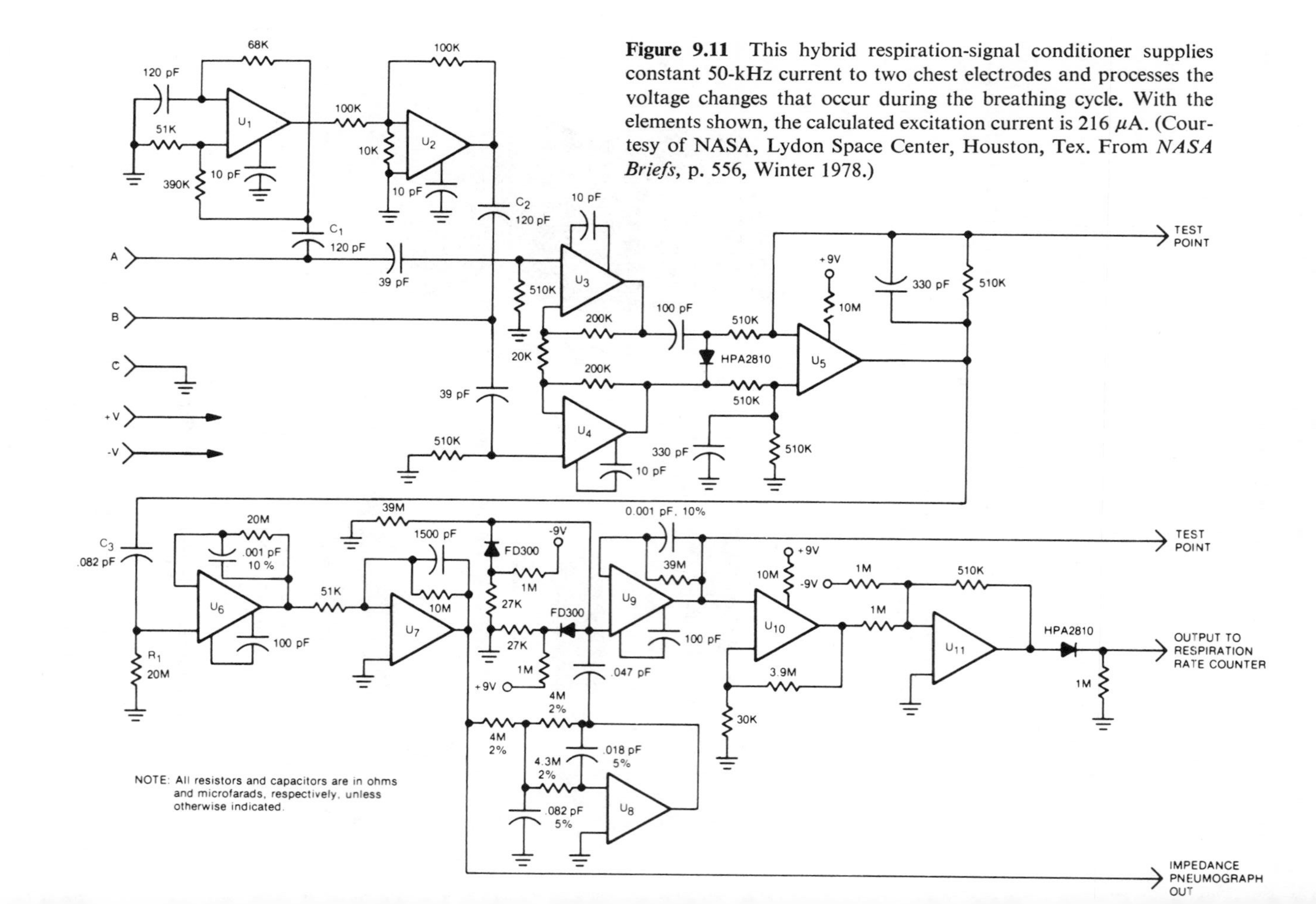

Figure 9.11 This hybrid respiration-signal conditioner supplies constant 50-kHz current to two chest electrodes and processes the voltage changes that occur during the breathing cycle. With the elements shown, the calculated excitation current is 216 μA. (Courtesy of NASA, Lydon Space Center, Houston, Tex. From *NASA Briefs*, p. 556, Winter 1978.)

These can be the same electrodes that monitor the ECG heart signal in the vital-signs monitor. The voltage developed across the chest impedance is amplified, rectified, and filtered. It also develops a signal for further processing.

9.8 REVIEW QUESTIONS

1. Draw a simplified human respiratory system.
2. Discuss how to make a spirometry measurement.
3. Discuss the meaning of respiratory minute volume, maximum ventilatory volume, and forced expiratory volume.
4. Discuss a digital processing spirometer.
5. Discuss automated pulmonary function measurements, such as ventilation, distribution, and diffusion, and the need for pulmonary function measurements.
6. Discuss a hemoglobimeter and an oximeter.
7. Discuss a cardiorespirogram.
8. When is a cardiorespirator used?
9. Discuss a hybrid respiration-rate signal conditioner which measures changes in the chest impedance during breathing.
10. Discuss the most important parameters needed to make a pulmonary function measurement.

9.9 REFERENCES

1. Strong, P.: *Biophysical Measurements*, 1st ed., 3rd printing, Tektronix, Inc., Beaverton, Oreg., 1973.
2. Thomas, H. E.: *Handbook of Biomedical Instrumentation and Measurement*, Reston Publishing Co., Inc., Reston, Va., 1974.
3. Cromwell, L., Weilbell, F. J., Pfeiffer, E. A., and Usselman, L. B.: *Biomedical Instrumentation and Measurements*, Prentice-Hall, Inc., Englewood Cliffs, N.J., 1973.
4. Jacobsen, B. A., and Webster, J. G.: *Medicine and Clinical Engineering*, Prentice-Hall, Inc., Englewood Cliffs, N.J., 1977.
5. Webster, J. G., ed.: *Medical Instrumentation: Application and Design*, Houghton Mifflin Company, Boston, 1978.
6. Morris, J. F.: Spirometry in the Evaluation of Pulmonary Function (Medical Progress), *Western Journal of Medicine*, Vol. 125, August 1976.
7. Blais, M. R., and Fenton, J. L.: Automated Pulmonary Function Measurements, *Hewlett-Packard Journal*, pp. 20–24, September 1979.
8. Ferris, B. C., Principal Investigator: Epidemiology Standardization Project, *American Review of Respiratory Disease*, July 1978.
9. *Chronic Obstructive Pulmonary Disease*, 5th ed., American Lung Association, New York, 1977.

10. Geddes, L. A., and Baker, L. E.: *Principles of Applied Bio-Medical Instrumentation*, 2nd ed., John Wiley & Sons, Inc., New York, 1975.
11. Edwards, R., Jr.: Evolution of Improved Capabilities in Spirometric Test Equipment, *CVP Journal of Cardiovascular and Pulmonary Technology*, March–April 1976.
12. Martin, D.: *Laboratory Experiments in Human Physiology*, Georgia State University, Department of Health and Physical Education, Atlanta, Ga., 1974.
13. *Clinical Pulmonary Function Testing*, Intermountain Thoracic Society, Salt Lake City, Utah, 1975.

10

CLINICAL LABORATORY MEASUREMENT

10.1 INTRODUCTION

The clinical laboratory is the cornerstone of modern medicine. This is due to the fact that most patients entering a hospital will have bolod counts and chemical blood tests to support or deny a physician's diagnosis or treatment. These blood tests may be repeated to monitor the process of the illness. To achieve this, most hospitals have their clinical laboratories automated and computerized to some extent. The effect of automation in clinical laboratory measurements has been an increase in the accuracy of tests, in the number of tests performed, and in the overall cost.

In dealing with clinical laboratory measurements, the laboratory instruments may be classified as either preparatory or analytical devices. Preparatory devices range from heaters to mixers and include all types of centrifuges and column and electrophoretic separation devices.

The analytic devices are usually more sophisticated instruments but retain identifiable characteristics in terms of their functions. They include photometric instruments, which use light energy to analyze the chemical properties of solutions. These include such devices as spectrophotometers, flame photometers, fluorometers, colorimeters, and densitometers. Photometers designed for visible spectral range are colorimeters that use nondispersing optics. Spectrophotometers have dispersing elements (prism or grating) and are built to have a higher order of spectral discrimination than that of colorimeters. In fluorimetry, a compound, usually in solution, is irridated with monochromatic light of a wavelength that it absorbs. In this manner, it is analogous to spectrophotometry. However, fluorescing materials have the ability to transform some of the absorbed light into emission at a longer wavelength. The amount of emitted fluorescence light is directly proportional to the concentration and is increased by electronic amplification. In colorimetry and spectrophotometry, zero concentration corresponds to the maximum signal and electronic amplification does not enhance the signal-to-noise ratio.

10.2 MECHANISM OF BLOOD

Blood is the transporting medium of the body through arteries, capillaries, and veins. The heart itself is the master pumping device which enables blood to reach all areas of the body.

Blood is composed of cells and/or plasma and makes up approximately 9% of the body weight. A newborn infant weighing 3.25 kg (about 7 pounds) usually has about 0.03 liter of blood. Blood is composed of blood cells and blood plasma.

There are two types of blood cells:

1. *Erythrocytes* or *red cells* appear as homogeneous circular disks with a magnitude of 4.5 to 5.5 million in about 1 mm^3 of blood. They have a yellowish-red tinge and consist of a stroma in which hemoglobin is deposited. Hemoglobin is made up of a protein called *globin* and non-protein portion called *hemotin.*
2. *White cells* are minute ameboid cells, variable in size, and classified as follows:
   (a) The *lymphocytes* arise from the reticular tissues of the lymph nodes of the body.
   (b) The *monocytes* include the mononuclear and transitional types. They are large cells.
   (c) The granular *leukocytes* show marked pseudopodial movement.
   The number of white cells in 1 mm^3 of blood is about 6000 to 10,000. White blood cells function to protect the body from pathogenic organisms and to promote tissue repair.

One of the newest devices used in automated clinical laboratories is the *differential white cell counter.* The counter picks up the images of slides under a microscope with a television camera. A computer analyzes the different shapes of the cells and gives a count of how many of each type of white blood cell are on the slide. This bioelectronic measurement is useful in diagnosing and monitoring blood diseases such as leukemia.

Blood platelets or thrombocytes are disk-shaped bodies. The sticking together of platelets seals small leaks in blood vessels, preventing loss of blood. In 1 mm^3 of blood, there are about 200,000 to 800,000 thrombocytes.

The blood plasma consists of a clear amber-colored fluid made up of water, protein (fibrinogen serum globulin and serum albumin), prothrombin (used in coagulation), nutrients, salts and other organic substances, gases, and antibodies.

The *hematocrit* is a measure of the red blood cells to blood plasma. The normal hematocrit is taken using blood from the arm. This is called the *limb hematocrit* and is the number of millimeters of packed red cells in 100 mm of blood. The *body hematocrit* is the total erthrocyte mass to the total blood volume.

The *hematocrit centrifuge* is used to separate blood into volumetric proportions (percentage) of corpuscular elements (cells) compared to total blood volume. Clinical centrifuges have a rotational speed of less than 20,000 (revolutions per minute). Centrifuges with rotational speeds in excess of 20,000 rpm are called ultracentrifuges and are used principally for reasearch.

When the hematocrit is spun down, the blood actually separates into three groups. At the bottom of the test tube are the packed red blood cells. Just above these cells is a white layer which consists of the white blood cells (leukocytes) and platelets. Above the white layer is the clear fluid called *plasma*.

The hematocrit process requires a fairly large amount of blood. The large sample size necessitates a low-rpm, long-spin-time centrifugation process. The handling of larger blood samples also requires the addition of an anticoagulant, usually versene, to the sample to keep it from coagulating and destroying the measurement.

To overcome these difficulties, the *microhematocrit* is used. The microhematocrit uses two heparnized or anticoagulating capillary tubes filled about two-thirds full with blood taken from a finger puncture. The capillary tube ends are sealed with molding clay and then placed into the head of the microhematocrit centrifuge. The blood-filled capillaries are then spun for 2 to 3 minutes at from 7000 to 12,000 rpm. The resulting separated components—plasma, erythrocytes, leukocytes, and platelets—are measured against a scale called a microhematocrit reader. The normal range of hematocritis is shown in Table 10.1.

TABLE 10.1 Normal Range of Hematocrit (%)

	Low	High
Infants	44	62
Children	35	37
Adult males	40	45
Adult females	36	45

The hematocrit method of measuring red blood cells content has an accuracy of about 98% (2% error rate). For this reason it is used as the screening methodology for anemia and polycythemia.

Many body disorders and diseases cause variations in the blood composition. In anemia, the red blood cell count is reduced. Other diseases cause changes in the chemical composition of blood serum or in other body fluids, such as urine. In diabetes mellitus, the blood glucose concentration is elevated.

In Fig. 10.1, a reflectance using a Dextrostrix reagent strip gives a measurement of the whole-blood glucose levels in the range 10 to 400 mg/100 ml. This instrument measures the reflected light from the surface of the reacted reagent area of Dextrostrix and converts this measurement by means of electronic circuitry to a reading on a calibrated meter scale.

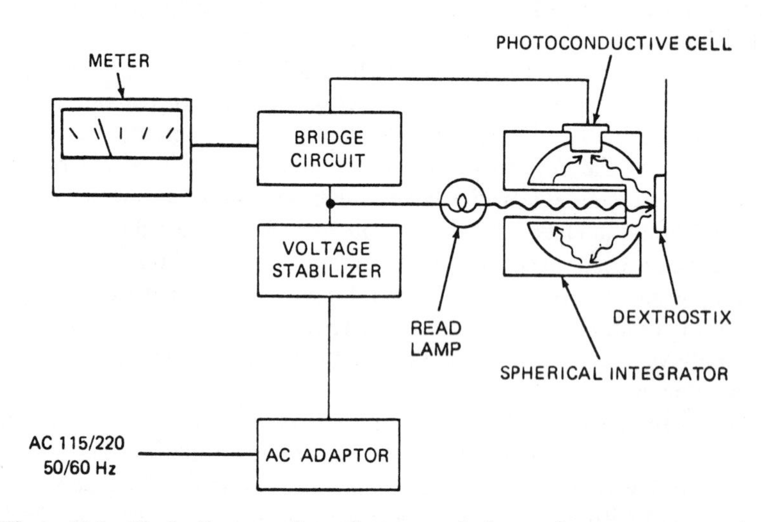

Figure 10.1 Block diagram of a reflectance colorimeter for measurement of blood glucose levels. (Courtesy of Ames Division, Miles Laboratories, Inc., Elkhart, Ind.)

10.3 CHEMICAL BLOOD TESTS

The blood samples most commonly required are listed in Table 10.2 and can be illustrated with an output sheet from an SMA-12 test done in the hospital.

Sodium and potassium may be measured using the flame photometry technique, which causes the normal colorless flame to appear yellow for sodium and violet for potassium when the solutions are aspirated into the flame. To determine the measurement of chlorides, a chloridimeter is used based on the coulometric electrochemical method. Other chemical blood tests include these for blood urea nitrogen and creatinine. Many electrolyte measurements, including the above, can be performed with automated clinical devices. The Radiometer Company of Copenhagen, Denmark, manufactures the FLM 3 flame photometer, which is automated to read out digital calcium, potassium, and lithium in blood serum and urine. When the flame photometer is connected

TABLE 10.2 Patient SMA-12 Test

Test	Normal Range[a]	Result
Calcium	8.6–10.4 mg/100 liters	10.5
Phosphorus	2.2–4.6 mg/100 ml	3.4
Glucose		
Less than 50 years old	72–127 mg/100 ml	
More than 50 years old	84–128 mg/100 ml	225.0
Uric acid		
Male	4.0–9.0 mg/100 ml	4.3
Female	2.8–7.7 mg/100 ml	
Cholesterol test		
20–29 years old	120–240 mg/100 ml	
30–39 years old	140–270 mg/100 ml	
40–49 years old	150–310 mg/100 ml	
50–59 years old	180–380 mg/100 ml	275.0
Protein test	6.6–8.2 g/100 ml	7.2
Albumin test	3.3–5.2 g/100 ml	4.3
Bilirubin test (adult)	0.1–1.2 mg/100 ml	0.7
Alkaline phosphatase (adult)	30–115 IU/liter	65.0
Lactic acid dehydrogenase (LDH)	100–225 IU/liter	250.0
Serum glutamic oxaloacetic transaminase (SGOT)	7–40 IU/liter	40.0

[a]IU, international units, refers to 1 micromole or substrate transformed or of product per minute. Units can be expressed per milliliters or per liter. The reaction temperature should be noted. Serum enzyme concentrations vary with age and with male and female and vary from test to test.

to an alphanumeric printer, the results can be printed out as fast as the samples can be diluted and numbered.

The Photovolt Corp. of New York City has introduced the Mastermind, a totally automatic electrolyte analyzer that measures blood urea nitrogen, glucose, sodium, potassium, chloride, and bicarbonate. The transducers are all electrochemical and the microprocessor provides both data handling and readout.

An acid–base laboratory can also measure pH, PCO_2, and PO_2 hemoglobin concentration and barometric pressure. With an accessory computer plasma bicarbonate, total CO_2, base excess, oxygen saturation, standard bicarbonate, and other parameters can be printed on a readout card or digitally displayed.

An automated hexiometer that can display the hemoglobin concentration, the oxygen saturation, or the carboxyhemoglobin fraction is shown in Fig. 10.2. when the inlet flap is open, the sample is either aspirated into the hemoximeter by the pump until the liquid sensor is reached or it is injected into the OSM2. In the latter case, superfluous blood passes to waste through the (upper) valve. When the flap is closed, the sample is ultrasonically hemolyzed into the cuvette

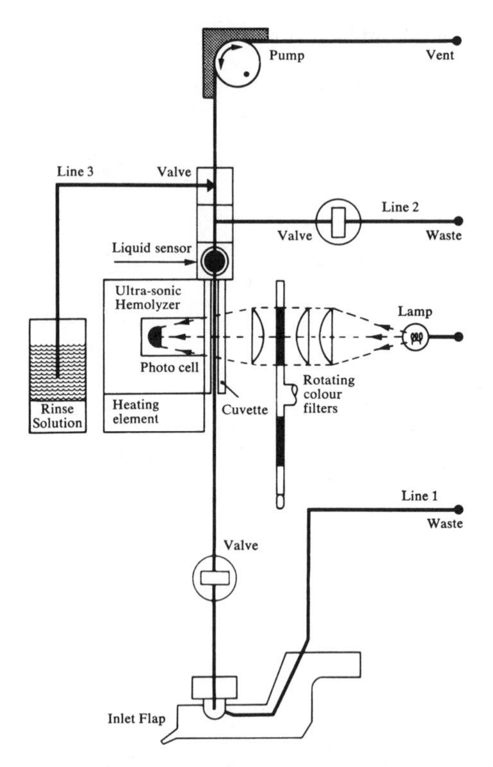

Figure 10.2 Hemoximeter which measures hemoglobin and percent oxygen saturation. (Courtesy of London Co., Cleveland, Ohio, Division of Radiometer of Copenhagen.)

and measurements are made at 505 and 600 nm. Results are then electronically calculated and displayed. The entire "wet" system, including the inlet flap, is rinsed with saline by means of the pump.

The hemoximeter shown in Fig. 10.2 determines hemoglobin concentration and oxygen saturation on 25-μl samples of whole blood or blood concentrate. Samples are automatically hemolyzed by ultrasound and then submitted to a photometric measurement.

10.3.1 Basic AutoAnalyzer Simple Principles[1]

The AutoAnalyzer is an automated system used to perform a variety of chemical analyses. Briefly stated, the AutoAnalyzer automates the normally time-consuming, tedious procedures of manual analyses—sampling, dilution, filtering, mixing, heating, and color measurement. It does all these procedures in a straightforward, logical way—automatically—and at a rate of speed and degree of precision once considered unattainable.

By preselected programming, sample fluids, reagents, and standards are brought together under controlled, sequenced conditions to cause a chemical

[1]Sections 10.3.1, 10.3.2, and 10.3.4 appear here courtesy of Technicon Instrument Corp., Tarrytown, New York.

reaction and, in most instances, color development. With a colorimetric procedure, the optical density of the fluid, being related to fluid concentration, is photoelectrically determined in a colorimeter. The colorimeter-output concentration values will appear as a series of curves or peaks on moving graph paper. Then, by means of graph overlay, the recorded peaks of the samples are compared with those of known standards.

The basic AutoAnalyzer (Fig. 10.3) consists of six separate units called modules, interconnected by plastic tubing, glass mixing coils, fittings, and attendant pieces. The six modules of the system are:

1. Sampler
2. Proportioning pump and manifold
3. Dialyzer
4. Heating bath
5. Colorimeter
6. Recorder

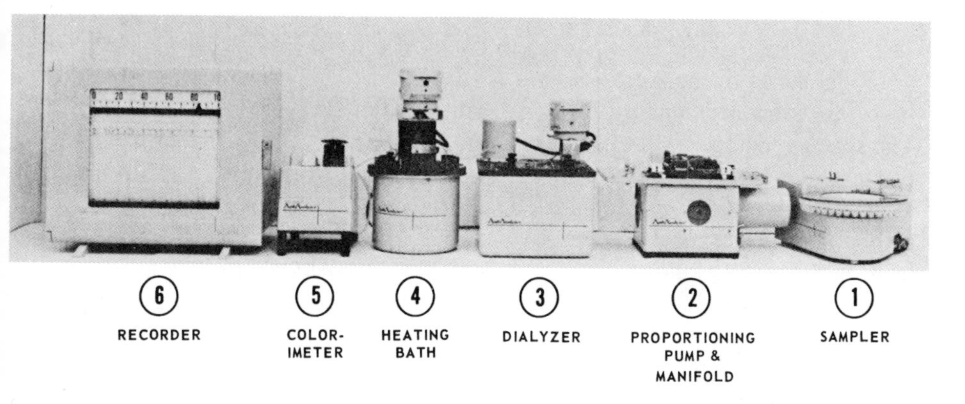

Figure 10.3 Basic AutoAnalyzer. (Courtesy of Technicon Instruments Corp.)

The number and types of modules required are dictated solely by the analytical determination. As other modules are required—a flame photometer, a fluorometer, a cell counter—they are simply added to, or substituted for, the basic six-module AutoAnalyzer system.

The principal function of each module in the basic AutoAnalyzer system is summarized as follows:

1. The sampler aspirates samples, standards, and wash solutions to the AutoAnalyzer system in a timed sequence.
2. The proportioning pump is used with the manifold to introduce and proportion the samples and reagents; it moves these fluids to other modules of the system at precise flow rates. The manifold is positioned

on the proportioning pump, and by pump action, aspirates, mixes, proportions, and advances the various fluids throughout system.

3. The dialyzer separates interfering substances from the sample material by permitting selective passage of sample components through a membrane.
4. The heating bath heats fluids continuously to effect a chemical reaction under controlled conditions of time and heat. It may be used for color development, enzymatic action, digestion, hydrolysis, and other procedures.
5. The colorimeter monitors the changes in optical density of the fluid stream flowing through a tubular flow cell. It converts values of optical densities into equivalent voltage signals for the recorder. A stabilizer unit supplies a regulated voltage to the colorimeter lamp.
6. The recorder converts the optical-density electrical signal from the colorimeter into an equivalent graphic display on a moving chart.

10.3.2 Theory of Operation

The heart of the AutoAnalyzer is the proportioning pump, where plastic tubes are stretched taut on a spring-loaded platen between two end blocks. The spring loading serves to keep each tube pressed hard against a roller which occludes the tube where it touches. A series of rollers, carried by a chain, pass along the tubing from end to end. There is never less than one roller occluding a tube at any one time. As the roller moves forward, fresh fluid is drawn in from behind while fluid in front is pushed forward. As the speed is fixed, the rate of pumped fluid is dictated entirely by the inside diameter of the pump tube. There are some 20 standard tube sizes available. A sketch of a typical proportioning pump is shown in Fig. 10.4.

The plastic *end blocks*, which serve to hold the tubes in a taut position on the platen, are provided with slots to hold as many as 15 tubes (23 tubes in the case of pump II).

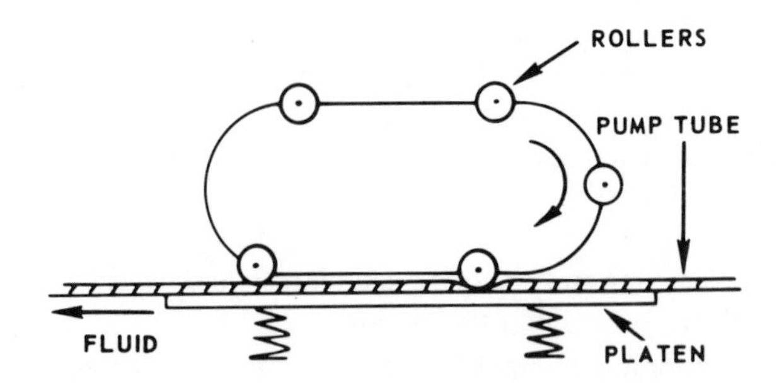

Figure 10.4 Proportioning pump: rollers occluding pump tubes on the platen. (Courtesy of Technicon Instruments Corp.)

The system of pump tubes, interconnecting tubes, mixing coils, glass, and plastic pieces is called the *manifold*. Each chemical analysis has a distinctive manifold, and each manifold is designed to give maximum efficiency (i.e., the fastest possible analysis rate consistent with the sensitivity and accuracy required). Since one AutoAnalyzer may have to perform several different analyses in a day, manifolds have been designed to be rapidly interchangeable.

Standard methods have been carefully studied and fabricated as *plattered manifolds*, where all components are rigidly fixed on a tray. Since distances are fixed and glass pieces correctly positioned, reproducibility is assured.

The automated analytical system shown in Fig. 10.5 is a simple example of a chemical analysis. A reagent added to the sample produces a characteristic color. The optical density of the final solution is used to quantitate the concentration of material.

Two pump tubes are selected: one having a delivery rate to pump the sample, and the other having a delivery rate to pump the reagent. At their outlets, the tubes are joined by a simple glass T fitting and the combined stream is then pumped to the colorimeter. The stream flows continuously through the flow cell and the recorder shows at any instant the optical density of the fluid in the flow cell.

Assume that the sample inlet tube is placed in water, representing a standard solution of zero content of the material to be analyzed. The combined mixture (reagent and water) is now passed through the colorimeter. At this point, the colorimeter control is adjusted to make the recorder read zero optical density, and the pen will trace out a steady line at this zero value. The equation representing the concentration of the reagent in the colorimeter stream is

$$C_R = \frac{b}{a + b} \tag{10-1}$$

where a = delivery rate to pump sample
b = delivery rate to pump reagent
C_R = concentration of reagent

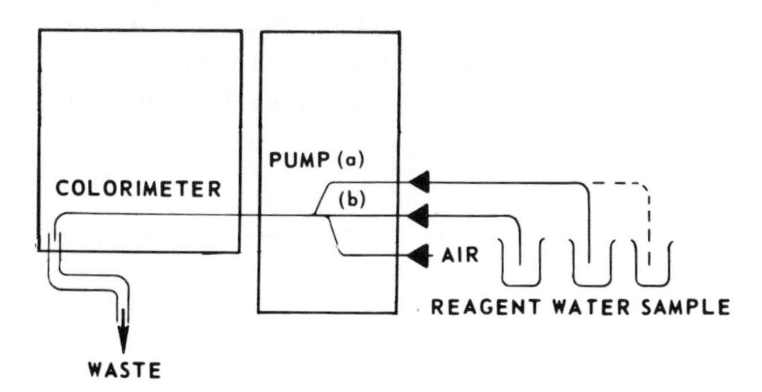

Figure 10.5 Example of an automated analytical system. (Courtesy of Technicon Instruments Corp.)

Imagine, next, that the sampling tube is taken from the water beaker and quickly placed into a beaker containing the sample material. The mixture of sample and reagent now becomes colored and, in time, the recorder shows an appropriate optical density. Note that the concentration of sample in the steam is given by

$$C_S = \frac{a}{a + b}$$

Even though the sample is introduced almost instantaneously, the recorder pen rises slowly (Fig. 10.6) and then continues to draw a line at the new level. A definite volume of new fluid was required to wash out any remnants of the previous fluid that adhered to the transmission tubing and flow cell. The system has a deffnite *wash time.*

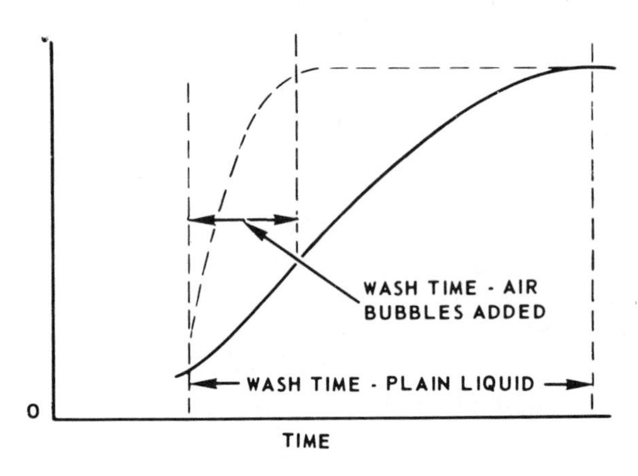

Figure 10.6 Wash-time curve. (Courtesy of Technicon Instruments Corp.)

As shown in Fig. 10.6, wash time plays a very significant part in determining the maximum rate at which samples may be analyzed. The art of automated analysis revolves around reducing the wash time; to this end, the first and probably the most important technique is *air segmentation.*

By introducing a series of air bubbles (Fig. 10.7) to segment the liquid stream, the wash time can be reduced from several minutes to as little as 30

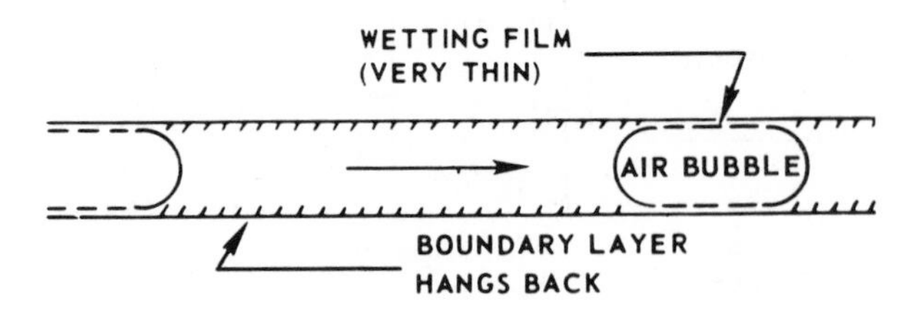

Figure 10.7 Air segmentation in a liquid stream. (Courtesy of Technicon Instruments Corp.)

seconds. A pump tube is used to inject air bubbles into the liquid stream through either a glass T fitting or a similar glass fitting.

In the absence of air bubbles, the boundary layer of a fluid clings to the wall of the tubing and hangs back to contaminate the liquid following.

When using air bubbles, the surface tension prevailing at the air–liquid interface reduces the stationary boundary layers to a thin film wetting the tube wall. The bulk of the fluid is thus caused to move forward and only the small amount wetting the wall will hang back.

When two liquid streams are thoroughly mixed, the proportion of the two liquids will be constant, not only throughout the length of any one segment, but also from segment to segment. In practice, this is not achieved perfectly: the proportioning is not quite constant and the variations show up as noise on the recorder. Provided that this noise is of sufficiently small amplitude, it will not have deleterious effects on the results.

Each extra reagent demanded by a chemical procedure is added, in turn, by using another pump tube, and is joined at the appropriate point by means of glass fittings. Whenever two liquid streams are joined, the solution must be mixed. Mixing is the function of the mixing coils.

In Fig. 10.8 a mixing fluid-inversion cycle is shown in which two or more liquid streams are joined together. Each segment of liquid in the combined stream comprises a mixture of fluids, the proportion of each being, on the average, the same as the ratio of the pumping rates. However, the fluids may not be mixed within the segment. To achieve proper mixing, a stream passes through a glass coil, or several coils, where each fluid segment is repeatedly inverted.

The following principles apply to these mixing coils:

1. A mixing coil operates efficiently only if its axis is horizontal: if the axis is vertical, the segments are never inverted.

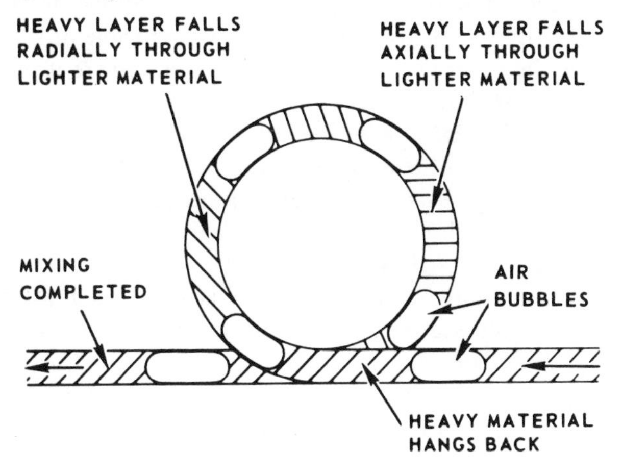

Figure 10.8 Mixing coil: fluid-inversion cycle. (Courtesy of Technicon Instruments Corp.)

2. To be effective, the length of the liquid segment must be less than a half circumference of the mixing coil; otherwise, the segment is not truly inverted as it goes up and over the coil.

Mixing within a liquid segment is also achieved, but at a slower pace, as the liquid flows through a tube. Thus, a long length of time-delay coil may in itself give sufficient mixing without the use of a mixing coil. However, long tube lengths achieve mixing only at the expense of wash time and are therefore to be avoided.

Chemical reactions do not always take place instantaneously. Often an appreciable amount of time is needed for a reaction to reach completion. If the reaction is fairly rapid, the mixing coils already mentioned (capacity 5 to 10 ml) may give enough time for the color to be developed, but in other cases the liquid must be held for a longer period. The "time delay coil" consists of about 40 ft. of glass tubing wound into a spiral coil. The fluid takes approximately 5 to 15 mins. to pass through this coil, depending on the volume of the coil and on the pumping rate. The "standard" time-delay coils have capacities of 28 ml, and the "large bore" of 88 ml. Standard coils are also available in half-length and quarter-length versions, with volumes of 14 and 7 ml, respectively. The delay time can be calculated when the coil volume and total pumping rate are known.

Once the fluid has passed through the heating bath or time-delay coil, the color is sufficiently developed to be measured. In the AutoAnalyzer system, color development need not be taken to a stage where it is relatively unaffected by time, as in most manual colorimetric methods. In manual methods, the time interval at which the color is measured may be difficult to control accurately, whereas in the AutoAnalyzer, time is controlled and constant for each reaction with standards subjected to exactly the same conditions as the unknowns.

Because the time element is so precisely reproducible, it is entirely acceptable to measure the color at a state of development where it might be on the steep part of the color/time curve.

Once the color has been developed, the liquid stream is pumped through the colorimeter flow cell to determine the optical density of the fluid. The flow cell is tubular and has an extremely small volume (0.05 to 0.1 ml) to speed the wash time. The tubular flow cell has an optical length of 15 mm, but other length are also available, as shown in Fig. 10.9.

The stream coming to the colorimeter is segmented with air bubbles. Since air bubbles would scatter the light in the colorimeter, they are removed by passing the stream through the horizontal leg of the *debubbler*. The air bubbles float upward and then flow to waste with excess fluid. The solution, free of air, is pumped downward through a fine-bore, nonwetting polyethylene tube to the flow cell.

The optical system within the colorimeter is shown in Fig. 10.10. The colorimeter lamp is fed from a constant voltage 6.3-V transformer. The light

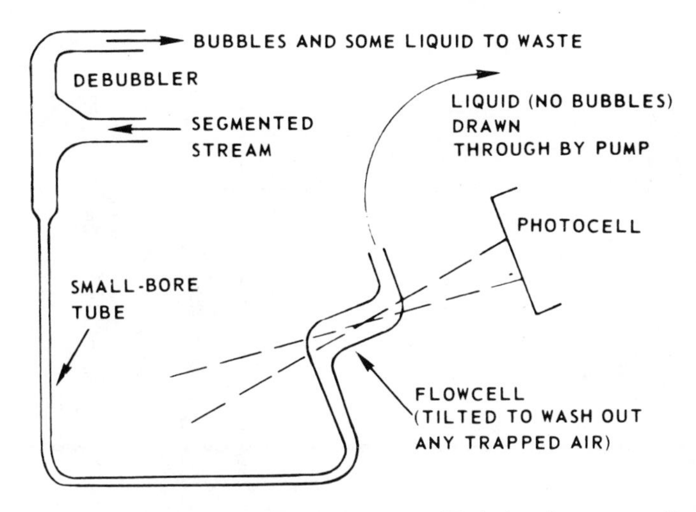

Figure 10.9 Flow cell and debubbler. (Courtesy of Technicon Instruments Corp.)

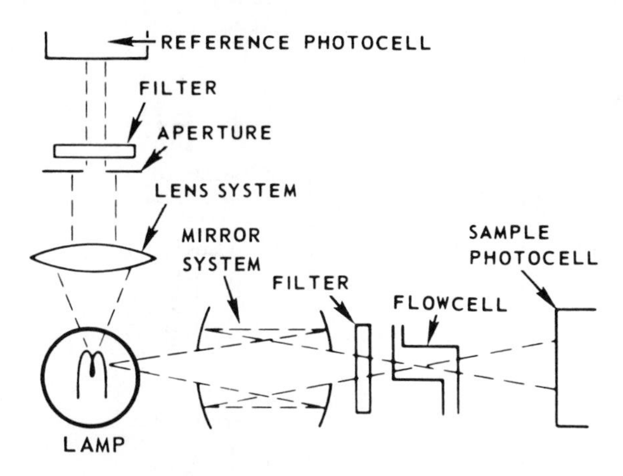

Figure 10.10 Colorimeter: optical system. (Courtesy of Technicon Instruments Corp.)

in one direction is collimated through a lens system, reduced to a single color band by an interference filter, and converted to electrical energy by the reference photocell. Another beam passes through a mirror system, an interference filter, and the flow cell to the sample photocell. A mirror system is used because it can collect a greater amount of light than a lens system, and thus increases the intensity of light passing into the flow cell. The electrical outputs from the photocells are applied to the recorder in the manner shown in Fig. 10.11.

The operation of the system is shown diagrammatically in Fig. 10.11. If the output from the sample cell, as it reaches the wiper of the slide wire, is not the same as the voltage being tapped off the slide wire by the wiper, a current will flow. This dc current is converted into ac by means of a chopper,

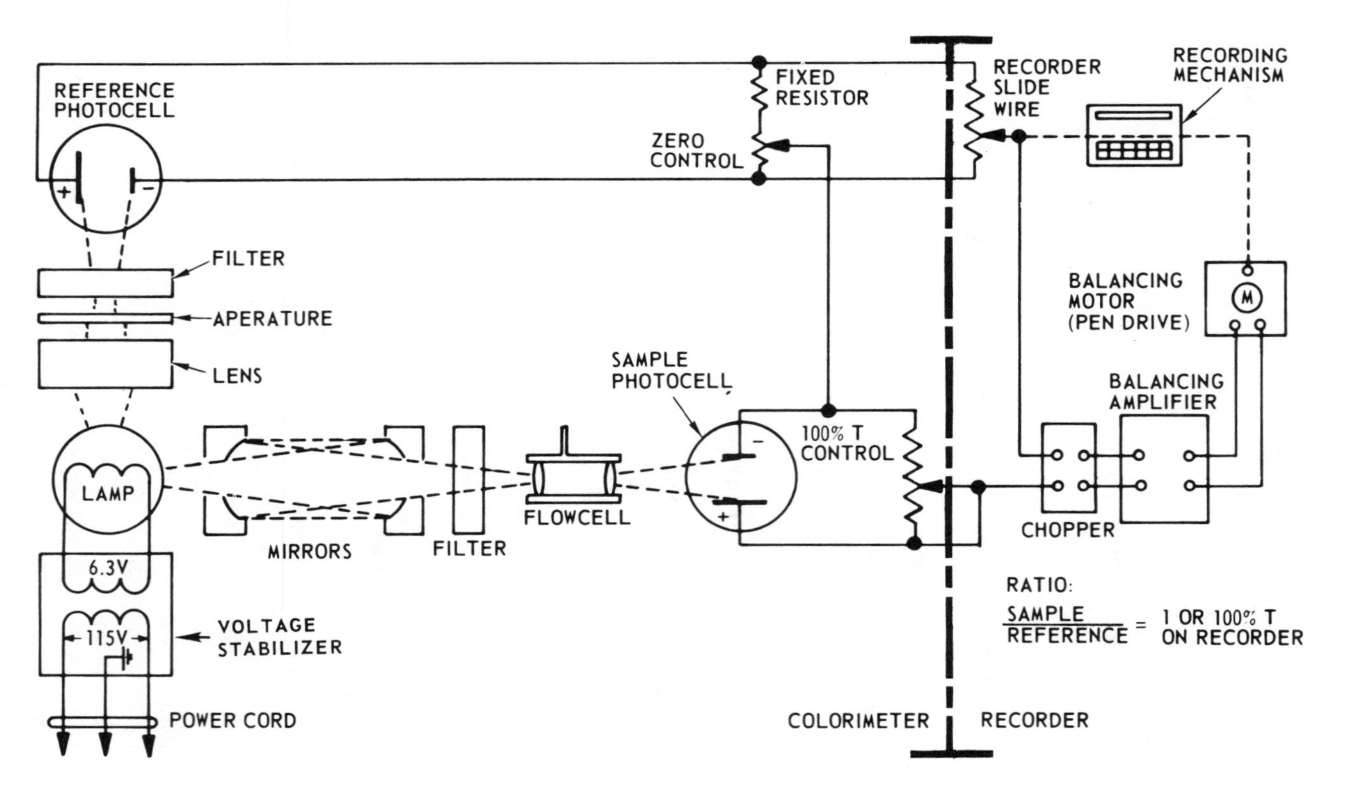

Figure 10.11 Colorimeter: electrical system. (Courtesy of Technicon Instruments Corp.)

turned into a voltage signal by a transformer, and fed to the input of an amplifier. The amplifier then generates power to drive a balancing motor, coupled to the wiper arm, so that the wiper arm moves in a direction to tap off a greater or lesser voltage, until the voltages balance. In this condition there is no current flowing through the wiper arm, and thus no power to move the pen. The pen takes up a position by the voltage output from the sample photocell as a fraction of the voltage output from the reference cell.

The instrument is set by filling the flow cell with water and then adjusting the outputs of the sample photocell and reference photocell to be equal so that their ratio is unity; the pen then reads 100% transmission (100%T) (optical density of zero). As colored liquids are brought into the flow cell, the amount of light reaching the sample photocell decreases and the pen moves to read in proportion to the percentage of transmission. If the light transmitted from the flow cell were zero (optical density of infinity), the recorder pen would read 0%T. This condition is simulated by placing a blank aperture in the sample side light path so that no light falls on the sample photocell, and then adjusting the zero control knob to make the pen move to the 0%T position.

In some chemical tests, the color change available over the desired working range is not very great. In these cases, the sensitivity of the colorimeter is improved by using a flow cell having a longer light path. Similarly, any test that produces a very dark color can be improved by using smaller flow cells. The sizes are 8, 15, and 50 mm.

The filters used in the colorimeter are specially selected for the narrow bandwidth of the light they transmit. Two filters are necessary and are labeled "reference" and "sample." The sample filter must be used only on the sample side. It is a filter chosen because the light transmitted has a wavelength which conforms accurately to that stated on the label. The filter used on the reference side, however, need not be so precise since it serves only to generate a voltage in the reference photocell approximately equal to that on the sample side.

The wavelength of the filter is designated on the filter holder. There are two other designations on the holder: one shows the bandwidth at half-height and the other shows the percentage of light transmitted. A designation reading 440-18-28 means that the filter will transmit light having a mean wavelength of 44 nm or 440 mμ, the amount transmitted in 28% of the input and the bandwidth is 18 nm at half-height (14% tramsmission), as shown in Fig. 10.12.

The function of the recorder is to display on a graph the variations of the electrical output ratio between the sample and reference photocells. The correct amplifier gain to use is close to the maximum that can be applied, short of having the pen oscillate.

In the system described so far, the sample has been mixed with reagents and eventually passed through the flow cell of the colorimeter. In many instances the sample may contain solids, or other interfering substances, which would invalidate the colorimetry. In conventional analytical chemistry, the sample

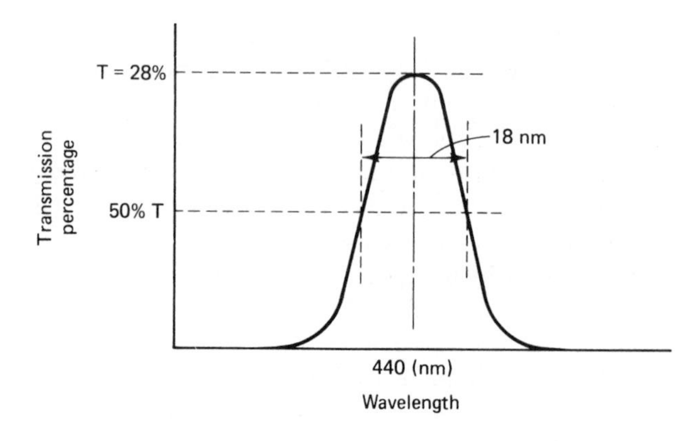

Figure 10.12 Colorimeter: interference filter curve. (Courtesy of Technicon Instruments Corp.)

would first have to be either centrifuged or filtered after addition of a precipitating agent. These operations are handled in the AutoAnalyzer by making use of the principle of dialysis.

The essential part of the dialyzer, shown in Fig. 10.13 is a matched pair of plastic plates having semicircular grooves. Tubing extensions are provided to give access to the grooves. The two plates are clamped together with a membrane between them. Think of this assembly as a tube with a $\frac{1}{16}$-in, bore, about 7 ft long, split down the center by the diffusing membrane. A stream carrying the sample (the donor) flows through one of these semicircular grooves while another stream (the recipient) flows on the other side of the membrane. Dialyzable materials contained in the sample dialyze through the membrane and are picked up by the recipient stream. On leaving the dialyzer, the donor stream usually goes directly to waste, while the recipient stream flows to the colorimeter, after first being subjected to further treatment as may be demanded by the analysis.

Particulate matter on the sample side passes over the dialyzer membrane and flow to waste, while dissolved material diffuses through. Macromolecules also flow to waste, since they cannot pass through the fine pores of the membrane.

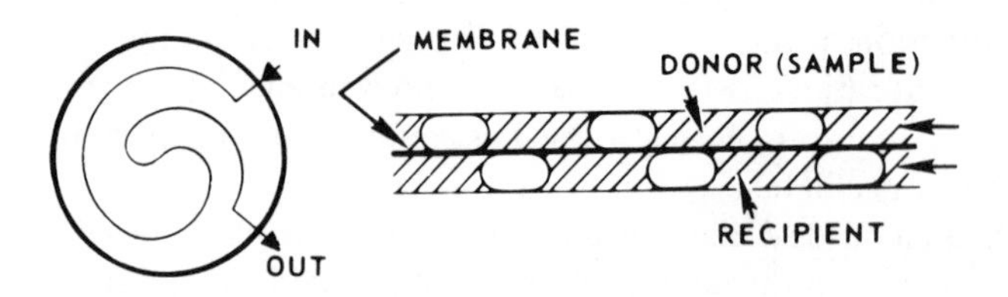

Figure 10.13 Dialyzer: sample and recipient dialyzing action. (Courtesy of Technicon Instruments Corp.)

The actual percentage of material that dialyzes across the membrane ranges from about 25% down to as little as 5%, depending on the time the stream is held in the dialyzer and on the dialyzing characteristics of the material concerned and the type of membrane. The amount of material dialyzing is strongly influenced by the temperature, so it is necessary that the streams (and the dialyzing element itself) be kept at a constant temperature. Therefore, the entire dialyzer is inserted into a constant-temperature water bath. Time-delay coils are submerged in the water bath to stabilize the temperature of the streams before they reach the dialyzer.

The functioning of the AutoAnalyzer as a system is more easily understood by studying the flow diagram of a typical Technicon determination of glucose as shown in Fig. 10.14. Figure 10.15 shows the Technicon continuous-flow AutoAnalyzer system.

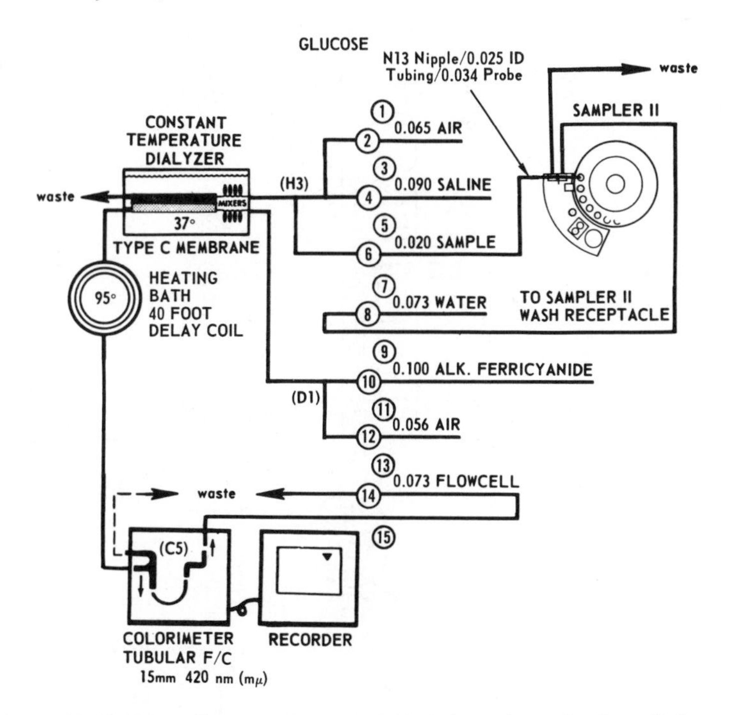

Figure 10.14 Flow diagram of a typical Technicon determination of glucose. (Courtesy of Technicon Instruments Corp.)

10.3.3 Continuous-Flow Technology

Continuous-flow analysis is the basis of the Technicon sequential analysis system. Samples and reagents, segmented by air bubbles, are aspirated by means of a proportioning pump to deliver the precise volume ratios needed for

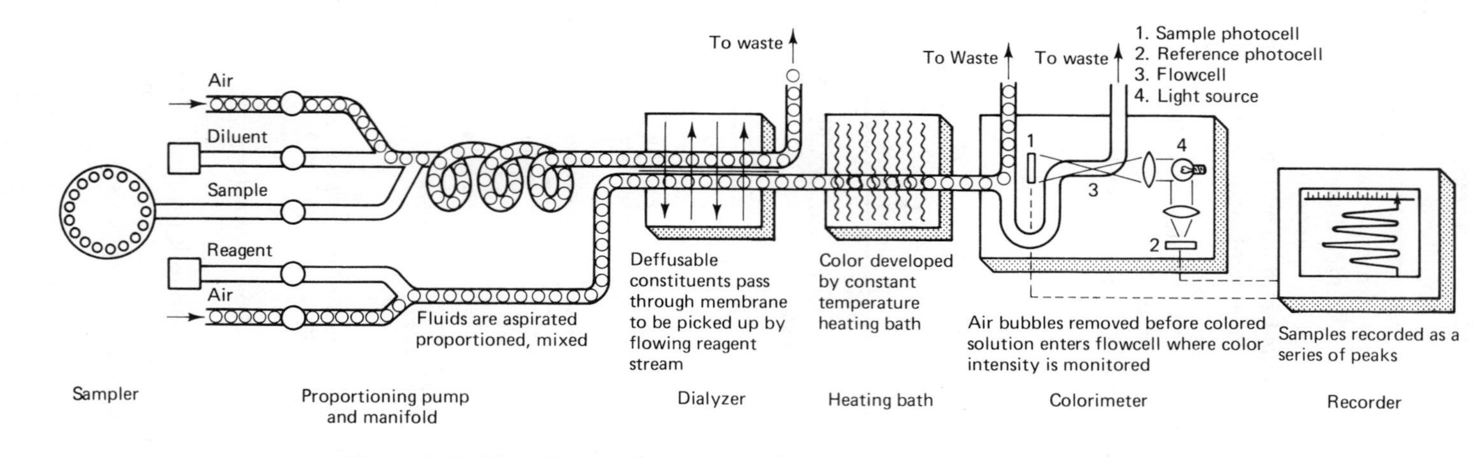

Figure 10.15 Flow diagram of a continuous flow Technicon AutoAnalyzer system. (Courtesy of Technicon Instruments Corp.)

specific analyses. The samples continuously flow through a hydraulic system and are brought together with reagents under controlled conditions, causing specific chemical reactions which are analyzed and measured.

The system continuously monitors the reaction, but the actual measurement is made only at the steady-state plateau after virtually all effects of possible sample reaction have been eliminated.

Figure 10.16 shows a continuous-flow scheme of the SMA-11 system which is a computer-controlled biochemical analyzer used for curve plotting.

The Technicon SMA II provides for continuous curve monitoring. The computer constantly checks all chemistry curves, comparing them to stored, ideal curve parameters for each specific chemistry, while continuously monitoring reference channel data. The computer determines, verifies, calculates, and prints out these results on one report form.

The Technicon SMA II provides for continuous curve monitoring. The computer constantly checks all chemistry curves, comparing them to stored, ideal curve parameters for each specific chemistry, while continuously monitoring reference channel data. The computer determines, verifies, calculates, and prints out the proper value for each test.

Phasing is automatic on the Technicon SMA II. No operator involvement is necessary. The system automatically stores all data from a patient sample until tests requested on that patient have been completed. The system then prints out these results on one report form. Figure 10.17 shows the SMA II traditional and digital outputs for convenient reporting.

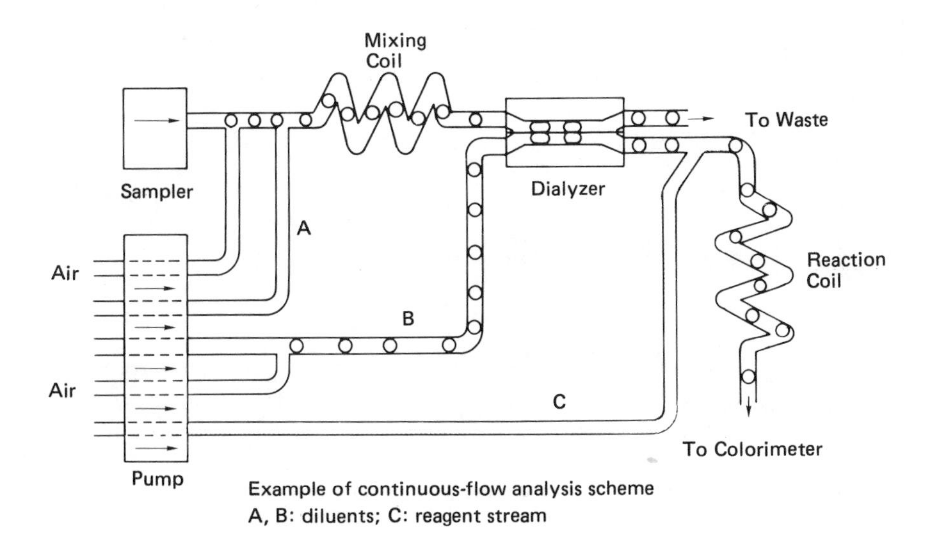

Figure 10.16 Continuous-flow Technicon SMA II system scheme. (Courtesy of Technicon Instruments Corp.)

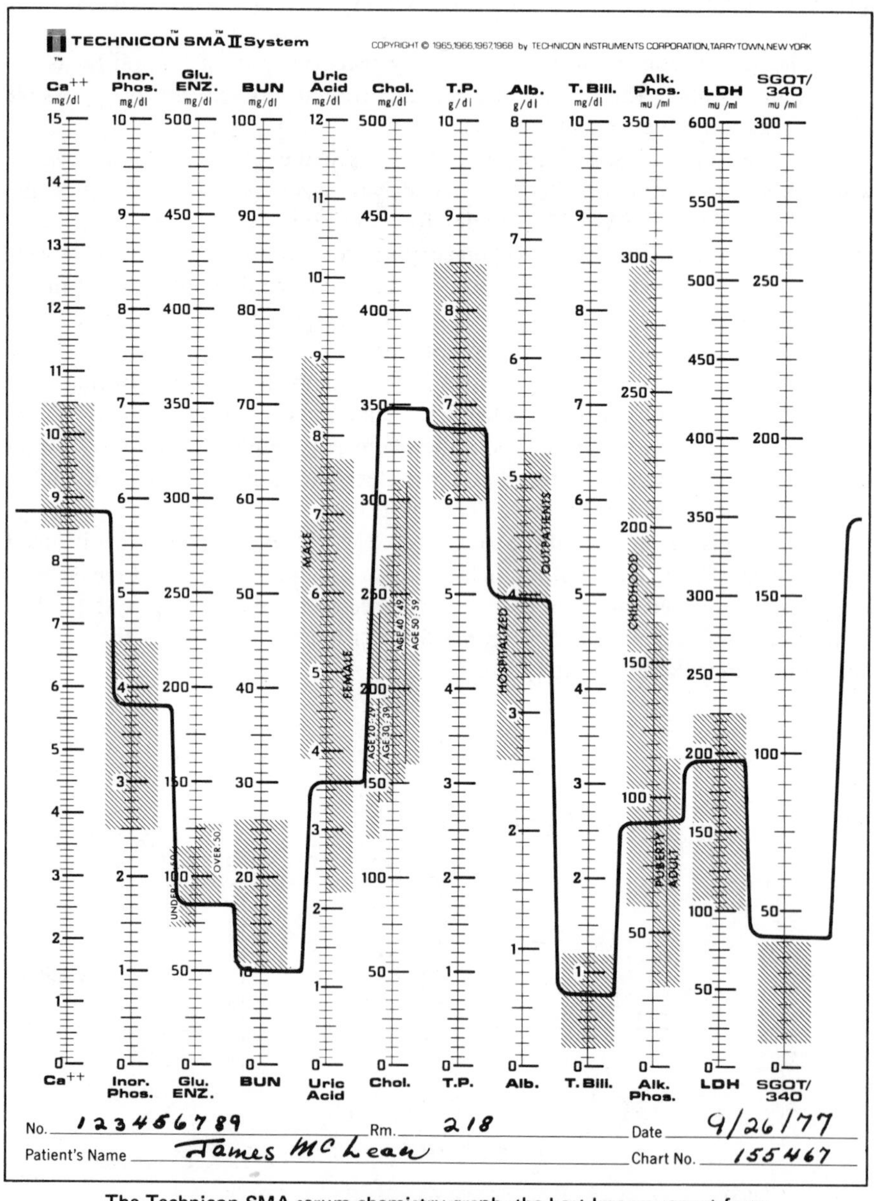

The Technicon SMA serum chemistry graph, the best-known report form in the clinical chemistry laboratory, is included for every patient report.

(a)

Figure 10.17 Technicon SMA II report. (Courtesy of Technicon Instruments Corp.)

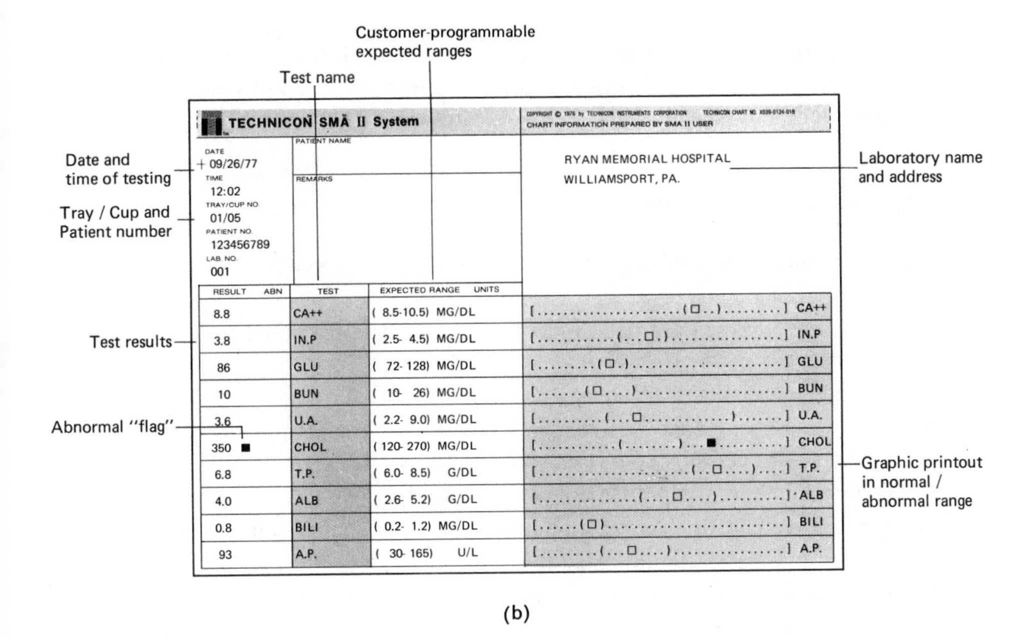

(b)

Figure 10.17 (*Continued*)

10.3.4 Other Automatic Analyzers

Besides continuous-flow automatic analyzers, discrete types of automated analyzers are now also being used with good results. These analyzers perform all tests in test tubes mounted on a carousel-type carrier or a chain belt, with the test tubes being rinsed once the analysis is completed. Another discrete sample analyzer called the DSA, manufactured by Beckman, uses disposable trays as reaction vessels. All automatic analyzers of this type use automatic syringe-type pumps to dispense the sample and add the reagents. After the incubation, the sample is aspirated into a colorimenter cuvette.

One other automatic tester has all the reagents for a given test to be contained in premeasured quantities in pockets of a plastic pouch. The sample is injected into another section of the pouch and the reagents added by destroying the separations between the pockets. This is accomplished by squeezing the reagent pockets. After the color reaction has been completed, the colorimetric determination is performed in the (transparent) plastic pouch with a special colorimeter.

Technicon has also developed the AutoSlide for achieving superior blood smears automatically for use in histology.

Automated clinical measurements are a rapidly growing field. To understnad the sophistication of this changing clinical field, the reader is advised to consult the latest journals and to contact manufacturers in the field in which he or she is interested. The bioelectronic measurements and microprocessors used in computers have truly made an impact on clinical analysis, and will make the devices described today obsolete tomorrow. Even in today's society, automated clinical measurement is truly complex.

10.4 REVIEW QUESTIONS

1. Discuss the meaning of photometry, colorimetry, and spectrophotometer.
2. Of what is blood composed?
3. Discuss the purpose of red and white blood cells. How many red and white blood cells are there in 1 mm^3 of blood?
4. Discuss the meaning of hematocrit.
5. Discuss a measurement for blood glucose levels. What is the significance of a blood glucose concentration value over 150 mg/100 ml?
6. Discuss the meaning of serum enzyme concentration, serology and histology.
7. Discuss a method to determine the measurement of sodium and potassium.
8. Why are automated clinical analyses used?
9. Discuss an automated continuous-flow clinical measuring system.
10. Discuss a method to measure oxygen saturation.
11. List five electrolytes that can be measured by automation.

10.5 REFERENCES

1. Thomas, H. E.: *Handbook of Automated Electronic Clinical Analysis*, Reston Publishing Co., Inc., Reston, Va., 1979.
2. Thomas, H. E.: *Handbook of Pharmaceutical and Clinical Measurement and Analysis*, Reston Publishing Co., Inc., Reston, Va., 1979.
3. Jacobson, B., and Webster, J. G.: *Medicine and Clinical Engineering*, Prentice-Hall, Inc., Englewood Cliffs, N.J., 1977.
4. *MED Equipment Buyer's Guide*, Medical Electronics, Pittsburgh, Pa., 1974.

11

ULTRASONIC MEASURING SYSTEMS

11.1 INTRODUCTION

Echo sounding has long been used for the measurement of distance by animals, such as bats, dolphins, and certain birds. It was employed by human beings shortly after World War I. After 1945, it found extensive use and applications in nondestructive testing of materials and in medical diagnosis. When an electric current is introduced into a piezoelectric crystal, the crystal vibrates and produces sound waves. If a very rapid, fluctuating current is employed, the sound produced is well beyond the range of human hearing in the realm of ultrasound. Each time a beam of this energy crosses the boundary or interface between two structures or tissues of different density, some of the energy is reflected at the interface and the echoes are picked up by the crystal and amplified as an electrical signal. The depth or position of a large number of reflecting structures can be plotted, as is done in charting the bed of the ocean. These echoes are immediately visible for interpretation on an oscilloscope screen and provide information for medical diagnosis.

Until 1952 the use of ultrasound in medical diagnosis has been confined to unidimensional echography. That is, transmission and pulse echo methods are comparable to a needle biopsy. By utilizing a 15-MHz pivotal crystal mounted in a water chamber closed by a rubber membrane, a simple linear movement of the probe at right angles to the beam produces the first two-dimensional echogram. Using this two-dimensional method, they examined many palpable tumors of the breast and were able to diagnose preoperatively 26 of 27 malignant tumors and 43 of 50 benign tumors confirmed by microscopical diagnosis.

This two-dimensional technique, although geometrically different, had a common limiting feature: that a beam reached any one point in a tissue from a single probe or transducer position. This defect was alleviated by Howry and coworkers when they developed the first practical instrument that employed a compound scan, called a *tomograph.* In 1952 an ultrasonic instrument termed a *sonoscope* was used in which a scanning method using lower-frequency, lower-power, lense-focused ultrasonic pulses produced echograms of high technical quality. The area to be examined was immersed in water to facilitate the propagation of sound and by scanning a complete 360° scan produced a picture that presented a cross section of excellent definition. Their ultrasound pictures of neoplastic breast masses and cross section through the forearm of one of the investigators were the first produced by the pulse echo method that showed the interior construction of solid human tissue.

Another ultrasonic examining technique developed within the last decade is based on the Doppler principle. By transmitting and receiving ultrasound, the instrument utilizing Doppler techniques detects the motion of organs and blood within the human body. A transducer containing both a transmitting and a receiving crystal is placed against the patient's chest or abdomen. A beam of low-intensity ultrasound is transmitted into the body as a continuous beam and part of that sound is reflected back from the internal structures. Ultrasound received from motionless structures has precisely the same frequency as the transmitted sound and is not heard. Ultrasound received from moving organs or flowing blood is slightly shifted frequency from the transmitted sound, and this difference in frequency is converted to an audible signal. For example, several distinctive Doppler sounds of clinical importance can be identified from the pregnant uterus. Most easily heard is that of the fetal pulse. It is produced by the moving heart or blood flow in the umbilical cord or a fetal artery. Depending on the structure through which the ultrasonic beam is directed, a distinctive placental sound is also easily identified and used to localize the placenta.

On the basis of the five major stepping stones in the historical development of diagnostic ultrasound—piezoelectric effect, transmission method, pulse echo method, scanning techniques, and Doppler techniques—many medical applications have been developed. Diagnostic ultrasound is thought to be a safe, painless, and riskless technique.

The power levels employed in ultrasonic testing are very low and on the basis of extensive clinical and experimental data, these tests are considered quite safe for the patient. Ultrasonic examinations are totally free of discomfort, completely external, repeatable as often as indicated, and, of course, do not expose the human body to ionizing radiation and therefore can be used in medical situations where x-ray examination cannot be applied.

The diagnostic ultrasonic techniques are a true interdisciplinary science that combines medicine and engineering in a way that assists the physician, bioengineer, and technologist in the struggle against disease by providing tools and techniques for precise and objective medical diagnosis. To name just some of the applications that the diagnostician has used in the past two decades:

1. In neurology, for example, ultrasonic testing is used to detect brain tumors, clots, and identify subdural hematoma within several hours after head injury occurs.

2. In cardiology, diagnostic ultrasonic techniques have been developed to detect obstruction or leakiness of the various heart valves; to study the motion of the various heart structures; to detect fluid accumulated in the pericardial sac, called pericardial effusion; to measure in liters per minute the blood output or cardiac output of the heart; and to provide an analysis of a cardiac valve motion by ultrasound that may help the physician detect abnormal performance or even clot material within the artificial valve. The routine use of ultrasonic techniques in cardiology is of particular value when the patient is too sick to be catheterized or when the information obtained by cardiac catheterization, as in a radiography, is erroneous or inconclusive or when the diagnosis remains obscure.

3. In ophthalmology, ultrasonic techniques have been very useful in detecting intraocular tumors, vitreous hemorrhages, and detached retinas. Many advances have been made in the use of ultrasound in opthalmology since 1960. The areas of endeavor include:

(a) Determination of axial length of the globe
(b) Diagnostic evaluations of introacular pathology
(c) Localization and extraction of intraocular foreign bodies
(d) The therapeutic and sometimes destructive use of ultrasound in the eye

The axial diameters of the eye globe, as a matter of fact, can be measured so accurately with ultrasound that these results can be used for diagnostic purposes in evaluating globe size in various clinical conditions such as myopia and glaucoma.

4. In the field of obstetrics, a Doppler ultrasonic instrument provides the obstetrician with a method by which the presence of fetal life can be determined conveniently at an earlier date than by many other means. A fetal impulse can be detected in all cases of live pregnancy of 12 weeks' gestation or more. Furthermore, the fetal impulse when present is always distinct. In some cases the heartbeat of a fetus can be heard at 10 weeks' gestation, particularly if the uterus is slightly elevated. After 12 weeks' gestation, a correct diagnosis of fetal life can be made in 94% of the cases. Doppler ultrasound, discussed in Section 9.3.2, allows the placenta to be localized at least 8 to 10 weeks before any other method of localization. Furthermore, an ultrasonic machine is capable of measuring the biparietal diameter of the fetal head and relating that to the weight of the fetus. This is important in evaluating normal fetal growth.

5. In the field of urology, diagnostic sound can be very useful in plotting an outline of the bladder and in computing residual urine volume in cubic

centimeters. This avoids the pain and risk of infection of catheterization, and comparing these results with several hundred patients where urine was withdrawn by catheter, the results of residual volume agreed within 10%.

6. Ultrasonic examination is also capable of locating foreign bodies in any part of human tissue, whether you have a piece of metal in the eye or a piece of glass in the leg. As a matter of fact, when it comes to glass, x-ray is not capable of seeing glass. Therefore, ultrasonic examination can be superior to x-ray in finding small pieces of glass: for example, if a person steps on shattered glass and has to have some pieces extracted from the sole of his or her foot.

7. Ultrasonic techniques can also detect and measure tumors in various parts of the human body, whether in the eye, the breast, the abdominal region or within the heart.

8. Within the medical field, diagnostic use of the ultrasonic echo method is becoming increasingly frequent. In the past, clinical diagnostic ultrasonic techniques had been used predominantly in brain work, detection of tumors, and cardiology diagnosis. However, significant progress has followed in the ultrasonic visualization of abdominal organs and application of this technique to obstetrics and many other fields. Developments during the last decade, especially in the last three to four years, have shown continuing refinement and advancement of ultrasonic techniques in all the areas previously cited, and point toward the ever-increasing potentialities of ultrasound in many other fields of diagnostic medicine.

9. Two-dimensional echocardiography and real-time measurements are never techniques which assist the clinician in diagnosis. Instruments are also being digitized for ease of operation and readout. Three-dimensional echocardiography showing a cross section of any part of the body is experimental and shows the physiological parameters, such as the heart valves, in motion.

11.2 ULTRASONOGRAPHIC TECHNIQUES

Ultrasonography is a technique by which ultrasonic energy is used to detect internal body organs. Bursts of ultrasonic energy are transmitted from a transducer through the skin and into the internal anatomy. When this energy strikes an interface between two tissues which have different acoustical impedance, reflections are returned to the transducer. The sound waves utilized are in the very high frequency range from 1 to 15 MHz. At these frequencies the sound waves obey all the laws of optics and can be focused into nearly parallel beams capable of penetrating body tissue.

The *transducer* is a piezoelectric crystal capable of both transmitting and receiving ultrasound energy. Pulses approximately 0.01 ms wide are emitted at a rate of 400 to 1000 per second. Between pulses, the crystal picks up reflected echoes, which are converted to electrical signals for display on an oscilloscope.

Although the basic function of ultrascan equipment is to measure distances between interfaces that separate body structures by timing echoes, this timed echo information is usually processed into different display modes. The forms of display are:

1. A-mode scan
2. T-M or M-mode scan
3. B-mode scan
4. C-mode scan

A-mode scan ultrasonography (Fig. 11.1) displays the amplified echo on the vertical channel of an oscilloscope, with the horizontal channel being deflected by a conventional sweep generator. This sweep generator is triggered from the impulse signal and the time delay between the beginning of the sweep and the echo appearing on the screen is proportional to the depth.

Actually, the A-mode scan is an amplitude of the echo as a function of distance, so that you would see the amplitude of the echo in the *Y*-direction and the distance in the *X*-direction corresponding to distance into the body.

An important factor in observing small structures is the resolution of the system. The higher the frequency, the shorter the wavelength and the better the resolution capability.

Since it is desirable to have like structures at different body depths appear similar, an electronic means of compensating for attenuation is used.

The time-motion scan is called the T-M scan mode or M-scan mode. Table 11.1 gives clinical sites and applications of the A-scan system.

Ultrasonography using the B-scan mode technique is referred to as *ultra-*

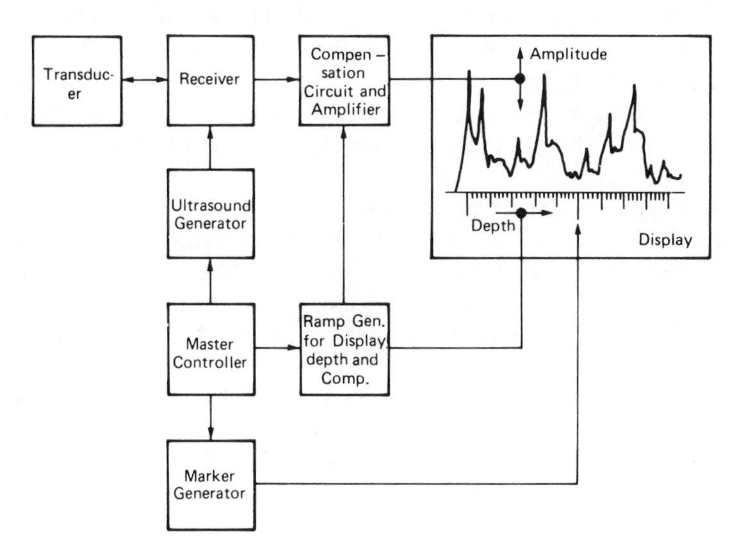

Figure 11.1 Block diagram of a simple A-mode equipment. (Courtesy of Brüel and Kjaer, Denmark.)

TABLE 11.1 Some Clinical Applications of the A-Scan Systems

Field	Applications
Mammography	Diagnosis and localization of malignant and benign tumors
Cardiovascular surgery	Aortic aneurism Pericardial effusion Pulmonary embolism Septal defects (using intracardiac probe) Mitral and aortic stenosis Mitral prolapse
Gastroenterology	Liver abscess, disease
Genitourinary surgery	Localization of renal calculi at operation Bladder dimension
Neurology	Brain midline localization: basis of method Intracranial pressure Others
Dentistry	Examination of the pulp Cavity localization
Obstetrics and gynecology	Fetal cephalometry Hydatidiform mole Placental localization
Opthalmology	Axial-length measurement of eye Diagnosis of disease Foreign-body localization and extraction Special probes
Miscellaneous	Disease at various sites

sonic scanning. It is a light intensity vs. time display. The result is a two-dimensional, cross-sectional presentation of the subject. Figure 11.2 shows the principle of a B-mode system with a scanning arm. The B-scan can be thought of a person cut in two and looking at the end portion or a sausage and looking at the cutting from the end of the sausage. Applications of the B-scan technique are outlined in Table 11.2.

The C-scan represents a cross-sectional display looking at a particular X-Y plane looking at the particular depth from the surface of the body. The term "real time" means that the picture is built up like it is in a cinema movie so that you get a real-life picture (i.e., seeing the heart really beating).

An example of the B-scan is shown in Fig. 11.3, which shows the liver and kidney in six frames.

A real-time portable hand-held ultrasonic device with the B-scan system for abdominal study is shown in Fig. 11.4. This system, called the Minivisor, is manufactured by Organon Teknika of Oklahoma City, Oklahoma. Pictorials with ultrasound images and corresponding anatomical drawings using the Minivisor are shown in Fig. 11.5.

Real-time B-mode scanning uses in cardiology as well as abdominal diag-

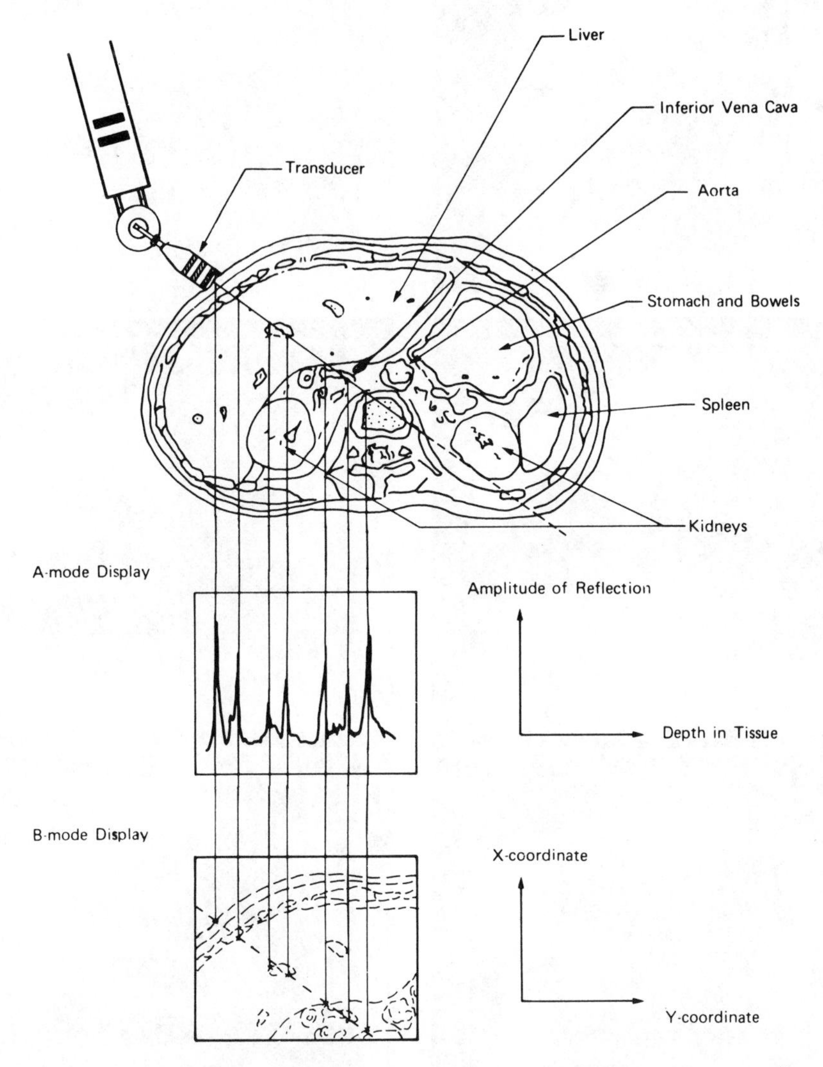

Figure 11.2 Principle of B-mode systems with a scanning arm. (Courtesy of Brüel and Kjaer, Denmark.)

nosis can be achieved by shifting the initial pulses along the linear array of transducers in a multielement transducer. The scan results in numerous parallel lines. By sweeping along the transducer array at a sufficiently repetition frequency, a real-time picture of moving organs is achieved (see Fig. 11.6).

A real-time two-dimensional image may also be obtained by sweeping one transducer mechanically or electronically over a certain area in rapid sequence. With sufficiently high sweep rate, the picture can be updated giving a real-time picture of the tissue examined.

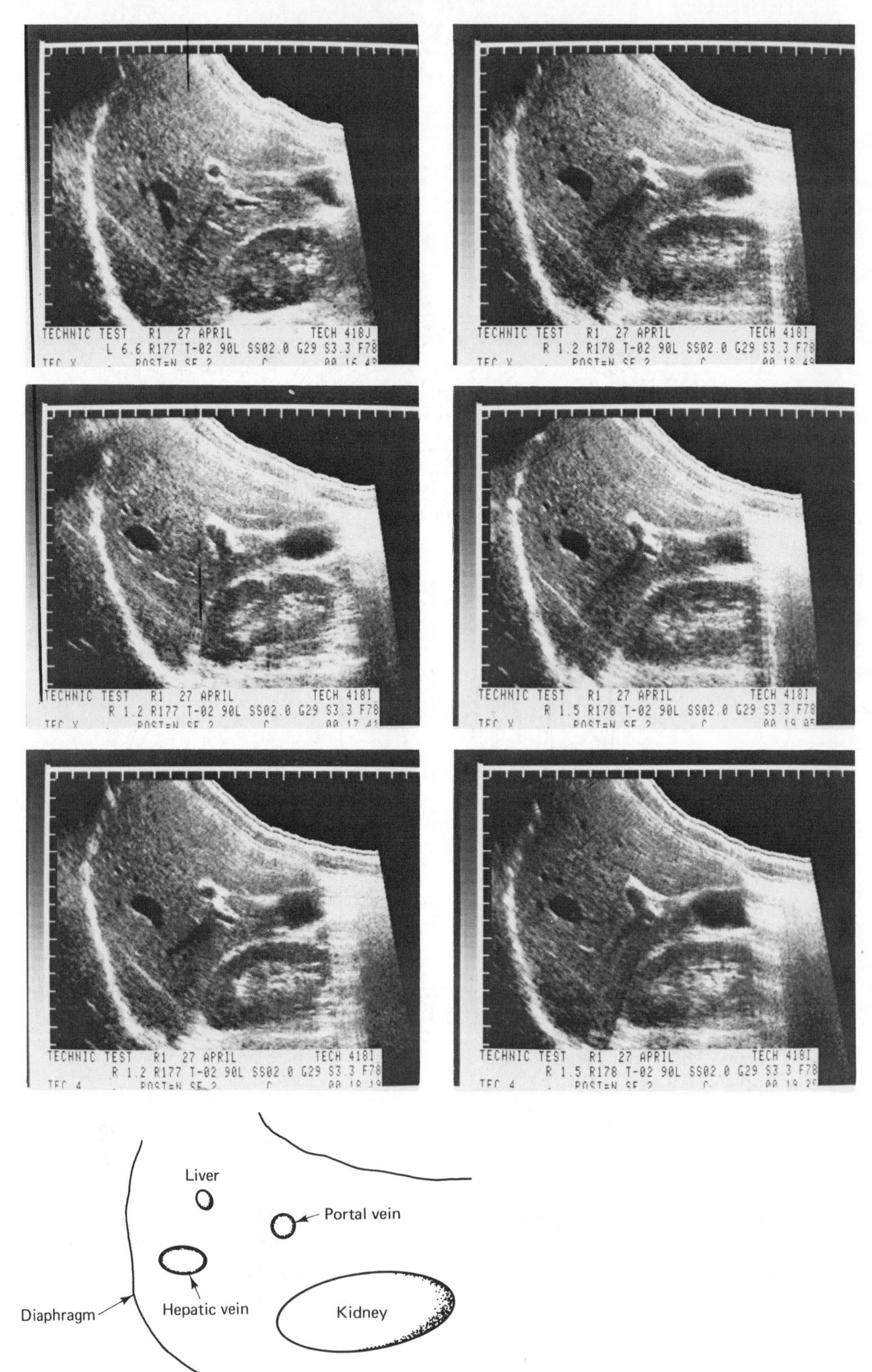

Figure 11.3 B-scans for liver and kidney.

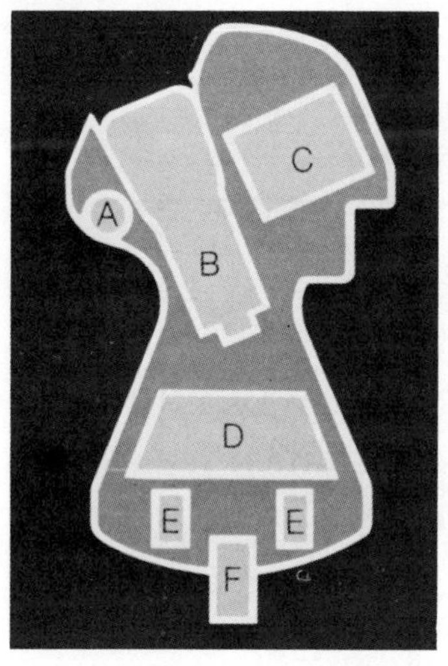

(a)

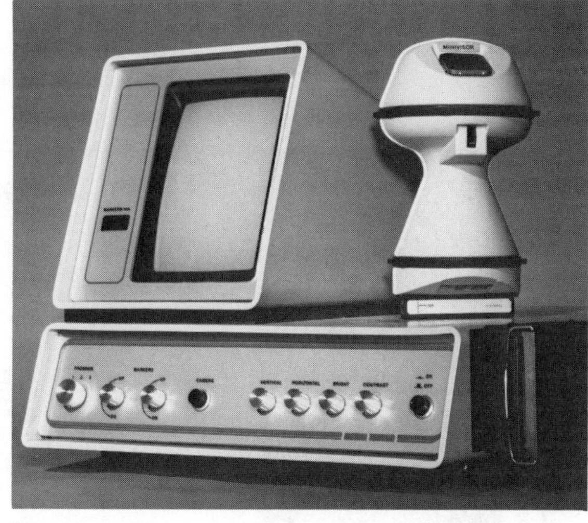

(b)

Figure 11.4 (a) Minivisor real-time portable hand-held B-scan system for abdominal studies: A, TGC switch; B, display tube; C, power supply; D, circuitry; E, batteries; F, transducer. (b) Hard-copy video display. (Courtesy of Organon Teknika Corp., Oklahoma City, Okla.)

TABLE 11.2 Some Clinical Applications of the Two-Dimensional B-Scan System

Field	Applications
Brain	Diagnosis and localization of malignant and benign tumors
Cardiovascular	Aortic aneurism
	Pericardial effusion
	Pulmonary embolism
	Septal defects (using intracardiac probe)
	Mitral and aortic stenosis
	Mitral prolapse
Gastroenterology	Liver abscess, disease
Genitourinary surgery	Localization of renal calculi at operation
	Bladder dimension
Neurology	Brain midline localization: basis of method
	Intracranial pressure
	Others
Dentistry	Examination of the pulp
	Cavity localization
Obstetrics and gynecology	Fetal cephalometry
	Hydatidiform mole
	Placental localization
Opthalmology	Axial-length measurement of eye
	Diagnosis of disease
	Foreign-body localization and extraction
	Special probes
Miscellaneous	Disease at various sites

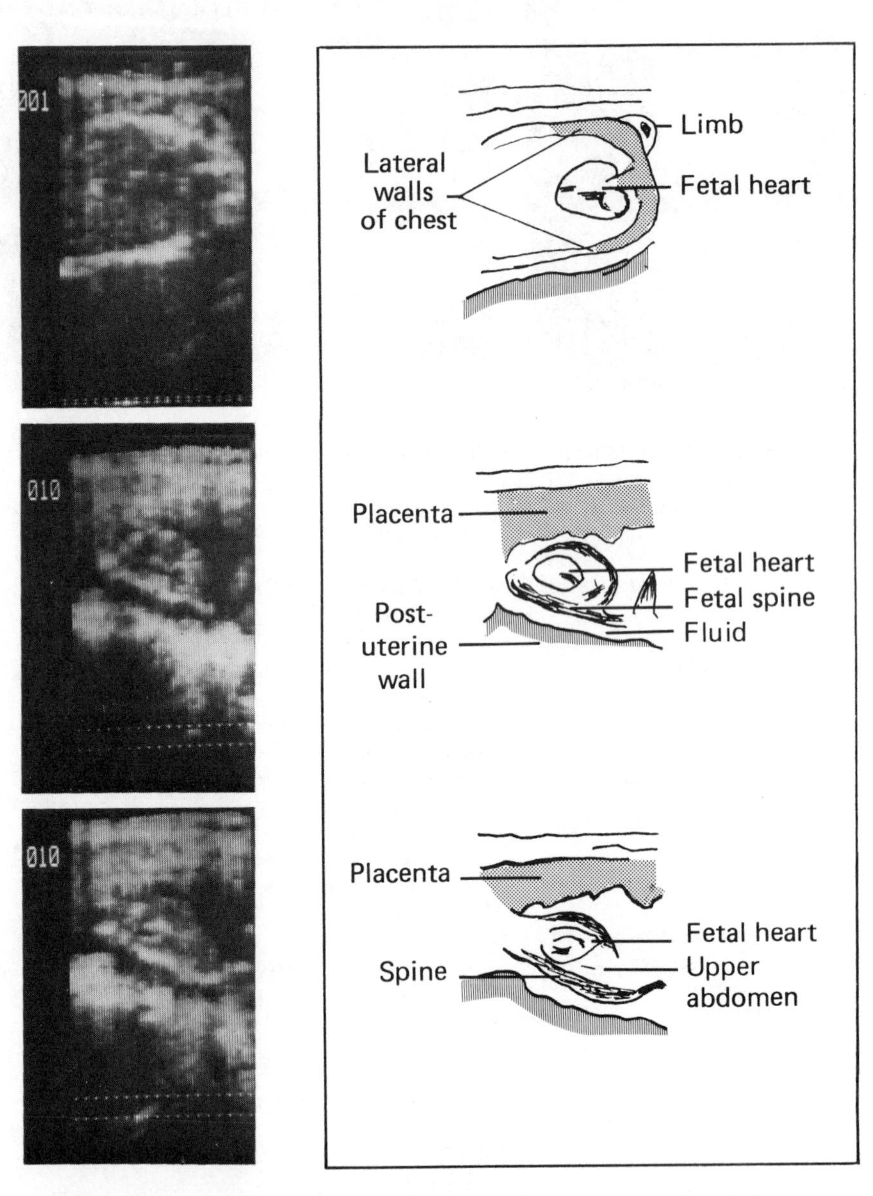

Figure 11.5 Ultrasonic images and anatomical images using the minivisor. (Courtesy of Organon Teknika Corp., Oklahoma City, Okla.)

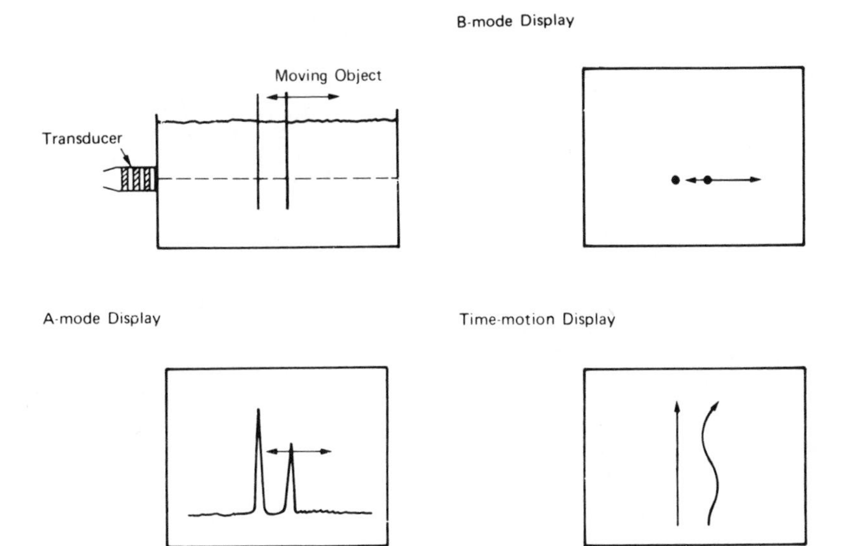

Figure 11.6 Buildup of a time–motion display. (Courtesy of Brüel and Kjaer, Denmark.)

11.3 THE MULTIELEMENT TRANSDUCER SYSTEM[1]

The scanning movement of a single transducer along a straight line in an upright position can be simulated electronically by electronic switching between numerous transducer elements placed in a straight line on the skin's surface.

On the display each transducer has its own line trace, echoes being shown in the normal way as small intensity-modulated dots. If the transducers are placed close enough to each other, the lines on the screen tend to merge together, in much the same way as the lines in a television picture. The display is no longer perceived as discrete echo traces, but as a two-dimensional picture of the structure underneath the transducer array. See Fig. 11.7 for a block diagram of a multielement transducer system.

If the electronic switching between the transducers is performed fast enough, the picture shown will be a flicker-free real-time two-dimensional display, giving living pictures of the reflecting boundaries. This allows the study of moving structures. Such as the heart and large abdominal vessels or fetal parts.

To give a compromise between a narrow, highly directional beam and a reasonable number of actual transducer elements per centimeter of skin surface,

[1]Sections 11.3 and 11.4 are excerpted from *Application of B & K Equipment to Diagnostic Ultrasound*, Bruel & Kjaer, Denmark, August 1978.

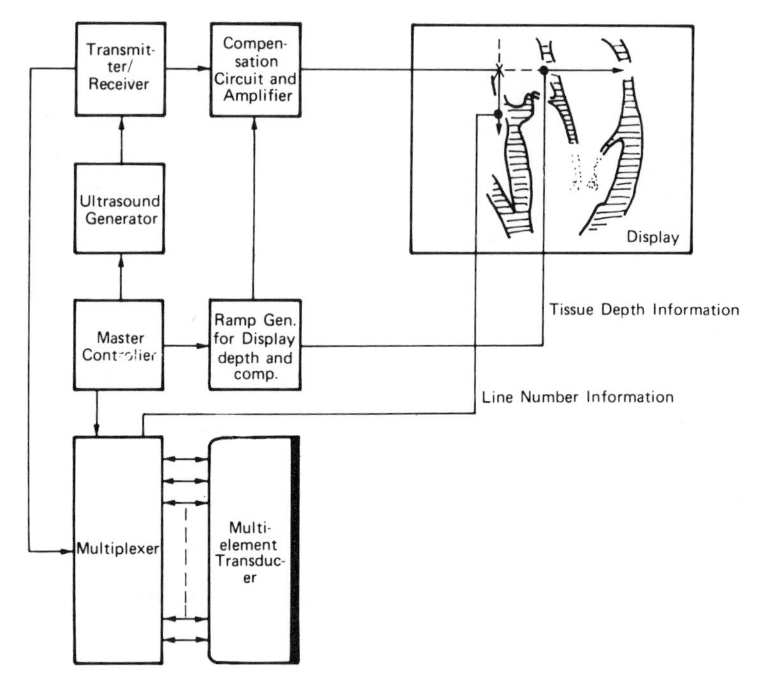

Figure 11.7 Block diagram of a real-time multielement transducer system. (Courtesy of Brüel and Kjaer, Denmark.)

elements are excited in pairs (1 and 2, 2 and 3, 3 and 4, etc.) by the system. This gives an active transducer area of a larger size, thus giving a beam of smaller opening angle; at the same time, the effective close-range resolution is not spoiled by having too great a distance between the transducer elements. Although the transducer beams will of course overlap, this in fact gives a substantial improvement in picture quality.

11.4 THE SECTOR-SCAN SYSTEM

On a sector-scan display, the echo traces appear to radiate from a single point, which is the hypothetical position of a fictive rotating or rocking transducer of very small size. The picture lines will form a sector of a circle, hence the name.

The actual number of lines in the sector will depend on the pulse-repetition frequency and the angular velocity of the sweep. In terms of depth in the examined tissue, there will be more lines per centimeter across the display close to the transducer than farther away with the same angular velocity. Thus, the resolution will gradually decrease with increasing distance from the fictive transducer point.

The block diagram of an ultrasonic sector-scan system of Fig. 11.8 reveals the same basic type of system as with the multielement transducer. However,

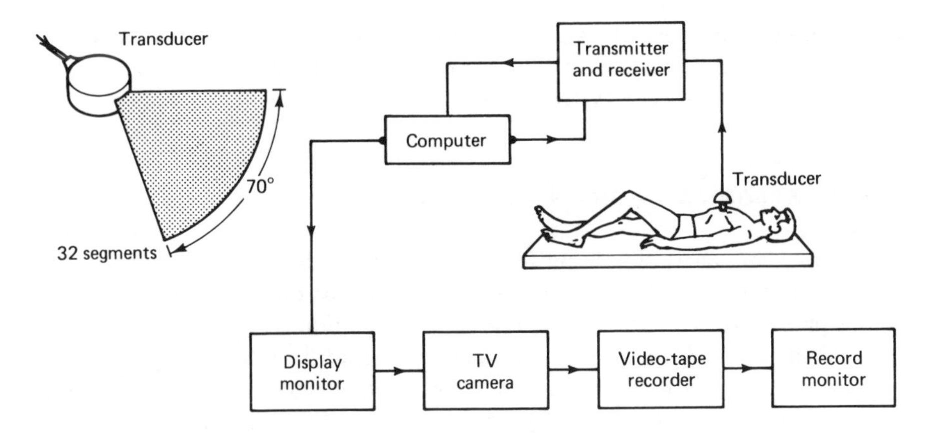

Figure 11.8 Ultrasonic sector-scan system. (Redrawn from Fig. 2, p. 80. Courtesy of *IEEE Spectrum*, January 1977, © 1977, IEEE.)

because of the synchronization system for the electronically synthesized or mechnically revolved or rocked "rotating" beam, the electronics have been complicated, whereas the transducer is simpler (and often cheaper and better) than a large multielement transducer.

A real-time sector scan of the right lobe of the liver with branching hepatic vein is shown Fig. 11.9. Simple electronically synthesized sector-scan systems

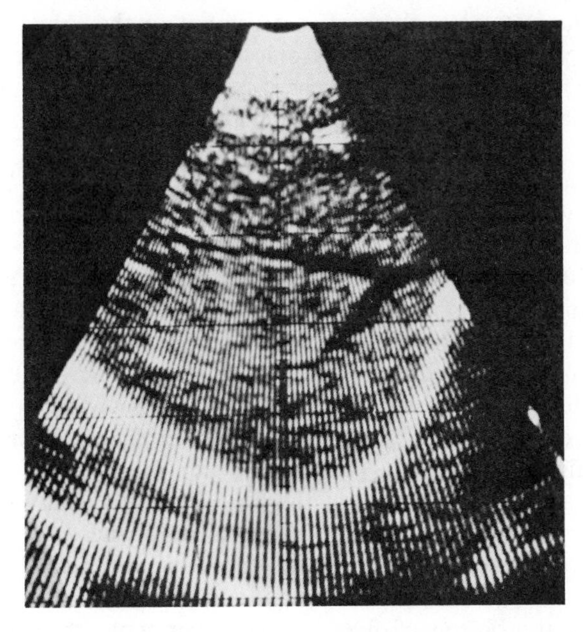

Figure 11.9 Real-time sector-scan of the right lobe of the liver with branching hepatic vein. (Courtesy of Brüel and Kjaer, Denmark.)

do not necessarily have the same directional characteristics when receiving as when transmitting, and will in this respect not be optimal when compared to other types of sector-scan systems.

11.5 ULTRASONICS IN BRAIN DISORDERS

For many years fairly simple pulse echo instruments have been used for determining midline shift of the brain following injury or disease. The name given to this procedure is *echoencepholography*. This technique is particularly valuable in accident cases, where there may be neither time nor facilities for angiographic studies, and the decision has to be made about emergency surgical intervention. In Japan such equipment now finds a place in specially equipped ambulances used for motor accident work, where emergency treatment, including surgery, is undertaken at the site of the accident because of the difficulties of getting serious cases back to the hospital through the intensely heavy traffic.

We can understand the principle of the echoencephalography procedure if we imagine a needle passing from one side of the skull through to the other side. The resistance that is felt when the needle penetrates the tissue surface is indicated on it by a mark corresponding to the amount of resistance. When the needle is pulled out, it is a one-dimensional report on the state of the path that the needle has followed through the head. This principle is realized when a sound pulse, instead of a needle, is sent into the head as a plane-wave packet. If a pulse hits interfaces, these offer resistance to the sound pulse and are partly reflected. The returning ultrasonic echoes are conducted to a display unit which converts the time differences between transmitter pulse and echo received so that they appear on a latitudinal axis, thereby creating a one-dimensional display which we call an A-scan.

The best frequency for routine echoencepholography is 2 MHz. In the brain, the wavelength of 0.75 mm is short enough to provide adequately accurate data about the site of the various interfaces of which the echoes originate (see Fig. 11.10). Furthermore, the area of amplification is large enough to pick up all reflections coming from the inside of the adult skull. A transducer at the frequency of 2 MHz and a diameter of 15 mm may be regarded as the optimum transducer because it is able to solve nearly every echoencephalographic problem. This probe produces a rectified ultrasonic beam of 80 mm in the direction of propagation so that a structure in the center of the brain can be easily localized.

Furthermore, manipulation of the lightweight transducer is very simple. A suitable application point in the temporal area where the surface of the 15-mm probe can make complete contact may be found in nearly every patient, whereas the 24-mm probe presents coupling problems.

For special problems of investigation, for example, the demonstration of epidural hematomas on the side of the hemorrhage of infantile hydrocephalue,

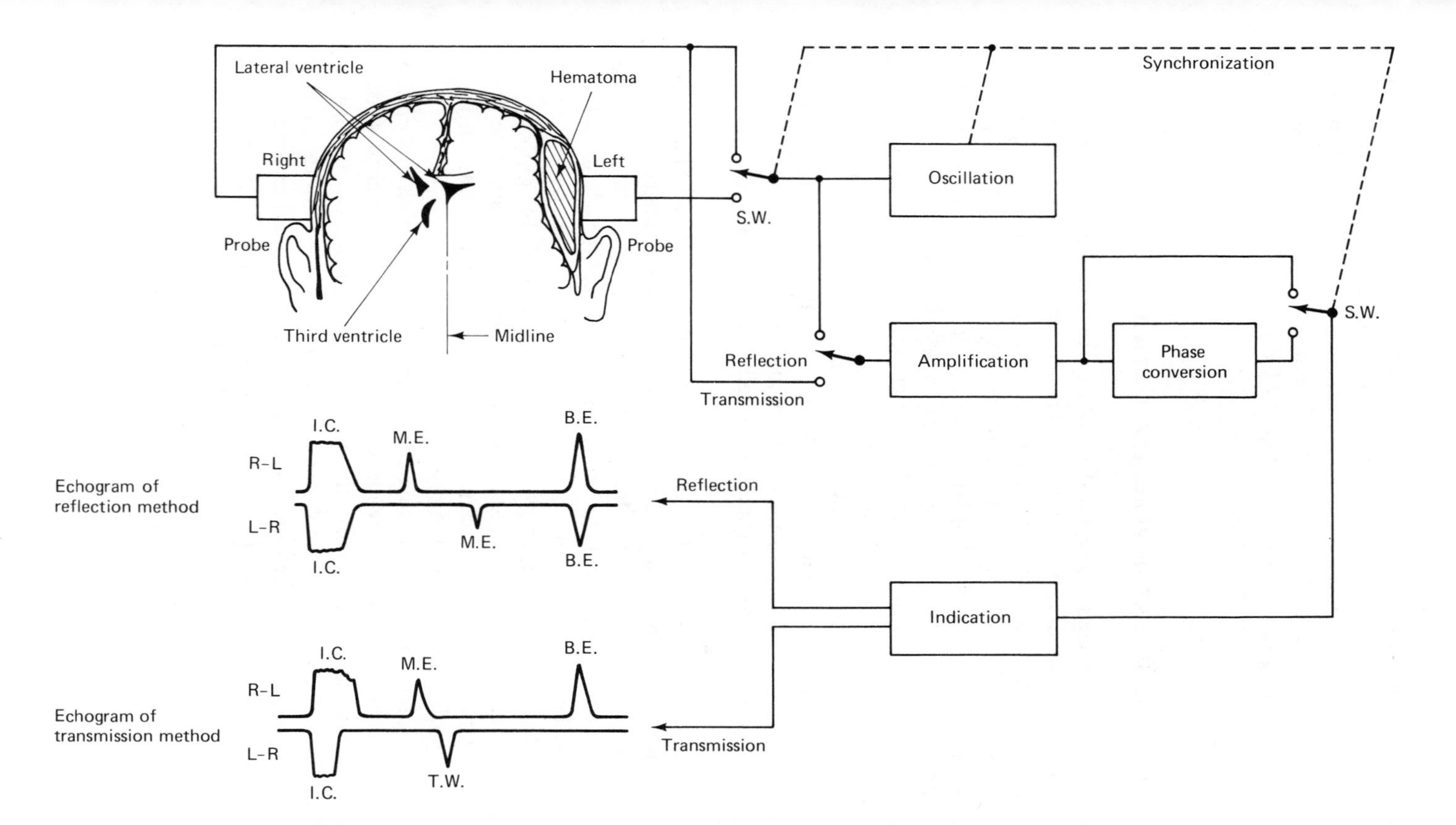

Figure 11.10 Block diagram of the echo encephalograph. (From J. M. Saba, Echo-Encephalography, *Medical Electronics and Data*, No. 5, pp. 96–101, September–October 1970.)

a transducer with the frequency of 4 MHz and a diameter of 10 mm has proved to be very useful because the depth of ultrasound penetration required in such cases is only slight. Since ultrasound of a frequency of 4 MHz is attenuated much more in bone, this condition is fulfilled. To achieve this with low-frequency transducers, special measures must be employed for the amplifier gain, time voltage controls, and reject controls.

11.6 ULTRASONICS IN CARDIOVASCULAR DISEASE

Since the advent of cardiac surgery, the demands on the accuracy of the cardiovascular diagnosis preceding the operation have increased tremendously. As a result, several new diagnostic techniques have been developed such as the measurement of the pressure on the heart chamber by heart catheterization and the various methods of angiocardiography. Nearly all of the methods are complicated and require a relatively large medical staff and involve some risk to the patient. They are also uncomfortable for the patient and therefore cannot be repeated frequently for control purposes.

In 1954, it was shown by two Swedish physicians, Edler and Hertz, that heart structures reflect ultrasound and that echoes could be obtained from the heart by placing a 2.5-MHz transducer of commercial diagnostic ultrasound equipment designed for material testing externally on the thoracic surface of a patient.

It has been shown that heart structures reflect ultrasound and that echoes can be obtained from the living human heart. An echocardiogram can be obtained from various parts of the heart. The pattern of movement of echo signals varies according to the location of the crystal on the chest and the direction of the beam. If the heart is enlarged, pulsating echo signals can be obtained over a significantly larger area than in normal cases. By applying the 2.5-MHz ultrasonic transducer to the third or fourth left interspace, 1 to 4 cm from the left sternal border, echo signals are obtained which pulsate in position and size synchronously with the heartbeat of the patient.

Since the lung tissue surrounding the heart has a very high absorption coefficient for ultrasound, with a frequency of about 2 MHz, no echoes from the body structures are obtained. On the other hand, the heart can be located by this method only from a relatively small area under the thorax of the patient where no lung tissue lies between the heart and the chest wall. Since the movement of these echoes represents the movement of heart structures, it was clear that this method could eventually be used for diagnostic purposes. In the diseased heart these typical movements are greatly altered.

Prior to identifying and analyzing the various echoes from the heart structures recorded on the echocardiogram, the basic function of the human heart must be understood. The contracting ventricles of the human heart constitute a unique type of pump. The source of energy is a repetitive contraction initiated through electrochemical reactions. As a pump, the human heart has two inlet valves, the mitral and tricuspid, and two outlet valves, the aortic and pulmonary. During a cardiac cycle, two heart sounds are generated that consist mainly of two separate energy bursts of vibrations occurring after closure of the two sets of valves. Partial obstruction of any valve or blood leakage due to imperfect closure results in turbulence of blood flow and concurrent audible sounds called *murmurs*. In heart disease diagnosis, such murmurs are often symptomatic of one or more malfunctioning valves.

Echocardiograms as measured by the ultrasonic echoing technique contain characteristic patterns for the valve motion and structural detail of normal and diseased valves. The ultimate objective of this analysis is to establish operational signatures of normal and diseased natural valves, and for normal and diseased structures in the heart, such as walls or implanted prosthetic valves. This includes the detection of particular valve faults either by interpretation of the echocardiogram itself or by analyzing the correlations between the ultrasonic echoes and heart sound recordings.

Obstruction in any heart valve is called *stenosis* and a leak in a heart valve is referred to in medical nomenclature as *insufficiency* or *regurgitation*. The echoes on the echocardiogram may be depicted in the two-dimensional A-presentation, or A-mode, which simply shows the depth of penetration and the amplitude of the reflected ultrasound energy. The depth of penetration scale is calibrated for an average speed of sound and tissue at body temperature of 1540 m/sec. On this scale, precise measurements of dimensions from the chest wall to the posterior wall of the left ventricle can be made, as well as measurement of the left ventricle and right ventricle at diameters or the thickness of the myocardium or heart muscle itself. A typical T-M or M-scan of the heart is shown in Fig. 11.11.

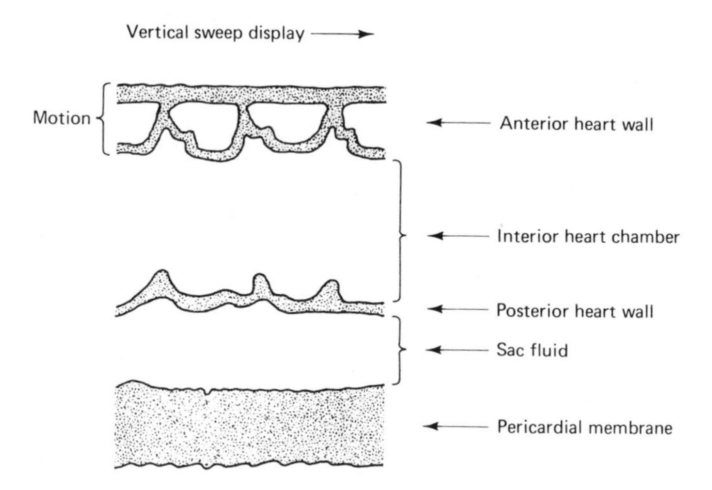

Figure 11.11 Typical T-M or M-mode scan of a heart.

The B-mode scan of Fig. 11.12 permits study of valvular or myocardial motion by depicting not only the depth of the structure but also the changing position of the echoes as though they were viewed from above. This presentation provides intensity modulation for the large number of echo amplitudes and, like the A-presentation, is recorded photographically on a polariod film.

Devices are available for converting the motion of the returning echoes into corresponding voltage variations which may then be displayed on a direct recorder. The term *echocardiogram* or *ultrasound cardiogram* identifies the graphic registration of ultrasound echoes from the heart. The first of these terms is perhaps more descriptive, shorter, and preferable. In this era of abbreviations, it is appropriate to recommend "echo" rather than "ECG," since the latter would obviously be confused with electrocardiogram.

The initial cost of the equipment is undeniably high. Adequate space and facilities are necessary to house the apparatus, the recording instrument, and the examining table. An experienced physician or technologist is needed to apply the transducer to the chest wall. A trained technician to record the reflected waveforms. The physician should be trained to discard artifacts, recognize loss of echo signals, and measure the velocity of leaflet closure of slopes and diastaline.

Once proficiency has been achieved, the echocardiogram of the interior leaflet of the mitral valve can generally be obtained within 10 minutes in most normal subjects and in nearly all patients with mitral valve disease. Figure

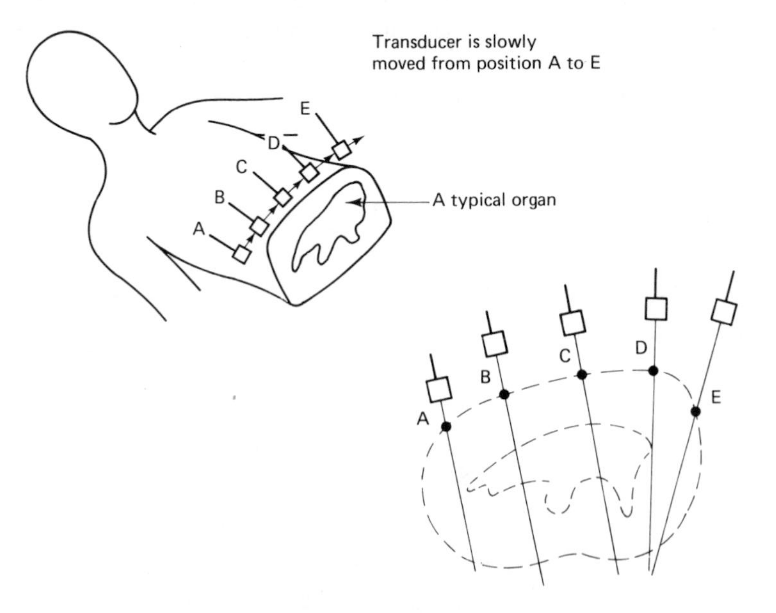

Figure 11.12 B-mode scan of the heart. The response may be recorded on a storage scope, and with all points interconnected the outline of the organ is determined.

11.13 shows the echogram of the right and left ventricles, interventricular septum, and mitral valve of a post-myocardium patient.

The clinical application of ultrasound in cardiovascular disease is well established and has gained widespread acceptance for the diagnosis of mitral or tricuspid valve disease. The waveform is specific for the scarred, contracted, and immobile anterior leaflet. The range of motion or amplitude of the recording by the B-scan, or direct, presentation is characteristically reduced. Echocardiography serves as an important guide in defining the precise structural abnormality of the mitral valve.

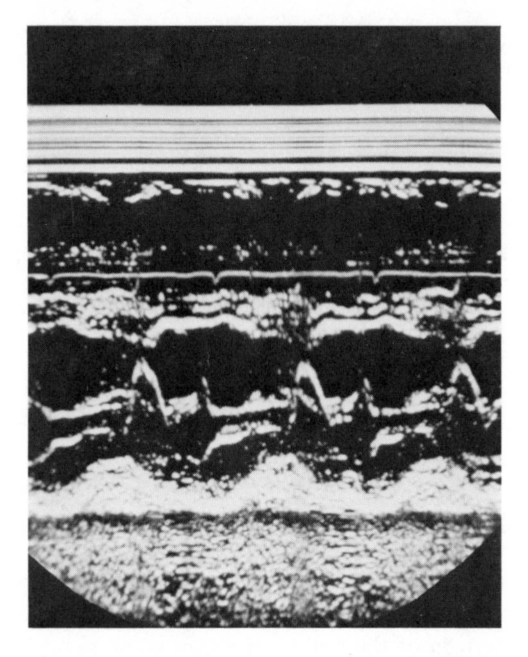

Figure 11.13 Echogram showing the right and left ventricles, interventricular septum, and mitral valve of a post myocardium patient.

A newer automated echocardiogram, using automatic gain control so that the person making the echogram does not turn any dials, is the Echomatic manufactured by Oreganon Teknika Corp. of Oklahoma City, Oklahoma. A normal heart scan without gain compensation is shown in Fig. 11.14. This instrument is used in the evaluation of mitral regurgitation.

Echocardiography has also been very valuable in detecting blood clots or tumors in the heart. Additionally it has been very helpful in detecting pericardial effusion or fluid accumulating in the pericardial sac, the sac that holds the entire heart. It has been shown that in normal healthy subjects a single echo is recorded from the posterial wall of the heart. In patients with pericardial effusion, the echo from the posterior wall of the left ventricle is separated from the echo of the pericardial sac. Observations have indicated that echocardi-

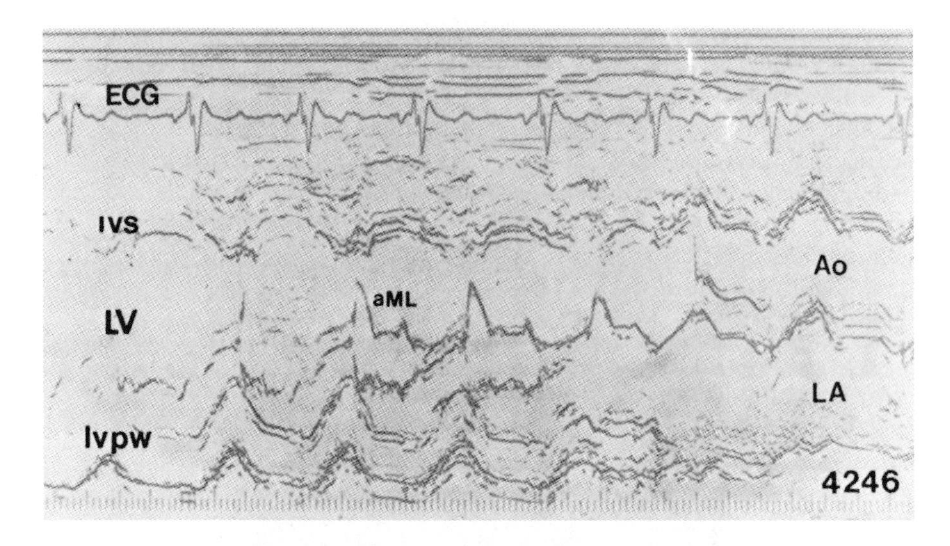

Figure 11.14 Heart scans using the Echomatic. ECG, electrocardiogram; IVS, interventricular septum; Ao, aorta; LV, left ventricle; aML, anterior mitral valve leaflets; LA, left atrium; Ivpw, interventricular posterior wall. (Courtesy of Organon Teknika Corp., Oklahoma City, Okla.)

ography seems to be a very simple, innocuous, and relatively accurate bedside technique for the detection of pericardial effusion.

11.7 ULTRASONIC MEASUREMENT OF CARDIAC OUTPUT

The advent of open-heart surgery, especially the latest technique for bypassing coronary arterial obstructions with a vein graft, has accentuated the need for more accurate evaluation of the ability of the heart to provide adequate blood flow. The single most important factor in evaluating the performance of the heart is the cardiac output measured in liters per minute. Cardiac output is a significant clinical entity in evaluating the performance of prosthetic cardiac valves, in monitoring the restoration of myocardial function in heart-attack patients, and in identifying the hyperkinetic or high-flow-state heart syndrome in youngsters with innocent murmurs.

The diagnosis of low cardiac output usually rests on recognition of the effects of low cardiac output rather than its direct measurement. Many techniques have been developed for the estimation of cardiac output, but most of these involve venous and arterial punctures and are not applicable to bedside monitoring. These punctures are not only painful but also involve some risk of infection. All cardiac output measurement techniques are indirect estimates based on various physical and mathematical assumptions. Therefore, each technique contains inherent errors that must be determined and minimized. Presently, the Fick principle, indicator dilution techniques, and thermal dilution techniques utilized in conjunction with intracardiac catheterization are the accepted standards for measuring the cardiac output during open-heart surgery.

The Fick principle is based on the fact that during its passage through the peripheral tissues, a certain amount of oxygen is taken up from the blood. The limitations of this method are the difficulty of obtaining accurate samples of expired air, the need for sampling in pulmonary artery blood, and the time-consuming analysis of expired air and blood oxygen concentrations.

Echocardiography is a simple noninvasive bedside technique used for measuring and monitoring changes in cardiac output.

11.8 ULTRASONICS IN OPHTHALMOLOGY

The human eye is an ideal organ for ultrasonic examination. It consists of a cornea, an anterior chamber, a lens, a vitreous space filled with homogeneous fluid, a retina, and a posterial wall. The ultrasonic transducer usually operates at a frequency from 5 to 15 MHz. Of course, the higher frequency is more desirable because it is associated with better resolution. However, since it has decreased penetration, the normal sensor used in eye work is usually a transducer at a frequency of about 10 MHz.

The transducer is placed either directly on the cornea after anesthetizing the cornea with a few drops of anesthetic or it is placed on the closed eye, directly on the lid, with some amount of ultrasonic jelly underneath the transducer to prevent air from generating noise in our ultrasonic beam. The ultrasonic pulses emanate from the transducer face and pass through the eye. Whenever a pulse meets an interface of different acoustic impedance, a portion of the energy is reflected in an echo. Residual energy continues to the next interface, which in turn reflects an echo, and these echoes establish a pattern. The anterior chamber, the lens, and the vitreous are normally acoustically homogeneous and produce no echoes.

The echogram of the normal eye usually shows an average diameter from the most anterior to the most posterior echo of 22 to 26 mm. The distance representing the anterior chamber is usually approximately 2 mm and the echo representing the lens normally measures 4 mm. The vitreous cavity has a diameter of about 18 mm.

One of the most significant areas for the A-mode application in eye work is the measurement of the dimensions of the dimensions of the globe. Changes associated in certain diseases, such as glaucoma or myopia, can be demonstrated. A particularly interesting application of diagnostic ultrasound is the analysis of the vitreous. Pathology in the vitreous cavity can present a variable acoustic response. Both the nature of the target lesion and the state of the surrounding vitreous are important in interpreting the echo configurations that are obtained. The greater the difference in their impedances, the more pronounced will be the echo response. Fresh hemorrhages in the vitreous differ from those of several days' standing in that they usually reflect clear sharp echoes. The latter frequently become quite acoustically homogeneous with the degenerative vitreous reaction and may not reflect any patterns at all.

If the hemorrhage is well separated from the surrounding vitreous, the echoes reflected may be of such caliber as to mimic a foreign body, and localization becomes quite difficult. However, frequently there is destruction of the normal vitreous architecture, which leads to eventual dispersion of the blood and prevents the formation of sharp interfaces with confusing patterns.

One of the most important applications of diagnostic ultrasound in ophthalmology is the detection of retinal detachment. The detachment can be fluid where the subretinal fluid is acoustically homogeneous and reflects no echoes. The detachment may be solid where echoes are reflected in terms of multiple blips distrupting the baseline and extending posterially to involve all of the globe. When a single high-amplitude spike is obtained, the sonar beam is probably striking the retina perpendicular to its surface. This is most easily demonstrated in an area of flat detachment. The evaluation of solid detachments may be carried our by ultrasonic techniques, even in the absence of a fundus view. The A-mode is capable of identifying the presence of a solid detachment. It may also identify tumors attached to the retina or located behind the retina in the choroid region.

Ultrasound is also useful in foreign-body localization within the eye, as well as extraction of this body by a surgical technique. Combined use of the A-mode and the B-scan technique can often provide more information about the eye and its diseases than either instrument alone. The goal of intensity-modulated B-scan ultrasonography is to display a two-dimensional cross section of the eye for diagnostic purposes. The principal advantage of the method is the ease with which specific areas of interest can be recognized and oriented in relation to known landmarks in the cross section. The optic nerve, for example, produces no identifying echoes, yet in intensive modulated ultrasonogram it is recognized as a well-defined void among the echoes from the retrobulbal fat pad.

The excursions of the optic nerve accompany eye movement even though echoes are not reflected by it. Special transducers have been developed for B-scan transducer. A suction cup transducer is a miniature transducer whose crystal is integrated in a corneal acrylic lens which acts as a suction cup.

11.9 OBSTETRIC APPLICATIONS OF DIAGNOSTIC ULTRASOUND

Whenever it is necessary, desirable, or convenient to terminate pregnancy prior to the onset of labor, the need for accurate estimation of the gestational age is a constantly recurring problem. It is of prime importance to avoid excessive prematurity. A significant error in the estimate of fetal weight or expected date of delivery imposes additional hazards on the fetus and increases the likelihood of prenatal death. Unfortunately, there are no clinical criteria that will sufficiently predict fetal maturity.

Additional evidence to support the clinical estimation of the gestational age can be obtained by diagnostic ultrasonic equipment. The measurement of the biparietal diameter of the fetal head within the uterus is a practical procedure. From this measurement one can make the usual prediction about birthrate and to a certain extent about gestational age. More important, the biparietal diameter indicates minimum birth weight. Armed with this information one can with greater certainty avoid the problems associated with premature delivery.

Ultrasonic fetal cepholometry is a painless, safe, and accurate method of measuring the biparietal diameter of the fetal head. Measurements correlate significantly with calibrated measurements made after birth. There is also statistically significant correlation between ultrasonic measurements and birth weight. These measurements have proved valuable to clinical obstetricians for a wide range of medical and obstetric complications that call for early termination of pregnancy. They are more valuable as indicators of the minimum birth weight than might be expected rather than the exact weight of a given infant. Their proper use is to supplement rather than to supplant the patient's history and clinical data. When so employed, they are useful in clarifying the question of fetal maturity.

The procedure (see Fig. 11.15) is simple, safe, and reliably accurate. After locating the fetal head by palpatation, the transducer of the diagnostic ultrasonic equipment is coupled to the abdominal wall either with water or lubricating jelly and then moved over the surface of the abdomen until a satisfactory echo pattern appears on the oscilloscope screen. If the transducer is perpendicular to the fetal skull at the biparietal diameter, three echoes will appear. Echo 1 is returned from the abdominal wall. Echoes 2 and 3 are returned from the near and far skull walls and are approximately the same height but higher than the first echo. The biparietal diameter measurement is obtained by counting the markers on a superimposed centimeter scale; posterior and anterior diameters of the fetal skull can produce a similar pattern. They can be recognized because these diameters are considerably larger than the biparietal diameter. Occasionally, it is possible to obtain an echo from the third ventricle, which is additional confirmation that the biparietal diameter has been located.

Once the fetal head is engaged, it is usually not possible to obtain a satisfactory measurement because of interfering echoes from the mother's pelvic bones. This is no disadvantage because the biparietal diameter is seldom needed following engagement. The pregnant abdomen has proved particularly ideal for pulse-echo ultrasonic techniques both because of its anatomical configuration and the fact that the pregnant uterus is filled with fluid, thus making good sonic contrast for depicting intrauterine structures. Pregnancy represents an area where a new diagnostic technique would be particularly valuable, since radiation

Spleen

1. Longitudinal Angled Intercostal Scans (Close to Horizontal in the Posterior Axillary Line)
2. Transverse Intercostal Scans

   The Spleen can also be Visualized in the Prone Position or with the Patient Lying on his Right Side

Gall Bladder

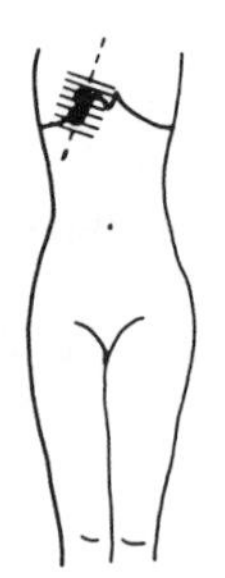

The Patient Should be Fasting

First the Long Axis is Determined by a few Transverse Scans

Serial Scans are then made Perpendicular to this Axis

Kidneys

Supine

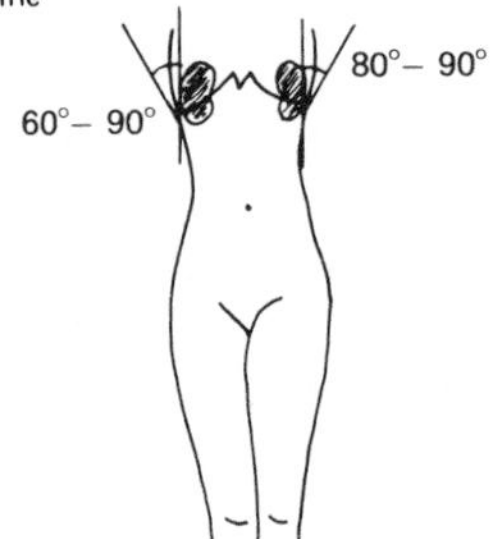

The Right Kidney is Usually Very Well Visualized Intercostally Through the Liver by Angled Longitudinal Scans as well as by Transverse Scans

In the Supine Position the Left Kidney is best Outlined by Almost Horizontal Longitudinal Scans Through the Spleen. Large Parts of the Kidney May, However, be Obscured by Gasfilled Intestines

Prone

For both Kidneys:

1. The Long Axis of the Kidney is Established by Transverse Sections of the Upper and the Lower Pole
2. The Slope of the Long Axis of the Kidney is Determined from a Longitudinal Scan Through this Axis
3. A Series of Longitudinal Scans Parallel to this Axis and a Series of Tilted Transverse Scans Perpendicular to this Axis are Produced

Figure 11.15 Suggested cross sections for abdominal, gynaecological, and obstetric scanning. (Courtesy of Brüer and Kjaer, Denmark.)

Urinary Bladder

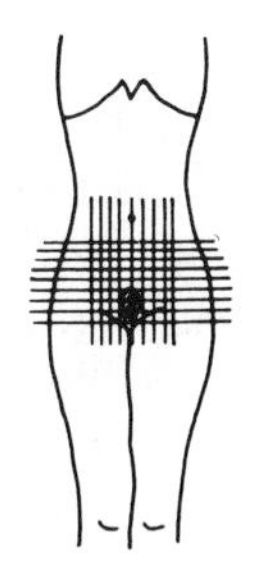

The Bladder Should be as Full as Possible

Non-Pregnant Uterus

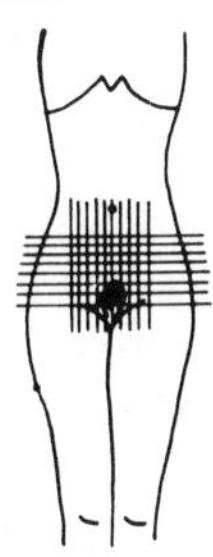

The Bladder Should be as Full as Possible

Pregnant Uterus

Early Late

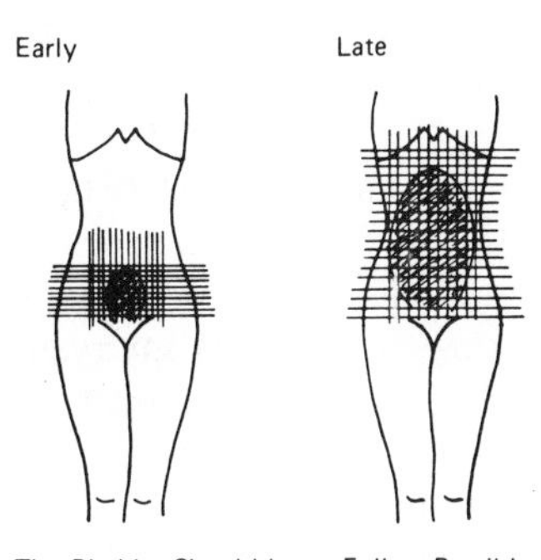

The Bladder Should be as Full as Possible, Particularly Necessary when Determining the Lower Margin of the Placenta if Previa is Suspected

Liver

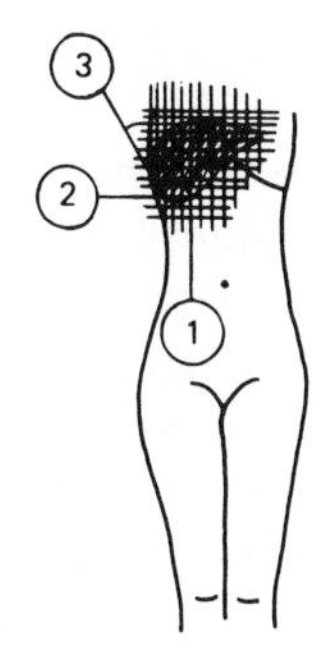

1. Serial Longitudinal Subcostal Scans with Patient Suspending his Respiration in Deep Inspiration
2. Serial Transverse Intercostal Scans
3. Angled Longitudinal (Including Horizontal) Intercostal Scans

Pancreas

The Patient Should be Fasting

1. Standard Longitudinal Scans Through the Long Axis of the Interior Vena Cava and the Aorta
2. Oblique Scans Through the Long Axis of the Pancreas (from the Hilum of the Right Kidney to the Hilum of the Spleen). The Superior Mesenteric Artery, the Portal and the Splenic Veins can be used for Orientation
3. Longitudinal and Transverse Scans Through the Left Kidney with the Patient in a Prone Position May Visualize the Tail of the Pancreas

Figure 11.15 (*Continued*)

hazards limit the use of x-ray and isotope techniques for study of various abnormalcies of pregnancy and for following fetal development over a long period of time.

Ultrasonic scanning, providing a B-scan presentation, can be achieved utilizing a compound contact scanner. In this method, the transducer, which contains a lead zurcinate crystal, moves mechanically 30° to each side of the perpendicular while the operator moves the transducer carriage across the pregnant abdomen. A storage oscilloscope tube stores on the oscilloscope screen the echo information reflected from the tissue interfaces. The skin surface is coated liberally with mineral oil to provide sonic contact between the tissues and the transducer. Each cross-sectional scan requires approximately 1 minute. The transducer carriage is then moved up and down the uterus in both horizontal and sagittal planes to provide cross-sectional presentations at approximately 30 minutes. Figure 11.16(a) shows the B-scan presentation of the cross section of a pregnant uterus, showing the fetal skull, including the bridge of the nose, two eye orbits, and the cross section of two limbs. The B-scan presentation of a cross section of a pregnant uterus showing the fetal skull and midline with the placenta above the skull appears in Fig. 11.16(b).

By reviewing the pictures taken at 2-cm intervals both horizontally and sagittally in a composite fashion, the viewer can plot out graphically the gross location and boundaries of the placenta. The placenta pattern should be seen in three horizontal and one to three sagittal films. Thus, ultrasound can provide specific localization in any part of the uterus, making it particularly useful prior to amniocentesio or intrauterine transfusion. In contrast to other diagnostic techniques, ultrasonic placentography visualized the placenta and its relationship to the uterine wall, the fetus, and the maternal abdomen. Echoes from the placenta form an ultrasonic pattern peculiar to that tissue and thus can outline precise boundaries for the placental margin.

The Doppler ultrasonic detector can often pick up the heart beat of a fetus before it can be heard by the stethoscope of a doctor. Thus, it is capable of differentiating the presence of a fetus from a tumor, and quite often the mother is happy to hear her baby's heartbeat as displayed on an oscilloscope or heard on a speaker.

Obstetricians and radiologists are relying more and more on diagnostic ultrasound in problem pregnancies. The biparietal diameter of the fetal head can be measured precisely in uterus. This measurement helps the obstetrician estimate fetal maturity and gauge the optimum time for surgical intervention when indicated. Also, by comparing a series of fetal head measurements with the growth curve in normal gestation, timely evidence of fetal distress may be elicited. A pictorial of a phased-array two-dimensional real-time section scan of patients with possible cardiac and abdominal disorders is shown in Fig. 11.17. This system is also used in obstetrics.

Diagnostic ultrasound is a new technique for localizing the placenta. The ultrasonic examination shows the placenta and its relationship to the fetus and

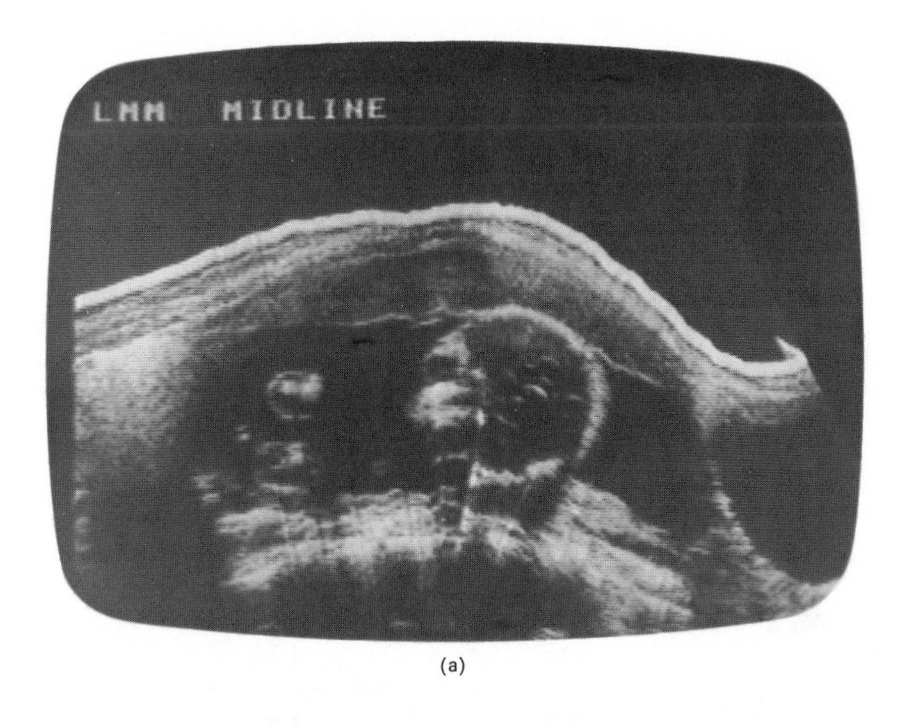

(a)

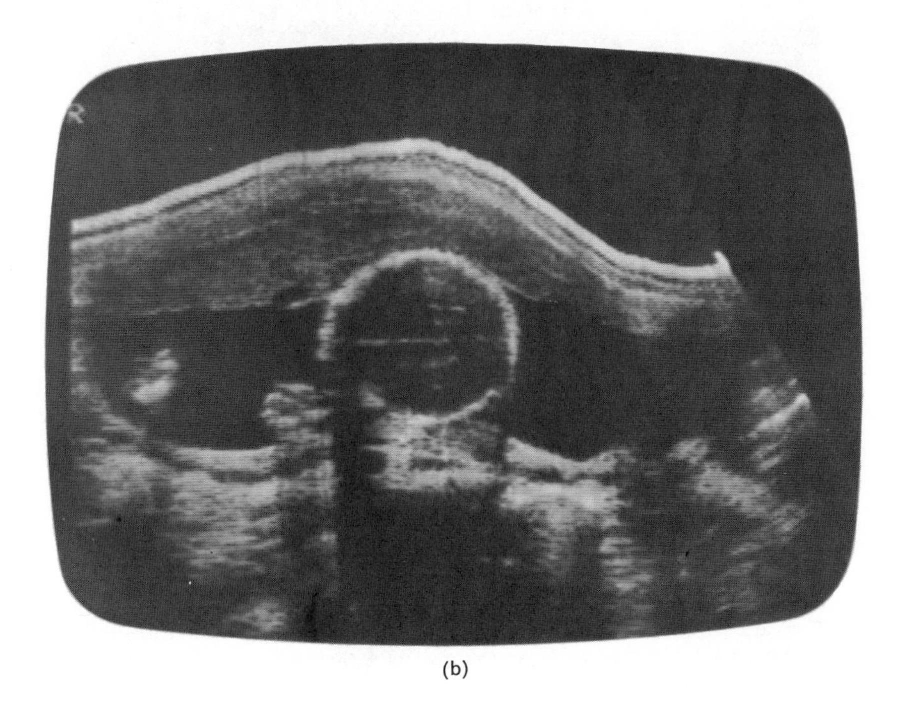

(b)

Figure 11.16 B-scan presentation of the cross section of a pregnant uterus.

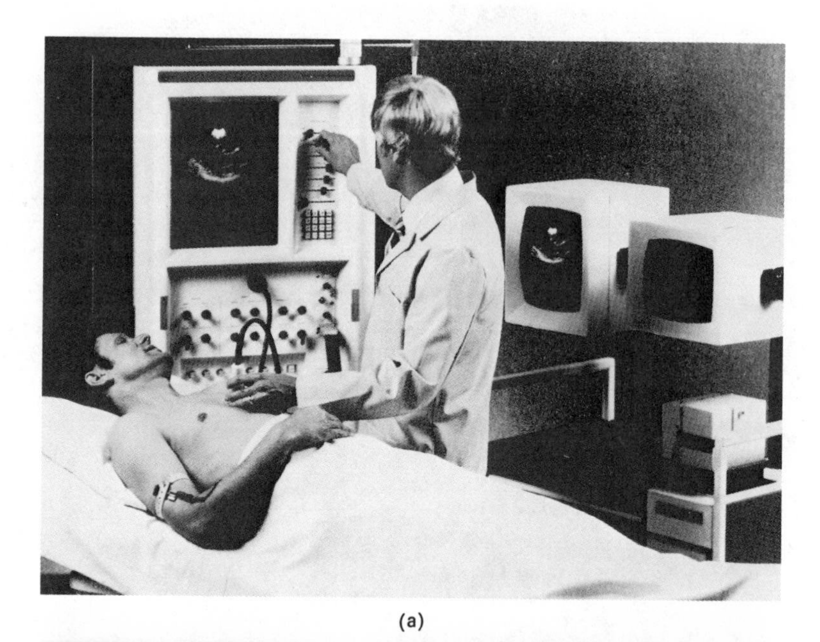

(a)

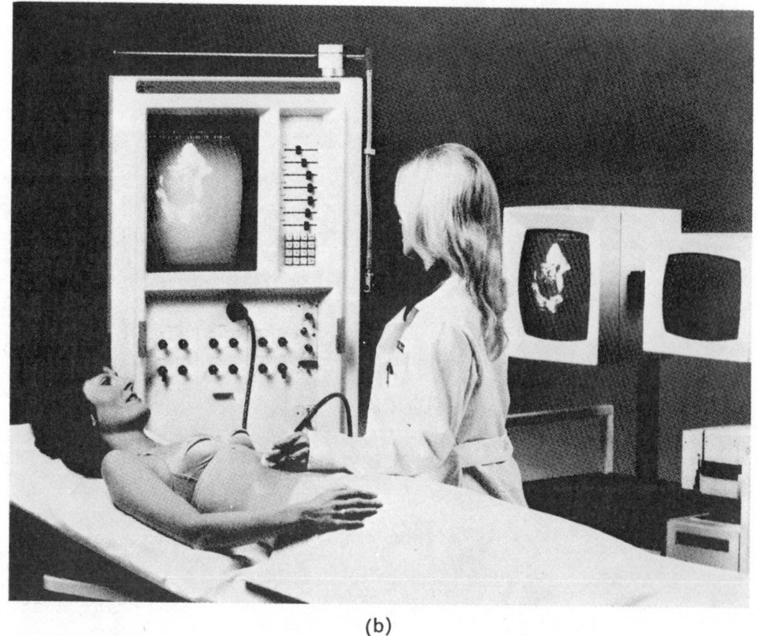

(b)

Figure 11.17 Two-dimensional sector scan for the heart (a) and abdomen (b). The Varian V-3000, phased-array ultrasonograph is used. The V-3000 is also used in obstetrics. Using microprocessor control, the V-3000 scans an 80° fan-shaped body cross section at the rate of 30 times per second. (Courtesy of Varian Associates Medical Group, Palo Alto, Calif.)

uterine wall. The peculiar ultrasonic patterns of the placental echoes permits a diagnosis of the boundaries of the placenta. Up to now the method of placental localizations have been physical exploration of the endocervix, x-ray, or isotope examination. The placenta is a soft tissue structure, and its margins are difficult to outline. Therefore, it is most desirable to reduce the ionizing radiation given to the pregnant woman as much as possible, and ultrasound does the job.

The power levels required for diagnostic ultrasound are approximately 0.004 W/cm^2, whereas those used for heat therapy and physical medicine range from 1 to 3 W/cm^2. Thus, diagnostic ultrasound represents a new and effective diagnostic approach for placental localization.

In cases where the uterus is large and distended, the differential diagnosis includes twins, a large fetus, or a pregnancy complicated by uterine tumor. In regard to the measures that can be taken to prevent premature delivery, the diagnosis of twins during pregnancy is important. The number of fetuses and their position in the uterus can be determined by ultrasonic examination. In such a case, the contours of two fetal heads can be seen side by side on the B-scan presentation.

Doppler ultrasonic techniques are also of great help and importance to the obstetrician. If a continuous beam of ultrasonic energy is directed at the tissue structure which moves relative to the sound source, some of the energy that is reflected back to the source will have undergone a slight shift in frequency, analogous to the apparent change in pitch of a train whistle as a train passes an observer. When the train is approaching the observer, the note sounds higher pitched than it would be if the train were stationary. When it passes, the apparent pitch drops below the stationary value. Of course, in the ultrasonic case, the ultrasound is by definition beyond the upper limit of audibility. However, it is fortunate that the difference between the transmitted sound and the reflected sound can easily be extracted electronically to be in the audible range.

This is the basis of the ultrasonic fetal heart detectors, which can be used to demonstrate the fetal heart beat as early as 12 weeks' gestation and to monitor the heartbeat, at any stage thereafter, including the period during labor.

11.10 REVIEW QUESTIONS

1. Discuss the principles involved in making an ultrasonic measurement.
2. Discuss three applications of clinical ultrasound.
3. Describe the transducer used in ultrasound.
4. Discuss the real-time display in ultrasonography.
5. Discuss the Doppler instruments used in ultrasound.
6. Discuss the multielement transducer system in ultrasound.
7. Discuss the sector-scan system in ultrasound.
8. What is an echoencephologram, and where is it used?

9. Discuss ultrasonics in cardiovascular disease.

10. List five applications of echocardiogram.

11.11 REFERENCES

1. Wells, P. N. T.: *Biomedical Ultrasonics,* Academic Press, Inc., New York, 1977.
2. Lees, S., Barber, F. E., and Lobene, R. R.: Enamel: Detection of Surface Changes by Ultrasound Science, *American Association for the Advancement of Science,* Vol. 169, No. 3952, pp. 1314–1316, September 25, 1970.
3. Saba, J. M.: Echo-Encephalography, *Medical Electronics and Data,* No. 5, pp. 96–101, September–October 1970.
4. Brascho, D. J.: *Diagnostic Ultrasound in Radiation Therapy Planning,* Unirad Corp, Denver, Colo., 1974.
5. Dreijer, N.: *Application of B & K Equipment to Diagnostic Ultrasound,* Brüel & Kjaer, Denmark, August 1978.
6. Soares, R. H.: *An Atlas of Echocardiography,* Unirad Corp., Denver, Colo., 1974.
7. Gramiak, R.: *Echocardiography,* Unirad Corp., Denver, Colo., 1973.
8. White, D. N.: *Ultrasonic Encephalography,* Medical Ultrasonic Laboratory, Queens University, Kingston, Ontario, 1970.
9. Lightvoet, C., Rusterborough, H., Kappen, L., and Bom, N.: Real Time Ultrasonic Imaging with a Hand-Held Scanner: Part I. Technical Description, *Ultrasound in Medicine and Biology, Med. Bio.,* Vol. 4, pp. 91–92, 1978.
10. Roelandt, J., Wladimiroff, J. W., and Boars, A. M.: Ultrasonic Real Time Imaging with a Hand Scanner: Part II. Initial Clinical Experience, *Ultrasound in Medicine and Biology,* Vol. 4, pp. 93–97, 1978.
11. Roelandt, J., Lima, L., Hajar, A., Walsh, W., and Kloster, F. E.: Clinical Experience with an Automatic Echocardiograph, *CVP Journal of Cardiovascular and Pulmonary Disorders,* Vol. 7, No. 4, pp. 27–30, June–July 1979.
12. Roelandt, J., and Kloster, F. E.: *Real Time Cross-sectional Analysis of the Heart by Ultrasound,* Medical Education Department of Oreganon Teknika Corp., Oklahoma City, Okla.

12

RADIOLOGICAL AND NUCLEAR MEASUREMENTS

12.1 INTRODUCTION

In the hospital environment, radiation monitoring is necessary for x-ray units and for monitoring the movement of radioactive tracer elements injected into a human subject.

Natural radiation may be alpha particles, beta particles, x-rays, neutrons, deuterons, and heavier particles. Alpha, veta, and gamma energy, which is emitted by radioactive isotopes, and x-rays are the most common types.

12.2 X-RAYS

X-rays, discovered by Roentgen in 1895, are nonluminous electromagnetic radiation of extremely short wavelength, generally less than 2Å or 2×10^{-6} m, produced by bombardment of one of the heavy metals by a stream of electrons moving at great velocity in a vacuum. X-rays can penetrate through body tissues and affect photographic plates and fluorescent screens.

X-rays are used in hospitals to study, diagnosis, and treat organic diseases, especially in the internal structure of the body. If you have a lung disorder, heart disorder, gallstones, ulcers, or tumor, the physician will order an x-ray. Even if the person is healthy, the x-ray may verify this.

Medical *radiology* is the application of x-rays and radioisotopes in the diagnosis and treatment of patients.

Radiography is an x-ray film of any part of the body. It is similar to a photograph except that it is a shadowgraph rather than a picture produced by reflected light. The energy source is an x-ray tube, and the black-and-white gradations depend on the differences of opacity in the tissues placed in the x-ray beam.

Tomography is an x-ray system that can display real-time and noninvasive cross sections of the heart as if it had been cut open with a surgical knife. A computer makes it possible to use a series of x-ray exposures from different angles to reveal the internal organs of the body in cross section instead of superimposing one on the other.

Figure 12.1 (a) shows a conventional x-ray picture generated by allowing x-rays to diverge from a source, pass through the body of the patient, and strike a sheet of photographic film. Figure 12.1(b) shows a tomogram made by having the x-ray source move in one direction during the exposure while the film moves in the opposite direction. With the projected image, shown in Fig. 12.1(c), one plane of the body remains stationary with respect to the synchronously moving x-ray source and detector. In Fig. 12.1(b), all planes other than the plane of choice are blurred.

A new x-ray tube used for microscopic study of living cells provides an exposure on x-ray sensitive paper that can be made in 100 nsec.

Photoroentgenography is an x-ray diagnosis tool for the detection of tuberculosis. X-ray machines can produce two chest films per minute. The chest image is projected on a standard-size fluorescent screen, where it is converted to visible light which is focused with optical lenses into smaller film. This older technique is being displaced by chemical studies.

Fluoroscopy is an x-ray examination of internal organs in motion, such as the chest wall, the heart, or the stomach and gastrointestinal tract. The images are shown in dark shadows against an illuminated background and are viewed in a darkened room on a fluorescent screen. The patient can be standing or lying, the x-ray machine is placed against the tissues to be investigated. The physician and patient wear special hoods for protection against x-ray radiation. In many instances, ultrasound diagnosis is replacing this technique.

An image-intensified tube, discovered by Irving Langmuir, converts x-rays to visible light which is used in various new ways. *Cinefluorograph* uses the imgae intensifier to take pictures of the fluoroscopic image. In cine techniques, the image tube and an optical system are used. A 16- or 35-mm motion-picture camera is placed in a patient position to record the output of the image-intensifier-tube phosphor. Closed-circuit television is another procedure to present the diagnosis of the disorder of the patient. It is remotely possible to integrate radiography, fluoroscopy, and tomography into one system with the recording techniques outlined.

A person who has an x-ray examination of the stomach will be given a radio-opaque material, a milk-like substance, to drink. The liquid contains barium sulfate, which absorbs x-rays to stop them from reaching the film or fluoroscopic screen so that the stomach and intestines will be outlined. Using

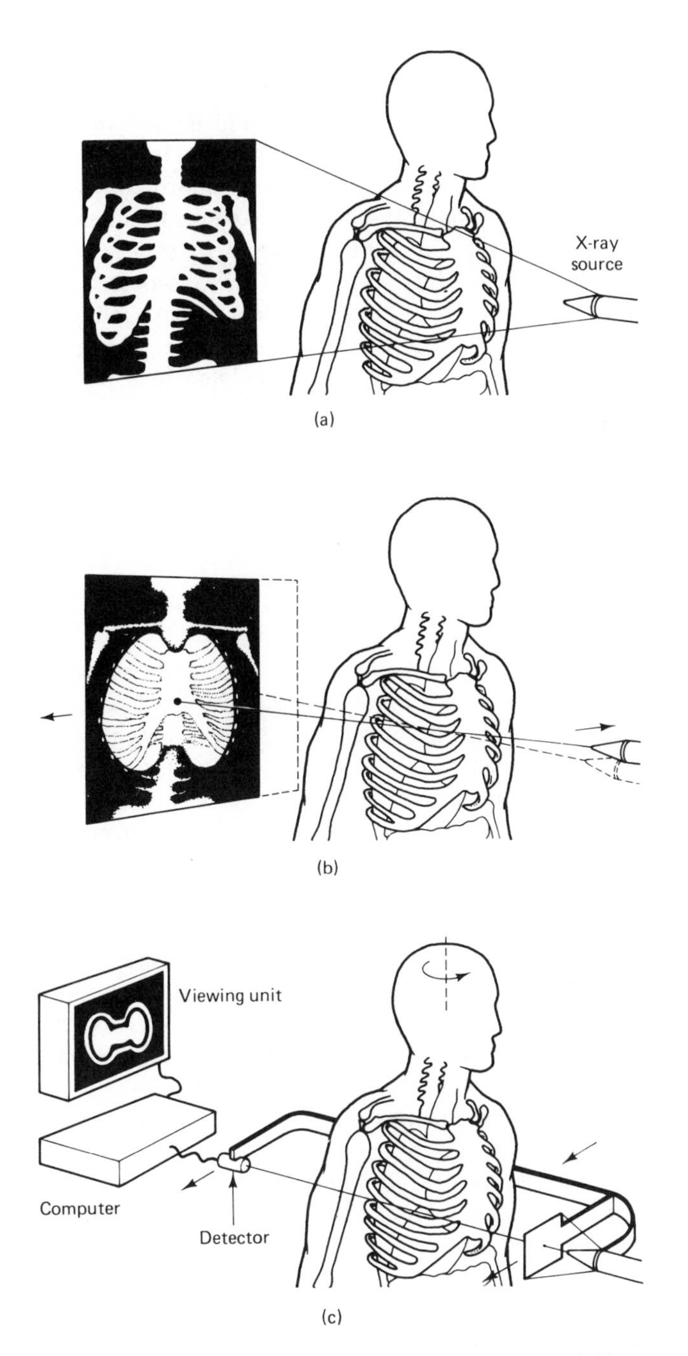

Figure 12.1 (a) Conventional x-ray picture, (b) tomogram, and (c) a reconstruction from projections. (Courtesy of *Scientific American*, October 1975, p. 58.)

this technique, the radiologist can detect an ulcer, tumor, or other pathology or study malfunctioning organs.

Radiation therapy is a therapeutic tool and is used in the treatment of cancer and other malignant and infectious diseases. Higher x-ray voltages are required than that for the diagnostic uses of x-ray fluoroscopy and radiography.

Nuclear radiology was first discovered by Henri Becquerel in 1896, when he noted spontaneous emission of rays similar to x-rays from uranium in its natural state. In 1897–1898, Marie and Pierre Curie discovered the rare element radium, in which radioactivity naturally occurs. Today, radioactive cobalt 60 and supervoltage x-ray units perform nuclear radiation. The radioactive tracers or isotopes injected into the body are accumulated in a particular organ or region. As an example, thyroid scanning uses iodide and sodium iodide; renal scanning uses chlormerodrin, and liver scanning uses rose bengal and colloidal gold radioactive materials. Imaging is generally performed by an external radiation detector. Resolution is still a problem, but scanning provides structural and functional information.

12.3 COMPUTER AXIAL TOMOGRAPHY

Ultrasonic devices as they exist today are used for soft tissues and are limited by the image resolution; they cannot be used with hard or bony tissue. For bony material such as the brain, conventional x-ray has been used, but is limited by the fact that the conventional x-ray image is a projection of information in a single direction. *Computer axial tomography* (CAT) was developed as a diagnostic-imaging technique in which anatomical information is digitally reconstructed on a screen from x-ray transmission data by scanning an area in many directions.

Figure 12.2 shows a block diagram of a computer axial tomography system. As shown in Fig. 12.2, the x-ray microbeam generator sends an electromagntetic beam of energy through the body of the patient and the x-ray beam feeds a scintillation detector. The x-ray beam is fed to a control unit which controls the microbeam generator and is also fed to a computer and an oscilloscope.

As a simple analogy, imagine that you wish to find the seeds inside an apple using a thin wire probe. By piercing an apple at many points along the "equator" you would hit a seed once in a while. Record each point of entry, and whether or not you hit a seed. This technique will eventually give you a profile of where the seeds are.

In computer axial tomography, a large number of very fine, low-level pulsed beams are used as probes. A xenon detector on the opposite side of the patient measures how much of each x-ray beam is transmitted through the body. Figure 12.3 shows a computer tomography pictorial of the head.

To make an image of a particular slice, rotate the pulsing x-ray source and detector around the patient. For each of tens of thousands of pulses and

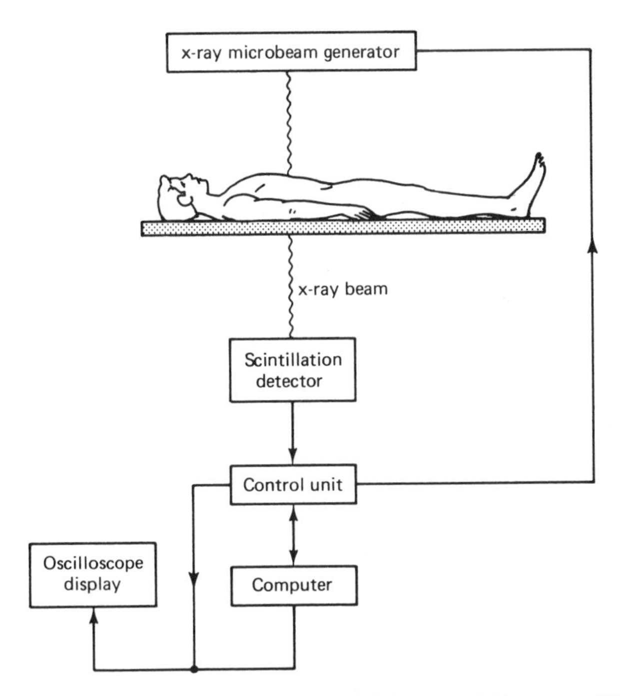

Figure 12.2 Block diagram of a computer axial tomography system. (Redrawn from Fig. 1, p. 79. Courtesy of *IEEE Spectrum*, January 1977, © 1977, IEEE.)

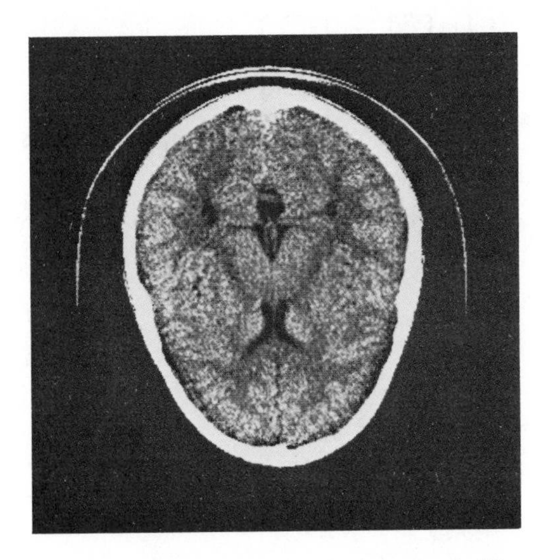

Figure 12.3 Computer tomography pictorial of the head. (Courtesy of General Electric Company, Medical Systems Division, Milwaukee, Wis.)

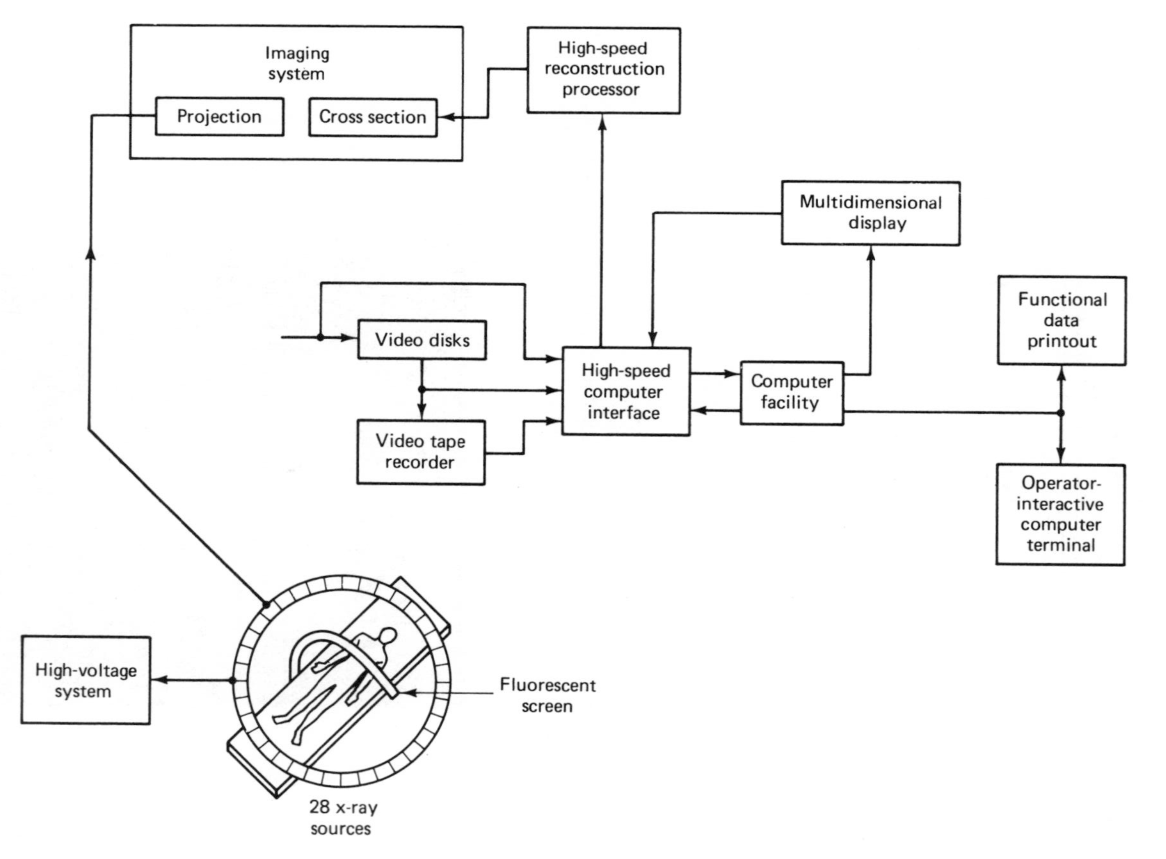

Figure 12.4 Newer computer axial tomography scanner. (Redrawn from figure on p. 77, courtesy of *IEEE Spectrum*, January 1979, © 1979, IEEE.)

positions taken, a computer records the information received, compiles the data, and reconstructs an exact image of the profile of the seeds. In human beings, an exact image of the organs or bones or tumors or other disorders can be distinguished by a clinician. The clinician can also differentiate fats from muscle, healthy tissues from diseased tissues, and at times, benign from malignant tumors.

A modification of the CAT scanner is called the whole-body scanner. In Fig. 12.4 a whole-body scanning system is used to take repeated accumulated "slices" of body section for analysis. In this system a single radiation source is directed at the body. Twenty-eight detectors are placed opposite the source on a single plane. The energy level detected by each head is fed into a computer, where it is logged in comparison to the source-energy-level control levels and the patient gantry position. A high-speed reconstructor extracts the signal from the computer, compares it to the level control setting and head position in relation to the gantry, and displays the resulting tomograph on one or more video consoles.

The current equipment for CAT scanning (Fig. 12.5) consists of a patient handling table, a scanning gantry which consists of a collimated x-ray source

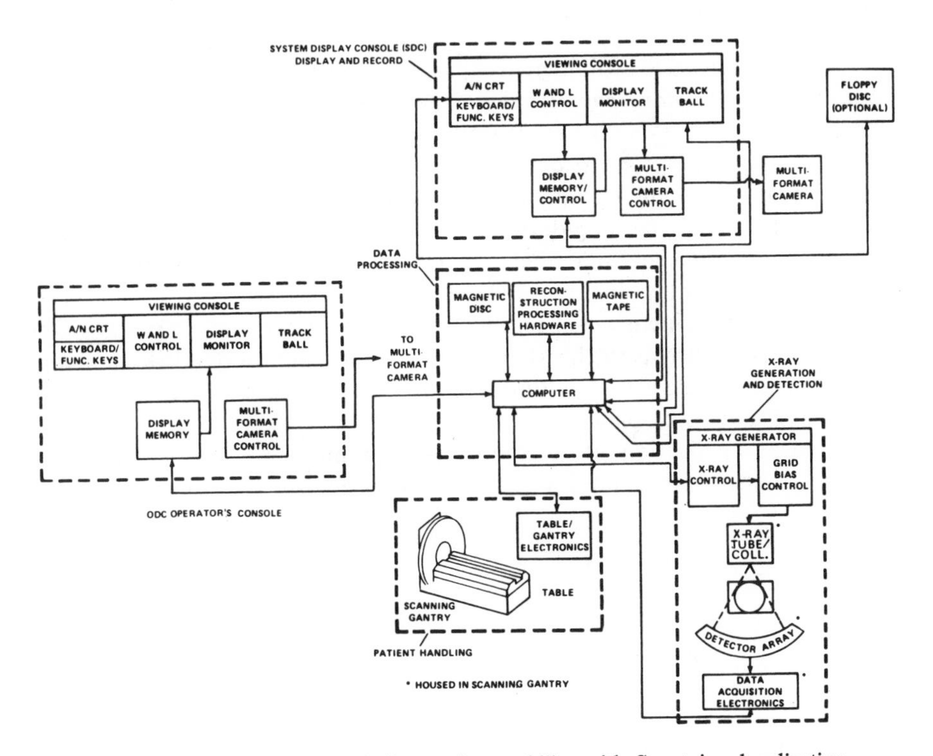

Figure 12.5 CT/T optional diagnostic capability with Scoutview localization system. (Courtesy of General Electric Company, Medical Systems Division, Milwaukee, Wis.)

and detectors, the data acquisition electronics, the x-ray generator, a computer, and the operator and viewing consoles.

Computer tomography was first discovered by Radon, who developed the theory of reconstruction in 1917. Oldendorf and Kuhl in 1961 and Edwards and Cormack in 1963 were the early investigators of image reconstruction for medical applications. The first brain scan used in the United States, at the Mayo Clinic and Massachusetts General Hospital in 1973, was really developed by Godfrey Hounsfield from the Central Research Laboratories of EMI Ltd., in England in 1967. It took six years to perfect the CAT scanning system.

Computer tomography, as used by radiologists, has been defined as the reconstruction by computer of a tomographic plane of an object or a slice. It is developed from multiple x-ray absorption measurements made around the object's periphery or a scan. The fidelity of the image depends on the x-ray source, the detectors, the number and speed of the measurements made, the details of the reconstruction technique or computer algorithm, the machine characteristics, and the methods of data display and interpreration.

12.4 NUCLEAR CARDIOLOGY

One of the newer basic radiologic techniques involves the principle of nuclear cardiology. *Radioisotope angiography* is a nondestructive technique (injection of radionuclide substances) that permits an evaluation of a wide variety of congenital and acquired heart disorders.

This measurement technique requires use of a gamma camera, which includes a magnetic recording device, a computer, and a short-time radiopharmaceutical isotope injected into the human body. The gamma camera detects, transforms, and displays the dynamic image.

Nuclear medicine is enhanced by use of a computer that simplifies study of human organs such as the heart. This newer procedure is often less traumatic to the patient, and the actual motion of the human organ can be used in dynamic playback on a video display.

Clinical application of a computer-based nuclear system with digital readout includes a procedure for cardiac analysis and the following:

1. Brain analysis
2. Lung analysis
3. Renal analysis
4. Functional analysis
5. Liver/pancreas analysis

Most of us do not know that the first diagnostic uses of radioactive tracers were in the study of the heart. In 1927, pioneers Blumgart and Weiss described their studies of normal patients and patients with heart diseases with radium salts as the tracer. The propsective of the use of nuclear cardiology has only recently become a reality with the development of safer and more effective tracers and detection systems. Two radioactive tracers used since 1975 are thallium-201 for imaging of myocardial perfusion and rechnetium-99M albumin for imaging of ventricular functions.

Measurements by radiologists that can now be made with intravenous injection of radioactive tracers include:

1. Right and left ventricular volumes
2. Right and left ventricular hypertrophy
3. Cardiac output at rest and exercise
4. Mitral valvular regurgitation
5. Wall motion of the right and left ventricles
6. Myocardial blood flow

A pictorial of isotope tracer information for the heart obtained with a gamma scintillation camera is shown in Fig. 12.6.

It will take years of further research to refine these measurements and define their precision, accuracy, and utility. Based on present technology, we can conclude that nuclear isotopes will probably play a major role in clinical cardiology and cardiovascular research. The population of patients taking the tracers is a limiting weakness of nuclear cardiology measurement. Nuclear cardiology cannot replace cardiac catheterization and contrast angiography,

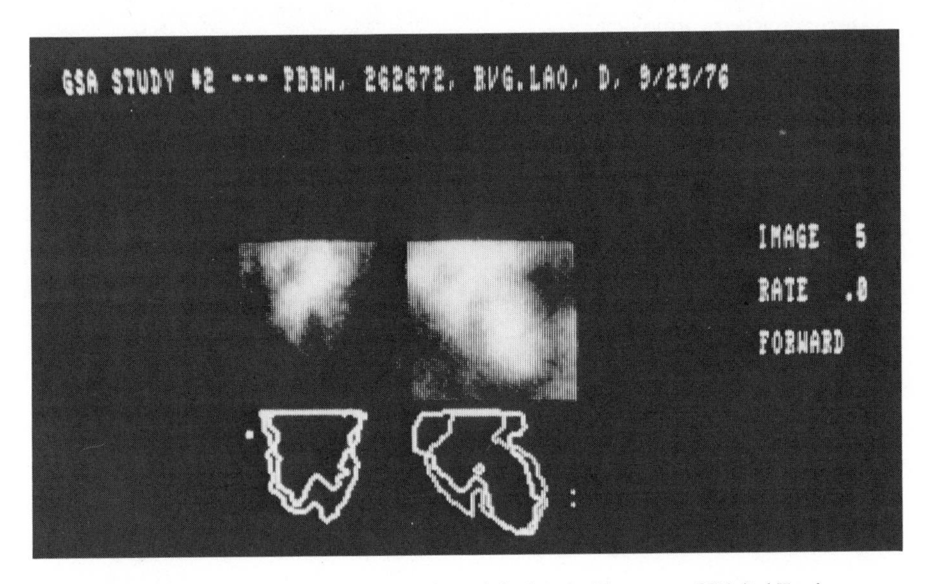

Figure 12.6 Isotope tracer information of the heart. (Courtesy of Digital Equipment Corp., Marlboro, Mass.)

which is used before, during, and after heart surgery. Each heart diagnostic measurement provides different types of information but usually complement each other.

The greatest appeal of nuclear cardiology, computer axial tomography, and echocardiography is their visual display of function. Human beings derive nearly all their sensory input through their eyes. People are better at perception than conception, and therefore like nuclear and echocardiographic images, especially when they are displayed as motion pictures. The greatest limitation of the nuclear studies at present is in quantification.

The radiation dose is low, and decades of experience with exposures in the level of nuclear cardiology studies have not revealed any deleterious effects. When we look at the quality of the first lung scans or myocardial perfusion studies performed with potassium-43, we can see how far we have come. Computers are now commonplace and are accepted as being necessary for first-class cardiovascular nuclear medicine studies. Cinematic display on color television is now widely used. New collimators are being introduced. Tomography is providing important improvements in sensitivity and quantification. At present spatial resolution is limited to about 2 cc with a precision of about 90% in quantitative assays using tomography.

12.5 REVIEW QUESTIONS

1. Define an x-ray.
2. Discuss computer axial tomography.
3. Discuss nuclear cardiology.
4. List two radioactive tracers.
5. What techniques are used to detect a congenital heart disorder?

12.6 REFERENCES

1. *The Story of X-ray*, General Electrical Company, Medical Systems Division, Milwaukee, Wis., 1978.
2. *Introduction to Computerized Tomography*, General Electrical Company, Medical Systems Division, Milwaukee, Wis., 1976.
3. *Radiological Health Training Resource Catalogue*, Bureau of Radiological Health, Rockville, Md., 1981.

13

THE INTENSIVE CARE UNIT

13.1 INTRODUCTION

In general, *intensive care* is assistance to critically ill patients 24 hours a day, so that further physiological complications are minimized. This care includes surveillance and treatment of patients with medical and surgical catastrophies, which include:

1. Patients with respiratory insufficiency and failure, including pulmonary edema and pneumonia
2. Circulatory emergencies:
   (a) Cerebral vascular accidents, which cover hemorrhage and thrombosis of the blood vessels
   (b) Acute heart failure, including cardiac arrythmias
   (c) Pulmonary embolism
   (d) Systemic embolism
3. Hepatic failure
4. Metabolic crises and diabetic coma
   (a) Glucose metabolism
   (b) Acute disorders of calcium metabolism
   (c) Acute adrenal cortical insufficiency
   (d) Disturbances of thyroid function
5. Fluid and electrolyte imbalances
6. Severe burns

7. Renal failure
8. Hyperbaric oxygenation
9. Organ homotransplantation
10. Drowning victims requiring serial study of blood constituents and serum electrolytes
11. Women patients with fetal problems
12. Critical pediatric problems
13. Life-threatening congenital defects in the newborn
14. Neurological problems

Intensive care units or wards have been developed and designed in the following categories:

1. Surgical and postoperative units (SICU)
2. Psychiatric units
3. Traumatic units
4. Neonatal units
5. Coronary units (CICU)
6. Mobile coronary units
7. Kidney dialysis units
8. Units treating addicts using drugs and narcotics, such as opium, cocaine, morphine, LSD, and many others
9. Pediatric units and neonatal units (NICU)

These units, which are not necessarily listed in the order of importance, are all called intensive care units. Every hospital, small or large, is an entity in itself. The number of intensive care units will, therefore, vary and so will the time the patient spends in the intensive care unit (generally, 3 days to 2 weeks).

In small hospitals one intensive care area may serve many purposes—coronary care, respiratory care, and postoperative recuperation.

There are basically two approaches now used to monitor critically ill patients in intensive care areas. The first is to provide a continuous presentation of the patient's condition by observing physiological variables. Bioelectronic measurements and the mechanics involved must be able to detect trends and changes which indicate whether prompt treatment is required. The other approach is to develop sophisticated monitoring designed to extend sensing capabilities by providing physiological facts that otherwise would not be available.

Development of an adequate monitoring system must include:

1. Development and validation of primary sensors.
2. Automation of measuring and controlling devices. There is a need for remote telemetry equipment to eliminate wiring.

3. The development and testing of pumps, pacemakers, and other artificial prosthetic devices.
4. The development of data-acquisition systems that include analog data-filtering techniques.
5. The elaboration of special-purpose data management techniques.
6. The provision for informational output procedures concerned with the condensation of data in a form that can be presented to physicians and nurses in a functional fashion for decision making.

Some intensive care units are awesome examples of poor electronic engineering, with wires running across the aisles and floors. Sensors and indicators may operate improperly or not at all; inadequate provisions are made at the bedside for standard medical measurements. Room planning and design are poor, including irregular mounting of terminals from bed to bed, inadequate nursing stations for monitors, insufficient isolation of patients, the need for the attendant to crawl over the patient's bed to attach instrumentation, and so forth.

Present-day intensive care units are still frightening to patients. Modern advances in electronics have created some new and serious psychological problems: tension, stresses, and depression in the hospital, as well as when the patient comes home. Some of these problems can be overcome by close doctor–patient and nurse–patient relationships, structural changes in intensive care design, and a minimizing of adverse sensory and/or sleep deprivation.

The purpose of the coronary intensive care unit (CICU) or coronary care unit (CCU) is to render the best medical and nursing care to critically ill patients with cardiovascular or associated malfunctions. This involves vigorous medical treatment to decrease the heart rate, blood pressure, heart size, and velocity of myocardial contraction.

Cardiovascular system failure may be due to myocardial infarction (40% of the population between 40 and 50 years old can get such an attack), pulmonary embolus, sepsis, hemorrhage, and many other heart diseases. The aim of treatment is to sustain the patient until the underlying disorders have been properly treated (i.e., sepsis, hypovolemia) or have been spontaneously (i.e., myocardial infarction, pulmonary emboli) resolved. For these seriously ill patients, it is important to detect any deterioration, so that appropriate treatment can be promptly instituted. Therefore, the following parameters should be continuously monitored or intermittently measured:

1. Heart rate and rhythms
2. Central venous pressure
3. Interarterial pressure
4. Cardiac output
5. Blood gases such as PO_2 and PCO_2

6. pH
7. Temperature
8. Transthorax impedance

A continuous monitoring scheme of four physiological parameters uses a computer system that incorporates analysis of each breath and heartbeat for changes in rate and rhythm, and measurement of systolic and diastolic blood pressure at intervals of 1 minute to 1 hour.

Electronic equipment now available can therefore display on an oscilloscope and record simultaneously the electrocardiogram, arterial pressure, venous pressure, dye dilution curves necessary to get cardiac output if this system is used, urine output, body temperature, and respiratory rate. An alarm system built into an electrocardiographic module is set off when the heart rate increases above or falls below certain rates. A module for the detection of premature ventricular beats and ventricular rhythms has also been developed. Such a module can detect any significant increase in the time duration of the QRS complex of the electrocardiogram. However, it does not differentiate a supraventricular beat with an aberrant ventricular conduction from a ventricular beat. Electronic equipment does not replace medical or nursing care, but rather increases the need for highly trained personnel.

13.2 ICU/CCU MONITORING

Cardiac monitoring is normally performed in intensive care units, including the coronary care units; operating rooms; and emergency wards. These devices are characterized by the amplification and display of cardiac activity and/or electrical events as they are detected through electrodes appropriately placed on a patient's body. The operational integrity of these devices is important because they are directly connected to the patient and extend the normal senses of medical and nursing personnel beyond their normal capabilities. A malfunctioning device can either present a serious hazard directly to the patient and/or operator, or can mislead the operator, who will then provide an inappropriate response that endangers the patient.

Cardiac monitoring began with the development of the electrocardiograph and has grown into our modern ICU and CCU complexes, which are begining to use computer assistance for analysis. Cardiac monitoring can be segmented into two sections: physiological monitors and electrocardiographs.

Physiological monitors can be separated into two groups, based on the signal-processing technique used in their design. These are:

1. Linear amplifier
2. Modulated carrier frequency amplifier

Both types can utilize identical input isolation techniques and can possess identical gains and bandwidths. Their basic circuits, however, are substantially different.

The linear amplifier physiological monitor processes the signal from the patient as it is detected, without modification of its component parts. This type of monitor is both relatively simple and direct. It does, however, require close control because there is little built-in patient protection in case of component failure.

To mitigate this situation, newer models of linear amplifier monitors (Fig. 13.1) use higher-grade power supply components as well as buffer amplifiers on the patient lead circuits. The combined effect of this change in design is the reduction of normal leakage current to a level which is more acceptable, while isolating the patient from the instrument through a higher-input impedance (on the order of 50 MΩ). The signal from the unity-gain amplifiers is fed to a lead selector network. The selected output voltage is then fed to a differential amplifier. A signal is obtained and fed to an ECG signal amplifier and then to an oscilloscope. The output of the ECG signal amplifier is then fed to a cardiotachometer with a low- and high-alarm-rate meter.

The physiological monitor that uses carrier frequency modulation techniques (Fig. 13.2) is somewhat more complex than the linear monitor. The input amplifier's lead selectors and patient leads are isolated from the high-level recorder circuits by insertion of a modulator in the signal flow path. The modulator is driven by its own oscillator operation at a carrier frequency, usually at or above 20 kHz. The patient's ECG signal is detected and amplified by the input amplifiers, which are powered by a low-level isolated power supply. The signal is then passed into the modulator, where it is chopped at the modulation oscillator's frequency. The output of the modulator is transformer coupled to a demodulator circuit, where the amplified patient signal detected from the carrier is amplified and coupled to the oscilloscope and rate meter circuits.

Carrier-frequency-modulated patient monitors usually possess a higher level of patient isolation than do linear amplifier units. They are, however, more susceptible to radio-frequency interference such as is generated by electrosurgical units.

Electrocardiographs are manufactured as either single- or multiple-channel devices. Two design principles are currently used based on linear amplification of modulated-carrier amplification. Because ECGs are seldom taken in areas where cauteries or diathemias are used, there is little difficulty with radio-frequency interference.

The single-channel electrocardiograph (either linear or modulated-carrier type) has the widest use in the clinical setting.

The multichannel electrocardiograph has been used in the research setting and is beginning to be found in increasing numbers in clinical use.

The basic electrocardiograph has the identical functions of the monitor

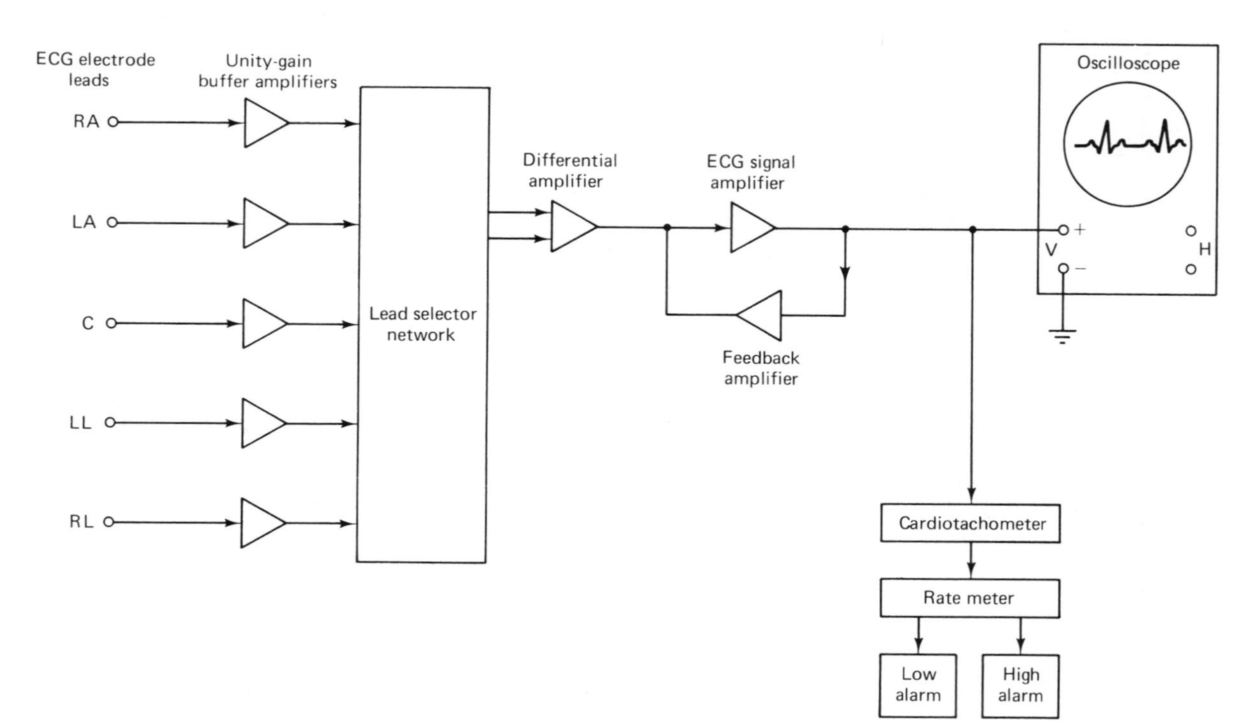

Figure 13.1 Simplified block diagram of a linear amplifier patient monitor.

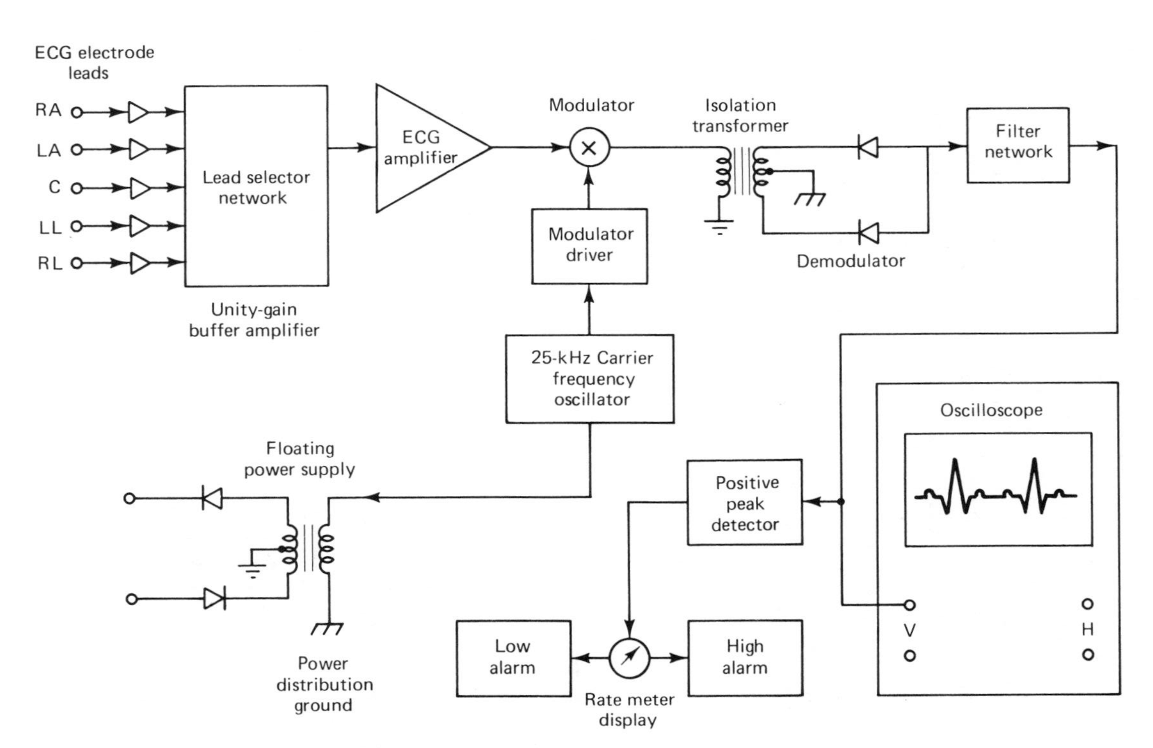

Figure 13.2 Simplified block diagram of a modulated carrier patient monitor.

except that its amplifier output is passed into a recorder assembly rather than into an oscilloscope display.

The creation of the concept of specialized monitoring facilities for critical classes of patients is still relatively new even though we are in the fourth generation of equipment. The original ICU installation (Fig. 13.3) carried all of the patient's ECG signals back to a central station, where individual rate meters and multichannel scopes were used. The second generation featured scope displays at the bedside with rate meters and an ECG recorder at the central station. The third generation of equipment featured rate meters and scopes at the bedside and a central station consisting of eight-channel scopes, tape memory loops, and an ECG recorder. The latest generation is a mixture of a return to the original—no bedside monitor and a complex central station rate-and-trace display—with a computer-based memory loop and a simple bedside rate-and-scope display coupled to a central station scope computer memory and ECG trace display.

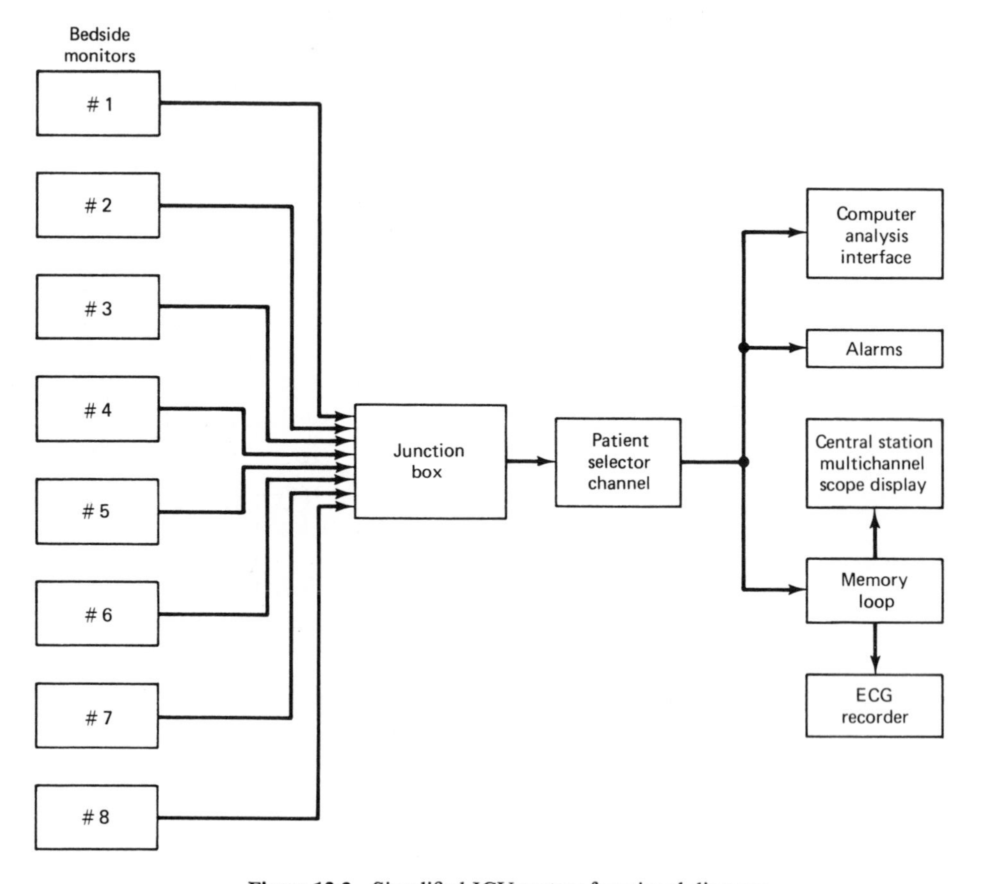

Figure 13.3 Simplified ICU system functional diagram.

Most large hospitals usually have a mix of the older and newer systems with telemetry thrown in. Smaller hospitals usually have either an older system or a newer system, but not both.

In any case, conceptually, an ICU system is a relatively straightforward converging/diverging branch network which allows multiple-function monitoring of several patients. In this system, up to eight (or more) patients are connected either directly or through bedside monitors by hard-wire cabling to a junction box (convergence). The signals are passed through the junction box to either a manully operated or automatic patient selector. The signals are all displayed on the central station scope(s). In parallel with the central station scope (divergence) are a rate meter alarm network, memory bank (tape or computer), and in some systems, the analog-to-digital computer interface. The memory loop feeds back into the scope display for delayed-trace display and also drives the ECG recorder for permanent-record generation. The ICU/CCU concept is, in reality, the natural extension of single-unit patient monitors.

Alarms visibly and audibly indicate the bed and parameter that require attention. Some systems differentiate between real ECG alarms and those caused by poor electrical contact. Modules include oscilloscopes, direct writers, tape records, panel meters, and digital readouts.

A delay tape recorder that continuously stores recent ECG data is particularly useful. If an alarm occurs, the recording process stops. Attendants may examine the record for the period immediately before the alarm. Stored records can run from 12 seconds to 5 minutes. One problem is that cartridge life is generally short.

13.3 BIOTELEMETRY

Telemetry instruments are normally used in intensive care units, progressive coronary care units, outpatient clinics, and research units. These devices are characterized by the transmission of a physiological function by either land lines (telephone) or radio-wave communication links (transmitters and receivers). In this context, telemetry is defined as the transmission of data over an arbitrary distance from a remote detection site to a central information retrieval and reduction center.

Telemetry can be either a hard-wire system or a radio-communications-link system. As used in the modern hospital, the hard-wire system is limited to mass screening operations from a central source such as electrocardiographic interpretation. The radio communication link, however, is becoming increasingly important in the daily monitoring of patients, especially where transient ambulatory arrhythmia is to be detected.

Radio-telemetry systems can be separated into two classes of equipment—those that modulate and transmit the signal and those that receive and demodulate the signal—or transmitters and receivers.

A block diagram of a typical biotelemetry system in the CCU is shown in Fig. 13.4. The signal from the human being via a sensor is sent to a transmitter through a signal conditioner. The signal is then picked up by a receiver, amplified, and fed to a readout device.

Telemetry can be applied internally or externally in the human body. It is possible to design a multichannel implantable telemetry system to monitor ECG and multiple pressures and temperature using a five- and eight-channel recorder. Biotelemetry is now popular in intensive care areas to monitor body temperature, ECG, and respiration rate.

For obtaining intracorporeal physiological data from outside the body, transduction systems such as ultrasonic blood-velocity detectors, ECG, EEG, optical oximeters, thermographs, electric impedance plethysmographs, and phonocardiographs are available. Portions of the body interior can be visualized through use of x-rays, neutron beams, and ultrasonic scanning.

When a satisfactory diagnosis cannot be made without probing body cavities or surgically penetrating the body, electronic equipment using fiber optics can be used.

Surveillance of cardiac pacemakers (Fig. 13.5) following implantation is a valuable tool in the management of patients with pacemakers using biotelemetry techniques. In Fig. 13.5 a sensing circuit feeds pacing circuitry fed to a telemetry computer transmitter. From an antenna, the signal is transmitted to an ECG telemetry receiver.

Dysrhythmia or pacemaker problems require continuous monitoring using biotelemetry. Efforts in the CCU must continue to reduce mortality after myocardial infarction. Factors that may influence prognosis include location and extent of infarction, presence of heart failure, shock, age, and the occurrence of intraventricular conduction defects. Deaths after acute myocardial infarction can be associated with congestive heart failure, pulmonary edema, cardiogenic shock, or arrhythmias.

The progressive coronary care unit (PCCU) or the intermediate coronary

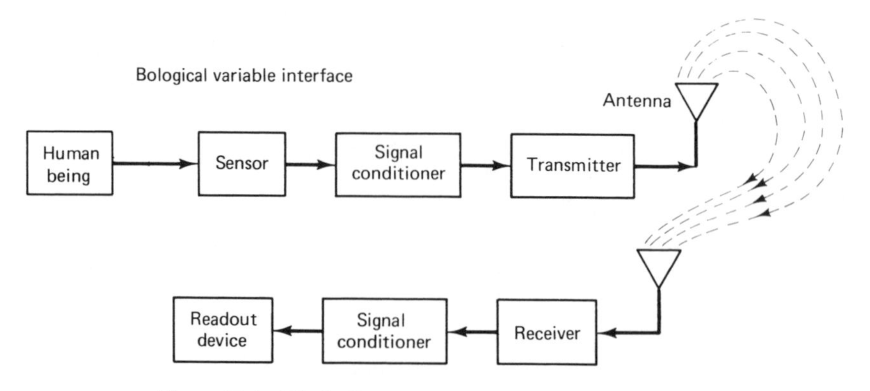

Figure 13.4 Block diagram of a typical biotelemetry system.

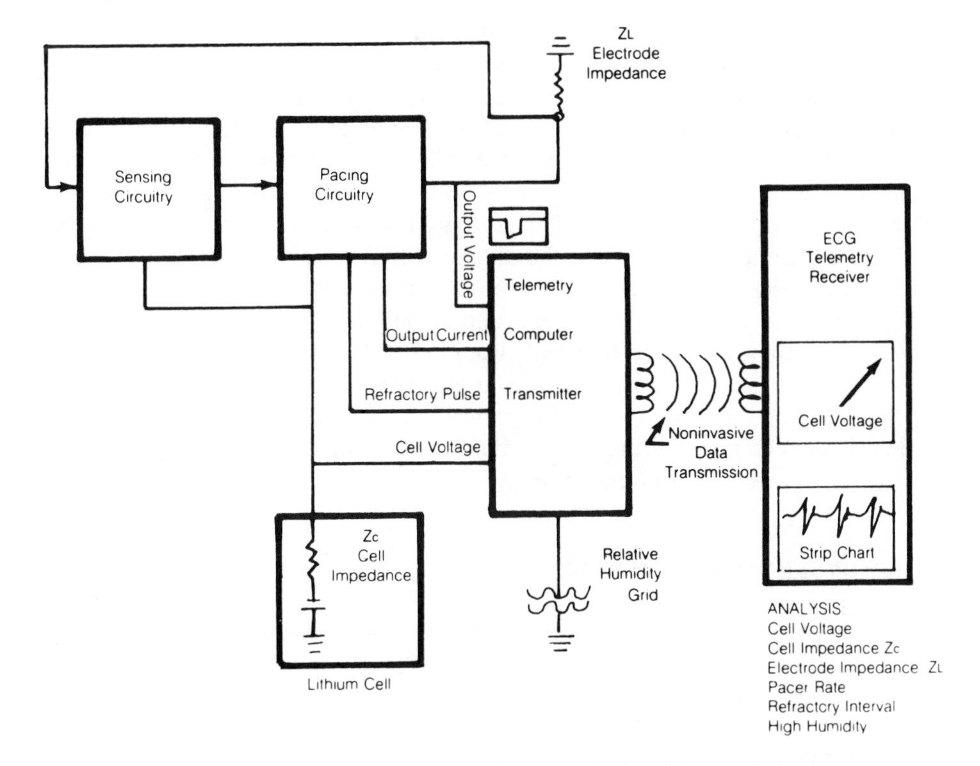

Figure 13.5 Pacemaker diagnostics system. (From C. J. Del Marco, Telemetry for Pacemaker Follow-up, *Osteopathic Annals*, Vol. 6, No. 10, October 1978, published by Insight Publishing Co., Inc., New York, N.Y. © 1978, p. 30.)

care unit also require the use of biotelemetry. The period of ECG monitoring should be extended beyond the traditional 4 or 5 days to provide maximum safety for the patient with acute myocardial infarction. Otherwise, the patient may die. The effective functioning of the PCCU results in early detection of new arrhythmia or extension of infarct with proper return of the patient to the CCU.

The PCCU is designed to do the following:

1. Reduce the incidence of post-CCU hospital sudden death by early detection and treatment of life-threatening dysrhythmias.
2. Enable the application of prompt resuscitation if required.
3. Provide rehabilitation by monitoring early ambulation.
4. Mobilize CCU beds by permitting earler transfer.

A separate monitoring system is provided for the PCCU, where the ECG is transmitted by an FM signal to a nearby receiver and then by hard wire to the monitor console bank in the CCU. The patient carries a small, portable transmitter, and movements are not inhibited. The bank of monitoring oscilloscopes is constantly observed by technicians or nurses, who also monitor the instruments of the CCU. Thus, the ambulatory patient is continuously monitored. Specific telephones connect the floors involved to the bank of monitors.

To avoid misinterpretations, the PCCU nurses notify the monitoring nurse or technicians when the patient is being bathed, or is in a nontransmitting area, or is disconnected, as when a routine electrocardiogram is obtained. Good communication between the patient care floor and the monitoring personnel is essential for the effective operation of this system. Patients in the PCCU are given bathroom privileges and are allowed to move about essentially at will as they gradually improve. Radio or television is permitted, but an earplug must be used to minimize interference with other patients. Use of the telephone is limited. Generally, the patient stays in the PCCU for several days up to a week or two. Readmission to the CCU is required if there is a recurrence of ischemic chest pain or arrhythmias.

13.4 DYSRHYTHMIA MONITORS

The prime purpose of acute coronary care is to prevent and treat dysrhythmias or arrhythmias. Microprocessing techniques have been developed in accurately generating a real-time graph in dysrhythmia monitors as well as other patient care monitors. The dysrhythmia monitor combines alphanumerics and waveshapes and color codes both by shapes. Outputs can be provided on graph paper and ORT screen and be constantly updated.

The waveshapes of an operating room monitor can be combined with alphanumerics as shown in Fig. 13.6. The chart recorder is shown in Fig. 13.7.

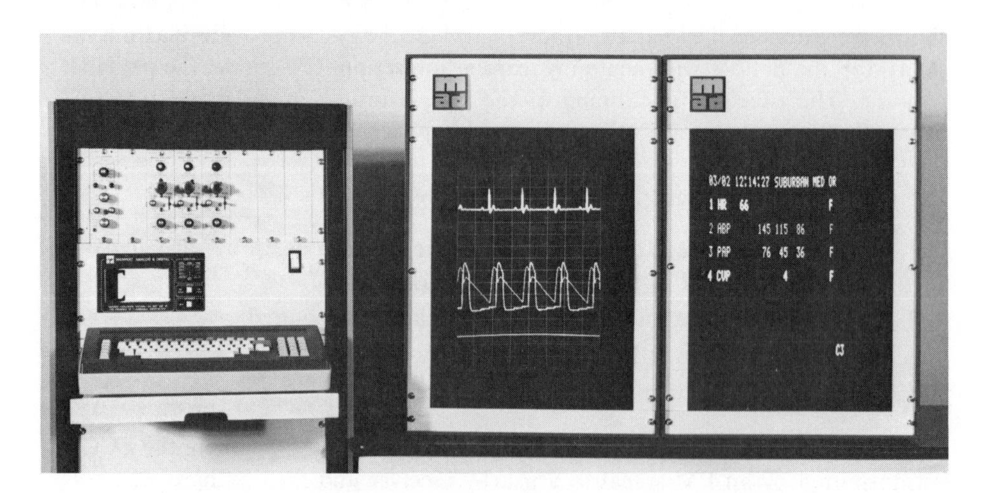

Figure 13.6 OR monitor: waveshapes can be combined with alphanumerics. (Courtesy of Midwest Analog and Digital Inc., New Berlin, Wis.)

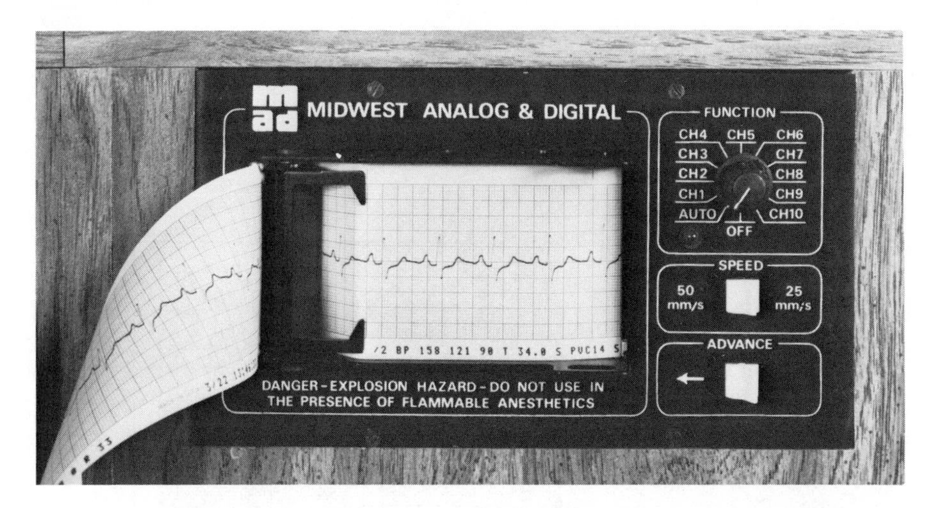

Figure 13.7 Chart recorder. (Courtesy of Midwest Analog and Digital Inc., New Berlin, Wis.)

Note that the edge printer prints out all parameters both automatically and manually. A sample recording is shown in Fig. 13.8. A sample recording from the pressure channel shows the systolic and diastolic pressure (158/118), and other pertinent information in the patient physiological status in the CCU is shown in Fig. 13.9. The microprocessor used in the CCU can be interfaced with other bioelectronic devices to obtain the required physiological parameters.

Dysrhythmia monitors must operate on every ECG complex presented to them, whether it be noisy, atypical, or having an R wave obscured by P and T waves. The monitor will obediently analyze everything, and if the ECG is unusual enough, it will probably make an incorrect judgment, either a false negative (undetected dysrhythmic beat) or a false positive (detected nondysrhythmic beat). To ensure correct operation of the device, it is necessary to know how it "thinks" and to provide data that are not likely to confuse it or to know when its judgment is probably too faulty.

The dysrhythmia monitor is essentially a QRS complex analyzer. Some consideration is given to P and T waves, but these are not a main concern. The reason is that the P wave is the first characteristic of the cycle to be obscured in noise and is not very reliably present. The decisions that are then made are:

1. Whether a beat is premature
2. Whether a premature beat is of ventricular or supraventricular origin

Each 30 seconds the average of each arrhythmia rate is stored for 1 hour. Each 15 minutes the average of each arrhythmia rate is stored for 24 hours. The results of this data storage may be displayed graphically in the form of

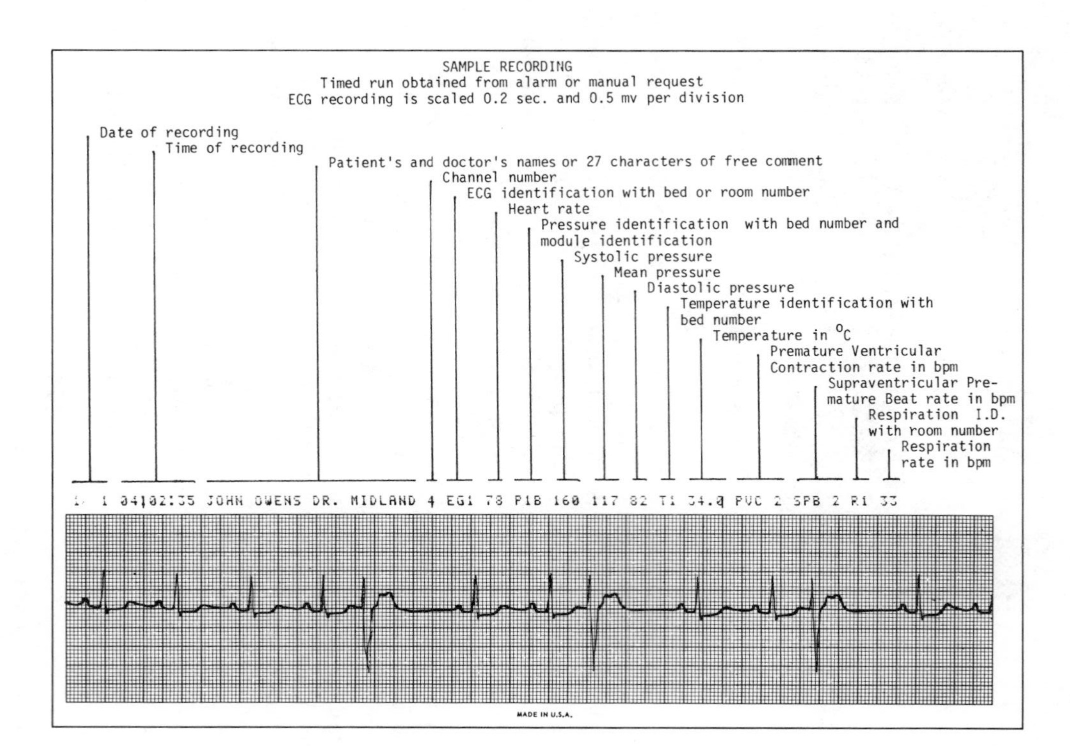

Figure 13.8 Sample recording. Timed run obtained from an alarm or manual request. The ECG recording is scaled 0.2 sec and 0.5 mV per division. (Courtesy of Midwest Analog and Digital Inc., New Berlin, Wis.)

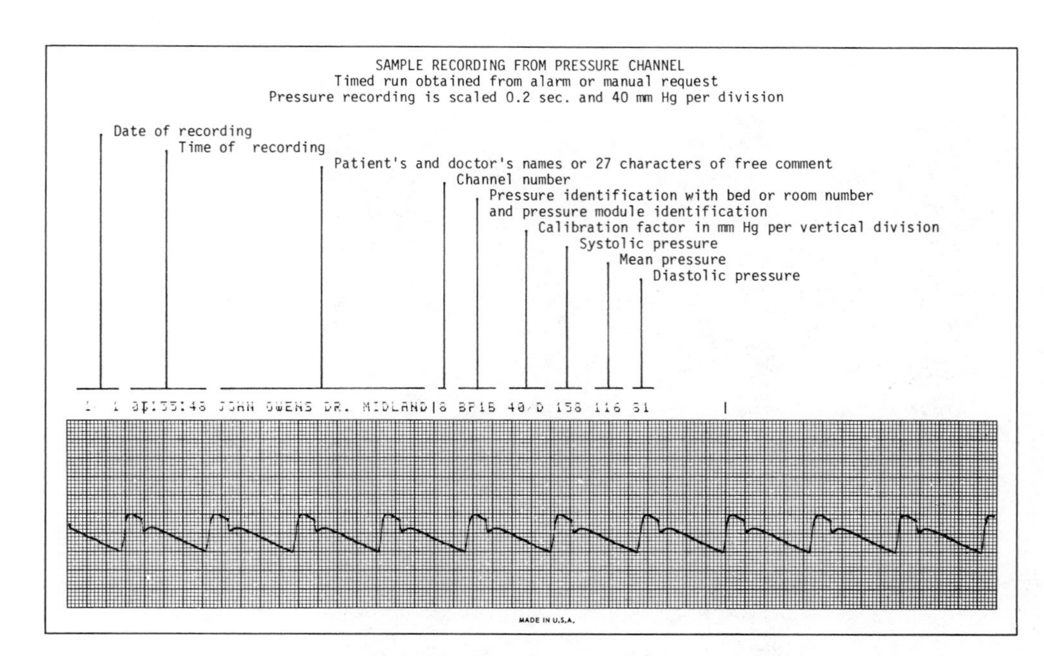

Figure 13.9 Sample recording from pressure channel. Timed run obtained from an alarm or manual request. The pressure recording is scaled 0.2 sec and 40 mmHg per division. (Courtesy of Midwest Analog and Digital Inc., New Berlin, Wis.)

a trend plot in the usual manner. Typical trends in physiological wave shapes are shown in Fig. 13.10. A grid can be placed on the back of the typical trends, and measurement of the patient's activity can be seen at any time.

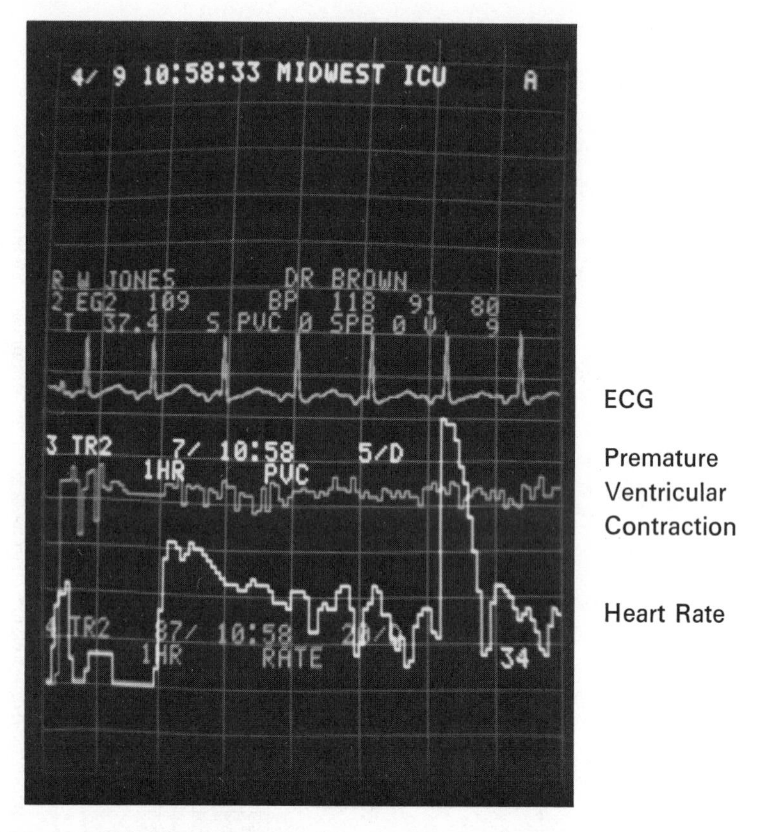

Figure 13.10 Typical trends in physiological waveshapes: (a) ECG; (b) 1-hr PVC (premature ventricular contraction) trend; (c) 1-hr heart rate trend. (Courtesy of Midwest Analog and Digital Inc., New Berlin, Wis.)

13.5 HOLTER RECORDINGS

A Holter recording is a continuous magnetic tape recording of a patient's electrocardiogram, customarily recorded for a period of 10 to 24 hours. Standard electrodes are attached to the patient's chest and leads are connected to a compact portable tape recorder. The recorder can be carried over the patient's shoulder like a camera while the person is active, or placed beside him or her during rest or sleep. The patient follows a normal or prescribed routine during the recording period and keeps a diary, noting the time of any symptoms experienced or any unusual activity performed. When the recording is completed, the patient returns the tape and recorder together with the diary. Special instrumentation operated by skilled personnel is utilized to examine the tape recording for arrhythmias, conduction abnormalities, and other morphological changes. Detected abnormalities are printed, with a time reference, on standard ECG chart paper for detailed analysis by the physician.

The performance of a Holter recording thus has three distinct and separable phases:

1. Obtaining the recording on the patient
2. Analyzing the recording for abnormal phenomena
3. Interpreting the significance of the detected phenomena within the context of the total clinical picture presented by the patient

In Fig. 13.11 the patient hookup and scanning system of a Holter recording are shown. In the patient hookup small, flat electrode disks are attached to the patient's chest, and three connecting leads are plugged into a special portable tape recorder. In the scanning system, when the recording is completed, the patient returns the tape, the recorder, and the diary. The recording is then scanned with special electronic equipment, and standard ECG strips are printed (with time notations) for all arrhythmias, conduction abnormalities, and morphological changes.

The Holter recording can be interfaced with the Analog and Digital Co. system so that the information can be easily stored for readout.

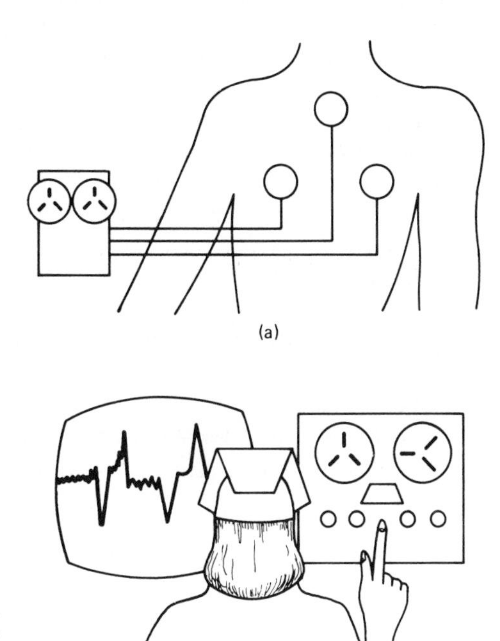

Figure 13.11 Holter recording as used in the patient care environment: (a) patient hookup; (b) scanning.

13.6 REVIEW QUESTIONS

1. What is meant by cardiac monitoring?
2. What is the function of a linear amplifier physiological monitoring process?
3. What is meant by a modulated carrier frequency amplifier?
4. List the lead selector position and measurement points and measurement type for an ECG.
5. List five places where patient care is used and define progressive care.
6. List five conditions for an adequate monitoring system.
7. What are the physiological parameters measured in the CCU?
8. Draw a block diagram of a simplified biotelemetry system.
9. Describe the purpose of an arrhythmia monitor.
10. What is a Holter recording?

13.7 REFERENCES

1. Jacobson, B., and Webster, J. G.: *Medicine and Clinical Engineering*, Prentice-Hall, Inc., Englewood Cliffs, N.J., 1977.
2. Cromwell, L., Werbell, F. J., and Pfeiffèr, E. A.: *Biomedical Instrumentation and Measurements*, 2nd ed., Prentice-Hall, Inc., Englewood Cliffs, N.J., 1980.
3. Webster, J. G., and Cook, A. M.: *Clinical Engineering: Principles and Practices*, Prentice-Hall, Inc., Englewood Cliffs, N.J., 1979.
4. Knoebel, S. B., Rasmussen, S., Lovelace, D. E., and Anderson, G. J.: Non-paroxysmal Junction Tachycardia in Acute Myocardial Infarction: Computer-Assisted Detection, *The American Journal of Cardiology*, Vol. 35, pp. 824–830, 1975.
5. Knoebel, S. B., Lovelace, D. E., Rasmussen, S., and Wash, S. E.: Computer Detection of Premature Ventricular Complexes: A Modified Approach, *The American Journal of Cardiology*, Vol. 358, pp. 440–447, 1976.
6. *Directions in Cardiovascular Medicine*, No. 8, Holter Recording (Continuous BCG Recording), Hoechst Pharmaceuticals, Inc., Somerville, N.J., 1973.
7. Guyton, A. C.: *Textbook of Medical Physiology*, 5th ed., W. B. Saunders Company, Philadelphia, 1976.
8. Summers, G. D.: Transducers for Bioimplantable Telemetry Systems, *IEEE Transactions on Industrial Electronics and Control Instrumentation*, Vol. IECI-17, No. 2, pp. 144–150, April 1970.
9. Clynes, M., and Milsum, J. H., eds.: *Biomedical Engineering Systems*, Inter-University Electronics Series, Vol. 10, McGraw-Hill Book Company, New York, 1970.
10. Mackay, R. S.: *Biomedical Telemetry*, 2nd ed., John Wiley & Sons, Inc., New York, 1970.

14

PERFORMANCE TESTING AND SAFETY

14.1 INTRODUCTION TO SIMULATION TECHNIQUES

Many medical centers around the country use innovative ECG/blood pressure simulators and other physiological simulators to improve patient monitor test procedures. Three primary objectives of patient monitor performance testing include:

1. Verifying that the physiological monitor provides clinically useful information.
2. Minimizing the impact of test procedures on the clinical environment.
3. Routine equipment performance testing, which reduces the cost of equipment repairs.

A total equipment control program gives us three important bits of information about a monitor:

1. The way the monitor is intended to work
2. The way the monitor actually works
3. The way the monitor is perceived to work

There is an initial evaluation and specification review which provides the description and technical details on how the monitor device is intended to work.

The incoming inspection and periodic performance testing provide data on how the monitor is working, and the in-service and repair activities give feedback on how the staff perceives it to work. Tests for acceptance differ from maintenance or repair tests.

Bioelectronic measuring devices that require performance testing include:

1. Cardiac monitors and heart rate meters
2. Blood pressure monitors
3. Temperature monitors
4. Respiration monitors
5. Arrhythmia detection and analysis systems
6. Holter cardiographs
7. Fetal monitors
8. Exercise/stress testing equipment
9. Cardiac catheterization systems
10. Electrocardiograph
11. Chart recorders

When developing procedures and equipment for performance testing, it is important to consider the essential characteristics of monitoring systems. For example, when evaluating the performance of heart rate displays, some characteristics to determine are as follows:

1. The ability to detect R-waves, including the response to R-waves as a function of shape, amplitude, and duration.
2. The response of the instrument to heart rate, both within and outside the specified range of the instrument.
3. The response to abnormal rhythms, such as bigeminy and ventricular tachycardia.
4. The response to noise and other unwanted signals, in particular 60-Hz interference, pacemaker spikes, and large T-waves.
5. Accuracy of alarm set points and the time required to activate the alarm after the set point has been exceeded.

A simulator that uses digital read-only memory to store physiological waveforms is useful in performance testing. The simulator produces a simulated ECG waveform in which the R-wave can be modified by shape, amplitude, and duration to test the ability of the monitor to detect R-waves. Simulated heart rates are adjustable from 30 to 375 beats per minute to test the monitor over the full range of rates that are likely to occur for adult and infant patients.

The response to abnormal rhythms can be tested by using a special mode of operation which allows the simulator to generate wide or normal waveforms at a variety of R-R intervals. A quick check of noise immunity of the cardiotach portion of the monitor can be obtained by a control on the simulator that injects 60-Hz voltage into the simulated ECG output. The accuracy and repeatability of a wide variety of heart rates permits the convenient testing and evaluation of monitor alarm systems.

Performance testing of dysrhythmia detection systems is becoming more important because of the increasing need for these systems. The requirements for testing dysrhythmia detection systems are similar in many respects to those for testing cardiac monitors and heart rate meters. Testing is simplified if dysrhythmia systems are considered to be advanced cardiac monitors. The primary differences are related to the dysrhythmia systems' ability to recognize and classify heart beats and patterns, to measure and analyze the interval between beats, to handle complex alarm criteria, and to report the results. The simulator also provides a powerful capability for testing dysrhythmia detection systems and incorporates a variable calibrated ECG output signal for performance testing of the recording and display devices associated with a patient monitoring module sequencer which stores the program in digital read-only memory. The flexibility and adaptability of this approach allow future ROM exchange as the sophistication of dysrhythmia systems improves to include the classification of supraventricular dysrhythmias. A simulator device can be also a valuable tool in testing blood pressure monitors, fetal monitors, and Holter cardiographic recording equipment.

Most monitoring systems now in use for cirtical care in CICUs, SICUs, NICUs, and labor rooms combine a multiparameter measurement system far more complex than earlier bedside ECG monitors. The incorporation of multiple test capabilities within one test instrument simplifies and accelerates the performance testing. An example of the effectiveness of the new simulator is demonstrated in testing fetal monitors. A previous test protocol used a conventional ECG simulator, a precision attenuator, an oscilloscope, and sphygmomanoperformeter. With both ECG and blood pressure incorporated in one instrument, the performance test time was reduced from an average 2.3 hours to an average 1.4 hours using the new simulator. For eight monitors being tested, this represents 7.2 person-hours saved per test cycle.

Figure 14.1 shows an ECG/blood pressure simulator, arrhythmia sequencer, respiration simulator, and temperature simulator. Figure 14.2 shows an ECG pattern sequencer for testing arrhythmia detection systems.

Introducing selected arrhythmias provides verification of the computation of instantaneous heart rate, also called heart frequency, in both fetal and neonatal monitors. Since the beat-to-beat rate is of primary importance in these patients, quantitative documentation of monitor performance provides confidence in the clinical data which directly affect treatment decisions. This is an example of an improvement in the quality of the performance test, which now provides information not easily obtained with conventional techniques.

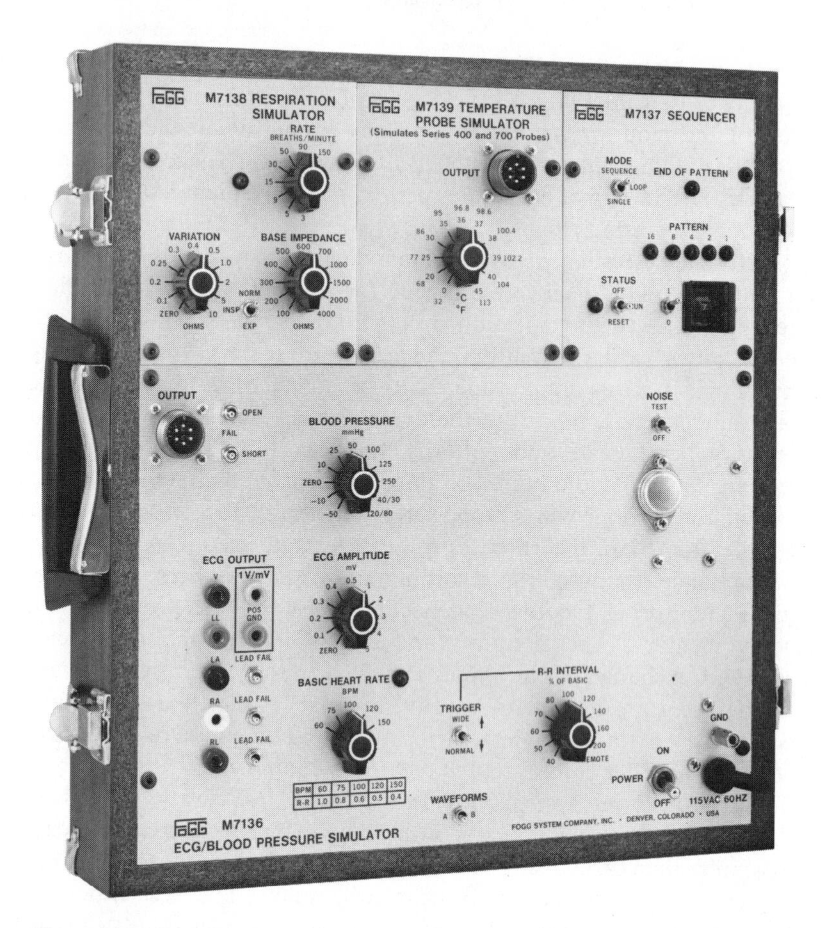

Figure 14.1 ECG/blood pressure simulator, arrhythmia sequencer, respiration simulators, and temperature simulator. (Courtesy of Fogg System Company, Inc., Aurora, Colo.)

14.2 TESTING THE PERFORMANCE OF BLOOD PRESSURE TRANSDUCERS

The need for testing blood pressure transducer performance includes:

1. Existing transducers represent a substantial investment, but most hospitals have little information about their performance characteristics. Calibration certificates have been misplaced, a testing program may not exist, and thus many transducers are returned to the manufacturer for repair or exchange. For these reasons many hospitals are planning to implement a systematic testing program to control expenditures in this area.
2. Performance testing of each transducer before and after use provides verification of the accuracy of patient pressure measurements. Practical

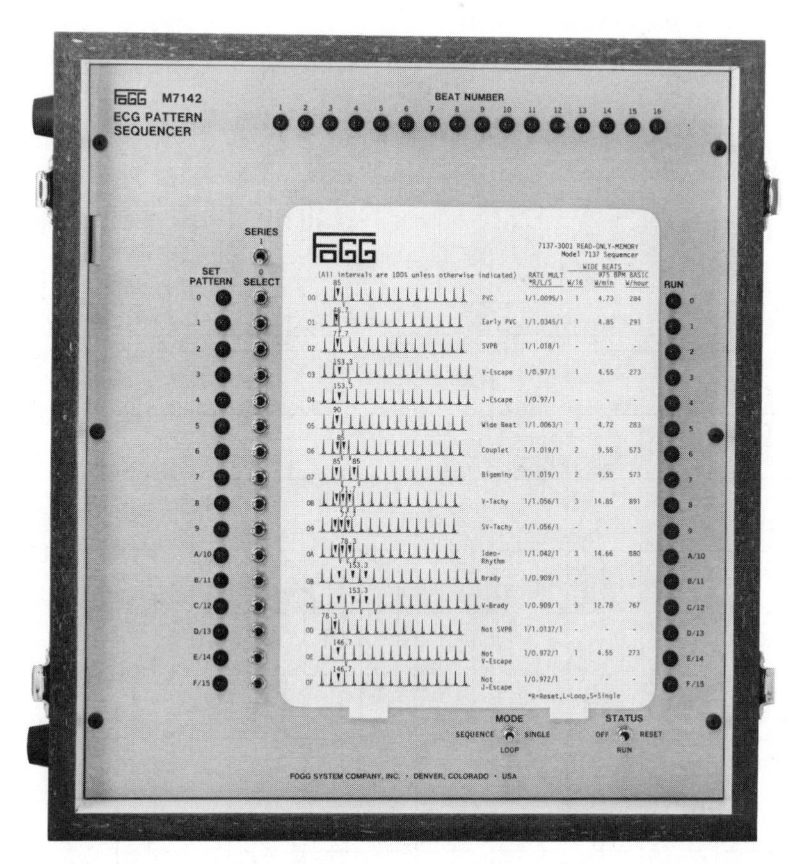

Figure 14.2 ECG pattern sequencer for testing arrhythmia detection systems. (Courtesy of Fogg System Company, Inc., Aurora, Colo.)

considerations may preclude this total approach, but frequent testing does offer a means of reducing the uncertainty in patient data.

3. If purchase of new transducers is planned, you need to determine if suppliers meet specifications established by the hospital. Later, when the supplier has been chosen and the new transducers arrive, there will be a need to check them against specifications before they are used.
4. The ICU reports a malfunctioning transducer. You need to determine if the problem is caused by the catheter, transducer, monitor, operator, or a combination of these factors.
5. Testing reveals that a transducer has failed. The transducer is sent to the manufacturer for repair or exchange. When it is returned to the hospital, you need to retest it to verify that it is performing to specifications.

A systematic program for on-side testing of transducer performance reduces maintenance and operating costs by minimizing service calls. It can

⑤ TEST STEP (TO BE COMPL)	⑥ MONITOR UNIT (CHECK ONE)	⑦ PRESSURE FUNCTION (CHECK ONE)	⑧ SIMULATED PRESSURE mmHg ZERO / 50	10 / 100	15 / 150	20 / 200	25 / 250	30 / 300	⑨ CHECK BOX (SEE LEGEND)	⑩ TEST STATUS	⑪ TEST BY DATE	⑫ APVD BY DATE
1	☒ BEDSIDE	☒ SYSTOLIC ☐ DIASTOLIC	ZERO: +1	10: +10	15: +15	20: +21	25: +26	30: +31	1; 2 300; 5 X	OK	RCR 8/19/77	H 8/22/77
	☐ REMOTE	☐ MEAN ☐ VENOUS	50: +50	100: +100	150: +151	200: +201	250: +251	300: +301	3 98; 4 4.02			
2	☐ BEDSIDE	☒ SYSTOLIC ☐ DIASTOLIC	ZERO △: +4	10: +14	15: +19	20: +25	25: +29	30: +34	1; 2 300; 5 X	△	RCR 8/19/77	H 8/22/77
	☒ REMOTE	☐ MEAN ☐ VENOUS	50: +53	100: +106	150: +155	200: +205	250: +256	300 △: +306	3 105; 4			
3	☒ BEDSIDE	☐ SYSTOLIC ☒ DIASTOLIC	ZERO: +1	10: +10	15: +15	20: +21	25: +25	30: +30	1; 2 300; 5 X	△	RCR 8/19/77	H 8/22/77
	☐ REMOTE	☐ MEAN ☐ VENOUS	50: +50	100: +100	150: +152	200: +199	250: +251	300 △: +281	3 101; 4 4.15			
4	☐ BEDSIDE	☐ SYSTOLIC ☒ DIASTOLIC	ZERO △: +7	10: +17	15: +22	20: +27	25: +33	30: +37	1; 2 300; 5 X	△ △ △	RCR 8/19/77	H 8/22/77
	☒ REMOTE	☐ MEAN ☐ VENOUS	50: +56	100: +106	150 △: +146	200: +192 (struck out)	250: +240	300 △: +280	3 108; 4			
	☐ BEDSIDE	☐ SYSTOLIC ☐ DIASTOLIC	ZERO	10	15	20	25	30	1 2 5 6 9 10			
	☐ REMOTE	☐ MEAN ☐ VENOUS	50	100	150	200	250	300	3 4 7 8 11 12			
	☐ BEDSIDE	☐ SYSTOLIC ☐ DIASTOLIC	ZERO	10	15	20	25	30	1 2 5 6 9 10			
	☐ REMOTE	☐ MEAN ☐ VENOUS	50	100	150	200	250	300	3 4 7 8 11 12			
	☐ BEDSIDE	☐ SYSTOLIC ☐ DIASTOLIC	ZERO	10	15	20	25	30	1 2 5 6 9 10			
	☐ REMOTE	☐ MEAN ☐ VENOUS	50	100	150	200	250	300	3 4 7 8 11 12			
	☐ BEDSIDE	☐ SYSTOLIC ☐ DIASTOLIC	ZERO	10	15	20	25	30	1 2 5 6 9 10			
	☐ REMOTE	☐ MEAN ☐ VENOUS	50	100	150	200	250	300	3 4 7 8 11 12			

CHECK BOX LEGEND		
1 ALTERNATE SIMULATOR S/N	5 ZERO CHECK	9
2 PRESSURE RANGE	6 NEG SIMULATED PRESSURE	10
3 MONITOR CAL SWITCH	7	11
4 GAIN CONTROL SETTING	8	12

① HOSPITAL MEMORIAL	
CLINICAL SERVICE	SURGICAL ICU
RESPONSIBLE DEPARTMENT	SURGERY
BUILDING	MAIN
FLOOR, WARD, WING, ETC.	3 WEST
ROOM, SUITE, ETC.	308

④ STATUS CODE
(TO BE COMPLETED BY USER)
△ OFFSET PROBLEM
△ LINEARITY PROBLEM
△ POSSIBLE SATURATION

③ SIMULATOR	
SERIAL NUMBER	184-48
MODEL NUMBER	BP-48

② MONITORING SYSTEM	
SERVICE ORGANIZATION / PHONE	XYZ 728-4900
WARRANTY	☐ 6 MO. ☒ 1 YEAR ☐ NONE
SER NO. / MODEL NO. BEDSIDE	389-265
SER NO. / MODEL NO. REMOTE	389-266
PURCHASE DATE	B 8-9-73 R 8-9-73
MANUFACTURER	ABC

Figure 14.3 Static test data taken on a blood pressure monitor. (Courtesy of Fogg System Company, Inc., Aurora, Colo.)

also reduce the number of "spare" transducers required to cover those returned for repair or exchange and nursing unit time spent changing transducers.

Figure 14.3 shows a static test blank form for taking simulated blood pressure on a monitor. Four tests are taken at the bedside and/or remote situations. Simulated pressure in mmHg are obtained from 0 to 300 mmHg.

14.3 CARDIAC MONITOR, HEART RATE METERS, AND ALARM STANDARDS[1]

The FDA has developed standards for cardiac monitors, heart rate meters, and alarms. The latest new standard (October 1978) establishes minimum performance and safety requirements for ECG heart rate monitors intended for monitoring critically ill patients. Subject to this standard are all parts of such monitors that are necessary for the following:

1. To obtain a heart rate indication via noninvasive ECG sensing from the patient's body.
2. To amplify this signal, transmit it, and display the heart signal and rate.
3. To provide alarms upon occurrence of cardiac arrest, bradycardia, and tachycardia which exceed preset alarm limits.

The instruments include the following:

1. Portable and battery-operated ECG monitors.
2. Operating room and intensive care heart rate monitors based on the ECG.
3. Intensive care and intermediate care ECG monitors using telemetry.
4. A subsystem of a more complex device (such as an arrhythmia monitor).

These standards include the requirements and test methods for devices used to monitor the heart rate of critically ill patients, so as to assure a reasonable level of safety and performance and provide the user with sufficient data to judge the effectiveness of the device and to operate it safely.

General requirements of instrumentation and test conditions of FDA standards include the following. Unless otherwise sepecified herein, all measurements and tests should be made within the range of standard operating conditions listed below:

Line voltage:	115 ± 5 V rms
Line frequency:	60 ± 1 Hz
Temperature:	20 ± 10°C
Relative humidity:	50 ± 20%

[1]Section 14.3 is based on the Final Second Draft of Standard for Cardiac Monitors, Heart Rate Meters and Alarms, FDA MDS-021-0043 and UBTL TR 1606-010. Support of the FDA of the Department of HEW is acknowledged in excerpting this material.

Actual conditions at the time of tests should be measured and reported. Measurement tolerances are ± 1.4°C for temperature and $\pm 5\%$ for humidity.

The waveforms needed for performance testing can be shown in terms of plotter curves of known amplitude and timing. To facilitate testing, however, these waveforms are also defined as test sequences on magnetic tape. The exact waveforms are stored in digital form on magnetic tape at 500 sampling points per second and an amplitude resolution of ± 2048 counts for an equivalent variation of ± 10 mV. This amounts to a resolution of better than $\pm 1\%$. This tape can be used with a computer system having a 12-bit digital-to-analog converter with a pacing rate of 500 samples per second.

The test circuit shown in Fig. 14.4 attenuates this signal to the low-millivolt level and filters the stepped voltage changes in the signal. The circuit also provides for the mixing of the ECG signal with a noise voltage and allows pacemaker pulses to pass through without being filtered or attenuated.

This circuit operates as follows. Amplifiers A_1 and A_2 comprise a deadband circuit that only passes, at unity gain, those portions of the input ECG signal that are greater than positive 5 V or less than negative 5 V. This ± 5-V limit is adjustable from 0 to ± 10 V. The purpose of this portion of the circuit is to allow pacemaker pulses, which are set to ± 5.5 V by the digital-to-analog converter to produce a 0.5-V output pulse without being filtered and attenuated. The input ECG also passes through a parallel signal path (shown just below A_1 and A_2) consisting of an *RC* low-pass filter (200-Hz cutoff frequency), a 1000 : 1 voltage attenuator, and a unity-gain inverting amplifier, A_4. This attenuated and smoothed signal is summed with the output of the deadband circuit and an externally generated noise signal in amplifier A_3. The output of A_3 provides the ECG signal for the cardiac monitors.

The test waveform amplitude at the input of the cardiac monitor is to be set at the specified values with errors no greater than $\pm 5\%$ or ± 15 μV, whichever is greater.

The R-R interval for the specified heart rate should be set with an error no greater than $\pm 1\%$ of the specified value or 5$\pm$ msec, whichever is greater. Durations are measured with standard universal counter.

Unless otherwise specified by the manufacturer, all tests will be performed with the left arm (LA) electrode connected to the signal source and all others, including any neutral electrode (N) connected to signal ground. The monitor in this mode should be switched to the standard lead I position.

Figure 14.5 shows the test signal simulating the QRS complex of the ECG. The QRS amplitude is $|a_r| + |a_s|$ and is found by following the procedures below:

1. Apply the test waveform specified in note 2 of Fig. 14.5 to the monitor leads.
2. Record the indicated heart rate.

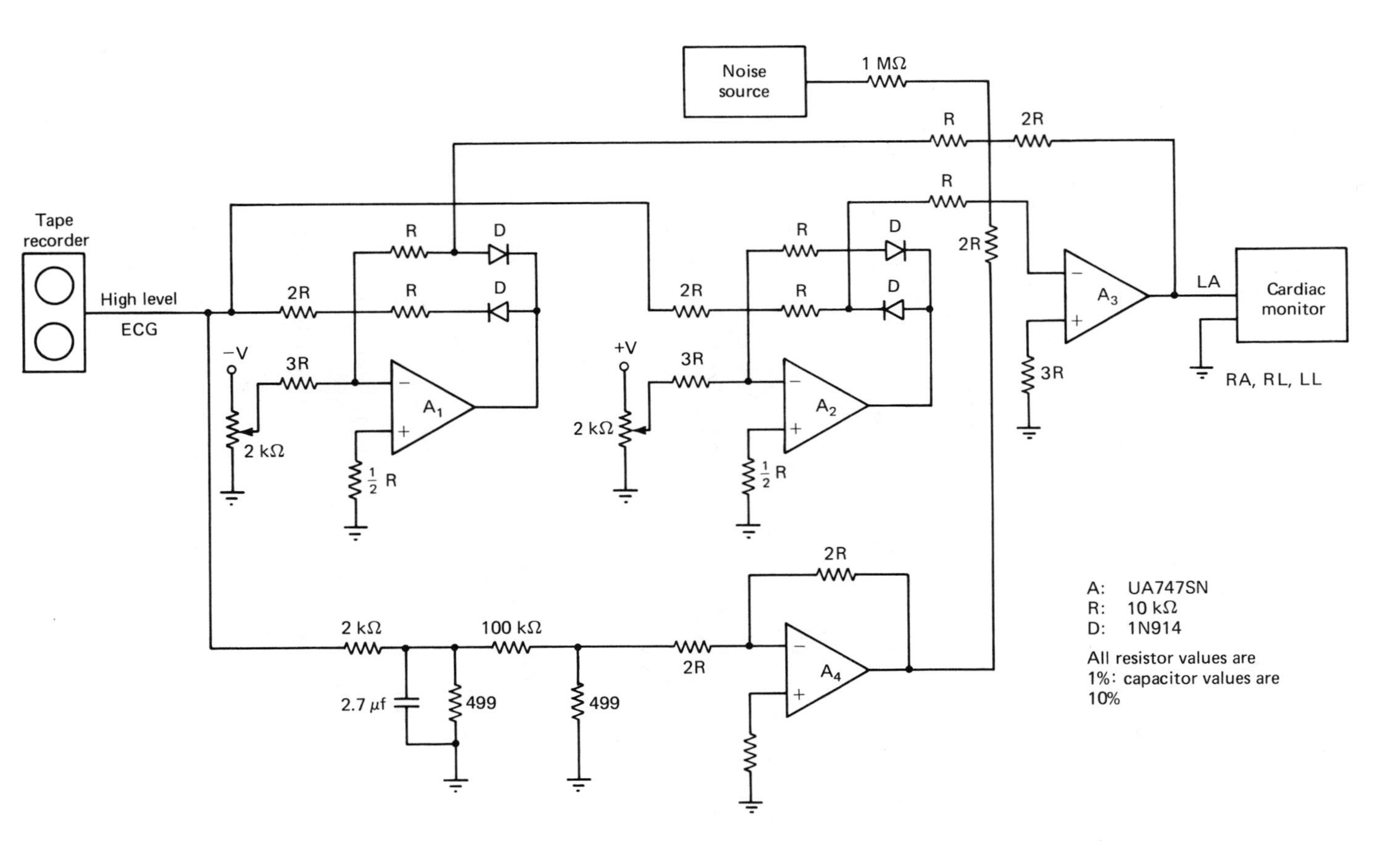

Figure 14.4 Cardiac monitor test circuit for attenuation, filtering, noise mixing, and pacemaker pulse transmission.

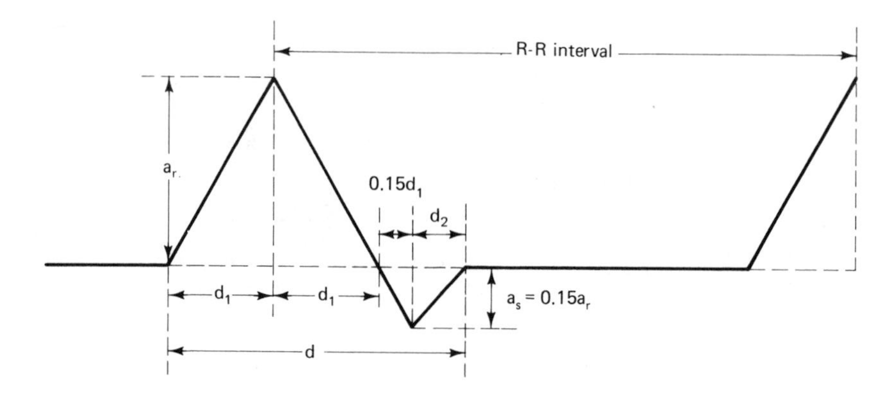

Figure 14.5 Test signal simulating the QRS complex of the ECG $|a_r| + |a_s|$ is QRS amplitude. *Note 1: Range and accuracy of the heart rate meter.* The minimum allowable heart rate meter range is to be 30 to 200 beats/min with an allowable error in readout of no greater than $\pm 10\%$ of the input rate of ± 5 beats/min, whichever is greater. Cardiac monitors which are disclosed to be for use with pediatric patients should have an extended heart rate range to at least 220 beats/min. *Note 2: Range of the QRS-wave amplitude and duration.* A cardiac monitor should meet the requirements of heart rate display and accuracy as specified in note 1 for a continuous train of simulated ECG pulses as shown in Fig. 14.4. The minimum range for QRS amplitude, $|a_r| + |a_s|$, is 0.5 to 5.0 mV and the duration of the QRS wave, d, is between 70 and 120 msec (40 and 120 msec for pediatric monitors). The slopes of the QRS signal should be in the range 6 to 300 mV/sec. The heart rate meter should not respond to signals where the QRS amplitude is 0.15 mV or less or where the duration, d, is 10 msec or less with amplitudes of 5 mV or less. The signal-to-noise ratio of the input signal should be greater than 10:1.

3. Run this test for all combinations of the following waveform parameters:
   (a) 0.5 and 5.0 mV QRS amplitude
   (b) 70 and 120 msec duration (40 and 120 msec for pediatric monitors)
   (c) 30 and 200 beats/min heart rate (30 and 220 beats/min heart rate for pediatric monitors)
4. Repeat this test with a waveform having intermediate values of amplitude, duration, and rate.
5. In all cases, the indicated heart rate should be within $\pm 10\%$ of 5 beats/min, whichever is greater, of the applied rate.
6. Apply the waveform with a QRS amplitude of 0.15 mV for all combinations of maximum and minimum duration and maximum and minimum rate.
7. The monitor should not respond to this waveform.
8. Repeat steps 6 and 7 for the waveform having a QRS amplitude of 5.0 mV and 10 msec duration at maximum and minimum heart rate.

14.4 ULTRASONIC PERFORMANCE TESTING

A simple performance testing technique developed by the American Institute of Ultrasound in Medicine, used with the scanning arm transducer, is to rotate the ultrasonic transducer around a pivot. If the equipment geometry is correct, a circle is displayed on an oscilloscope. If the circle display is distorted, an elliptical shape will be seen. Sensitivity and compensation can be rechecked by attaching a 16-mm-thick sheet of plexiglass to the transducer using scanning gel and noting any reflections that can be seen on the A-mode display for a given control setting.

Resolution and geometric distortion can be checked by use of a water bath with suitable reflectors such as metal rods and thin piano wires with varying spaces or other reflecting structures.

Figure 14.6 shows a test block used in a static B-scan. A test block with scanning gel or oil gives a performance test check that the depth base and scanning arm coordinate separation system are correctly functioning in a static B-scan system. Correct adjustment of one single scan along the edge of the test block displays the three sides of a square with a cross in the center of an oscillo-

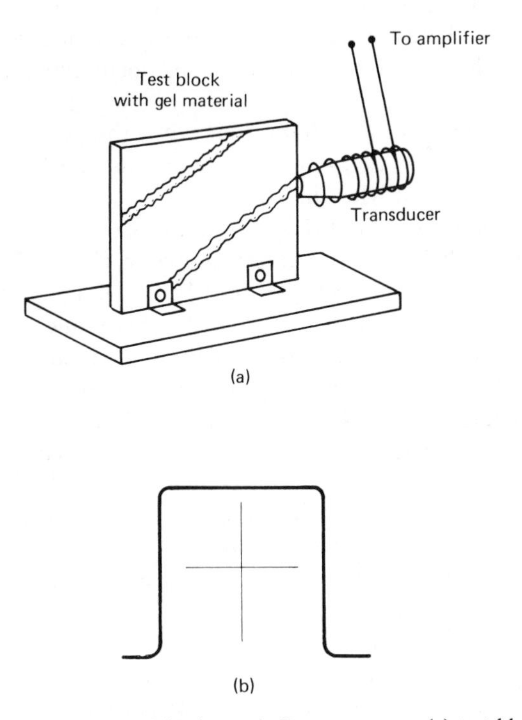

Figure 14.6 Test block used in the static B-scan system: (a) test block with the transducer touching the block; (b) display showing correctly functioning equipment.

scope. The three sides of the square seen on the display after a scan come from the transmitting pulse. The cross in the center of the display is put together from individual indirect reflections via the test block material to the surface of the test block and back to the transducer via the same path. Incorrect or faulty adjustments give distorted displays on the oscilloscope.

A 100-mm test object, which consists of stainless steel rods mounted parallel to each other in a trapezoidal grid pattern, has been adapted as a standard by the American Institute of Ultrasound in Medicine, which has provided a standard procedure for its use. This test object measures the depth calibration, axial resolution, dead zone depth, lateral resolution, depth-compensation circuitry characteristics, display characteristics, and others. A gel or oil material should be used with the test object.

14.5 SAFETY HAZARDS ASSOCIATED WITH CARDIAC MONITORS AND BIOTELEMETRY

All cardiac monitors and electrocardiographs are passive devices in that they are not supposed to generate any stimulus applied to the patient. When operated normally, these devices are usually safe. When defective, operated in conjunction with other devices or operated incorrectly, they can become hazardous.

A defective device becomes hazardous when the power supply or chassis is lifted from ground. (Battery-operated monitors and electrocardiographs are not normally grounded when in use).

Figure 14.7 shows stages of the ECG monitor. Early ECG monitors used a ground-referenced differential amplifier [shown in Fig. 14.7 (a)] in which the right leg was wired with an electrode directly to minimize power-line interference. With internally conductive leads in the body and open ground wires on the patient or any electrically operated device in conjunction with the setup of the patient, lethal currents may be produced. A solution to this problem is the right leg amplifier [Fig. 14.7(b)]. With indwelling electrodes for cardiac catheterization, a complete isolated input amplifier is necessitated, as shown in Fig. 14.7(c).

Fautly line-cord attachments are the major cause of this ECG monitor failure. The ground can fail at the chassis, however, resulting in the identical hazard. When the ground is lifted, there is no path to drain off the leakage currents present in all electrical devices.

If the chassis is not grounded, it is possible to raise the chassis's potential substantially aboveground by reversing the line-cord plug's polarity, contact with an active device, or contact with a second instrument, which is also defective. In any of these instances, if either the patient or operator touches the case while in contact with ground, they will probably receive an electrical shock.

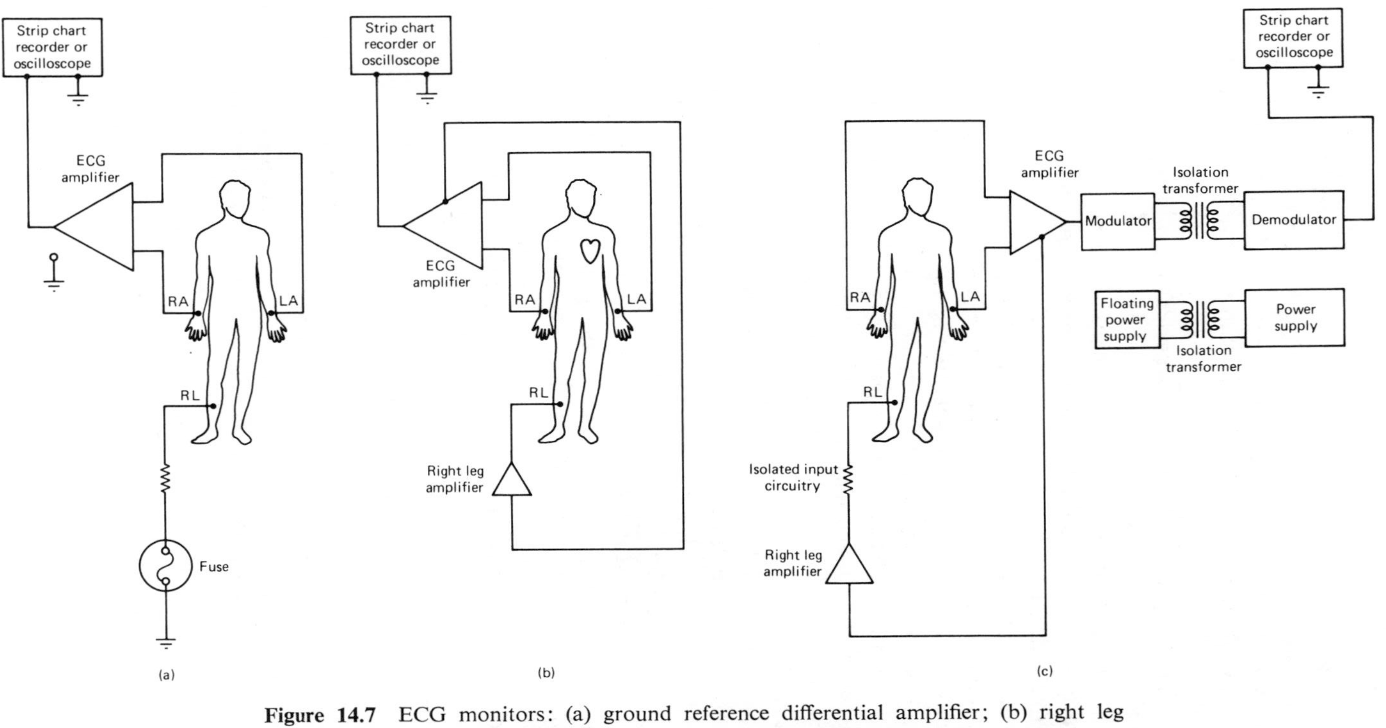

Figure 14.7 ECG monitors: (a) ground reference differential amplifier; (b) right leg amplifier; (c) isolated input amplifier. (Courtesy of Hewlett-Packard Co., Palo Alto, Calif.)

When the cardiac monitor or ECG is operated with other devices, such as a ventilator, electrosurgical unit, or defibrillator, special care must be taken to ensure that their chassis and electrical circuits are properly grounded. Furthermore, the operation of these additional devices should not detract from the function of the monitor of ECG.

Ventilators, for example, sometimes utilize large, electronically controlled motors which may generate line transients that can affect the monitor. If the ventilator's ground is defective, the leakage will appear as a voltage above-ground on the chassis.

An electrosurgical unit can cause interference on the ECG or monitor trace if its power supply grounds are even partially defective. If the patient ground circuit of this device is defective or improperly placed, its radio-frequency (RF) energy can pass through the monitor's ground system, causing, in some instances, burns on the patient's skin at the monitor's electrode positions. If the level of energy generated by the electrosurgical unit is high enough, damage to the monitor's input circuit is also possible.

In all instances, some temporary RF-energy-caused deflections are usually present when the monitor is used in surgery. But a bad ground will cause wild deflections of the monitor's trace and recovery is slow.

Carrier-frequency monitors are also susceptible to feedthrough interference from electrosurgical units, diathermias, and ultrasonic nebulizers. This occurs when the device's RF energy couples directly into the tuned modulator/demodulator circuit's transformers. When this occurs, it is usually not possible to rectify the situation in the field.

A defibrillator can also represent a serious hazard when used with a monitor or ECG. If the monitor is an older unit, does not have a defibrillator synchronizer circuit, or is not defibrillator protected, it must be disconnected (the patient cable removed from its socket and placed at the foot of the bed) from the patient prior to either cardioversion or emergency reversion. If this is not done, serious damage can be inflected on the monitor. A second possible ground path is available to the defibrillator's output energy. The chassis of the monitor can rise, under defeated ground conditions, to the level of the defibrillator output.

Electrocardiographs and monitors are *not* simple instruments. They are, in fact, complex electronic and electromechanical devices which require proper maintenance and calibration. This erroneous opinion of the simplicity of these devices is a consequence of familiarity with the ECG. Everyone who works in a hospital sees this device, usually on a daily basis. Most nursing and medical personnel can connect it to patients and record their ECG traces. In most instances, this device becomes as familiar in operation as a television set. Unfortunately, familiarity breeds contempt. Contempt leads to harsh or foolish actions where reason and caution should prevail. The result is, at best, a broken machine. The ultimate disaster is the possible electrocution of patient and/or operator.

One can observe personnel transporting ECG and monitoring equipment through the hospital, bumping into doors, walls, into elevators, out of elevators, and into beds. Their patient cables are rolled over, pulled on, and frequently left covered with redux cream. Their line cords are constantly plugged into and yanked out of wall sockets and, in some instances, there is a two-to-three-prong adapter permanently affixed to the line-cord attachment.

The fact that these machines are constantly receiving shocks is compounded by the fact that they can be used on a patient who is already connected to a wide range of devices. Any one of these pieces of apparatus (electric beds, reading lamps, respirators, ultrasonic nebulizers, hypo–hyperthermias, or volume-controlled ventilators) in conjunction with the cardiac monitor or ECG can create situations of extreme hazard to both patient and operator.

Electrocardiographs and monitors are not normally considered to be fire hazards. Three conditions can occur, however, which may develop into a potential fire hazard.

All ECGs and most central stations have chart movers, with their associated electromechanical drive elements. These mechanisms are usually partially exposed. Even though they are temporarily stored on the instrument, they can spill down into the chart drive, shorting out the electric motor and its control circuits. This will cause a localized instrument fire, requiring expensive repairs. No material should be stored on top of the ECG or any other instrument.

The electromechanical drive unit can also burn up if it is not kept clean, aligned, and clear of foreign material, such as paper clips, fountain pens, and so on. Any one of these causative factors can jam the drive, resulting in overheating and damage.

All ECGs use a heated stylus to trace the patient's cardiac-cycle potential on waxed-paper tape. The stylus can become hot enough to burn through the paper. If the ECG is used in an explosive atmosphere, such as surgery or an emergency ward, the stylus's heat can exceed the flash point of the explosive agent.

The ECG and patient monitor is so mobile that, even though it is supposed to be used at the heart station (ECG office) or in the ICU and operating room, it can, at a moment's notice, be moved to an area that is not closely supervised. The resulting electrical hazard condition is undefined.

Operators of ECGs and monitors should learn to:

1. Ground the instrument when no U-ground outlet is present.
2. Check to see that other devices in contact with the patient are also grounded.
3. Operate the instrument without touching the patient, metal structural surfaces, or other conductors.
4. Transport the ECG by pulling it through doorways and onto and off elevators.

5. Clean the patient electrodes when they are covered with redux cream.
6. Disconnect the patient cable prior to defibrillation.

All telemetry transmitters are active devices in that they generate an RF energy signal that is radiated into the instrument's immediate environment. This energy, although not hazardous to the patient being monitored by the telemetry system, *can seriously endanger a patient who is being paced* by either an external or internal pacemaker. The radiated energy can interface with the demand, inhibit, or rate modes of the pacemaker.

Although the telemetry transmitter is usually quite compact, its cord set should be dressed neatly and the unit should be secured to the patient via a strap harness.

Silk, nylon, rayon, or woolen garments should not be worn by a patient using a telemetry transmitter because of the excessive noise that is caused by static-charge buildup when the environment is dry (low humidity).

Most telemetry system components are relatively small and thus quite portable. If they are not mounted securely or are handled carelessly by personnel, they can be dropped. If a unit has been dropped or struck a sharp blow, it must be removed from service immediately and subjected to a preventive maintenance inspection. A telemetry transmitter or receiver may appear to be operating correctly after being subjected to a drop and yet have sustained sufficient damge to create a serious patient or operator hazard.

A final hazard situation is possible because of a common misconception of the use of telemetry. For example, attempts have been made to use telemetry as an extension of the CCU. When a patient has recovered from initial cardiac distress, he or she is placed on telemetry and lodged in an area remote from the CCU. The nursing staff is then responsible for monitoring the patient. Unfortunately, in most instances, the nursing staff cannot leave the ICU to respond to telemetered alarms and cannot locate the "patient" fast enough to rectify the situation. When false alarms are frequent, rapid deterioration of staff morale and effectiveness sets in.

14.6 REVIEW QUESTIONS

1. List two primary objectives of patient monitor performance testing.
2. List seven types of equipment that require performance testing in a hospital.
3. Describe a simulation technique for testing an arrhythmia detection system.
4. Describe the need for testing the blood pressure monitor and transducer performance.
5. Describe an important cardiac monitor standard and safety hazard.

14.7 REFERENCES

1. Sinai Hospital Uses ECG/Blood Pressure Simulator to Improve Patient Monitor Performance Test Procedures, *Medical Electronics and Equipment News*, May–June 1979, pp. 25–26.
2. Final Second Draft of Standard for Cardiac Monitors, Heart Rate Meters and Alarms, FDA MDS-021-0043, October 15, 1978.
3. *Multiphasic Health Testing*, American Society for Hospital Engineering of the American Hospital Association, Chicago, 1979.
4. American Institute of Ultrasound in Medicine Standard 100 Millimeter Test Object, Including Standard Procedure for Its Use, AIUM Meeting, Seattle, 1974. AIUM is located in Oklahoma City, Okla.
5. *The 1981 NEC Code* (NFPA-70-1981), *Safe Use of Electricity* (NFPA 76B), and *Inhalation Anesthetics* (NFPA 56A), National Fire Protection Association, Boston, Mass.
6. *JCAH Standards*, Joint Commission of Accreditation for Hospitals, Chicago.

GLOSSARY OF MICROPROCESSOR-BASED CONTROL SYSTEM TERMS[1]

access time: The interval between a request for stored information and the delivery of the information; often used as a reference to the speed of a memory.

accumulator: A data-word holding register for arithmetic, logical, and input/output operations.

accuracy: The deviation, or error, by which an actual output varies from an expected ideal or absolute output. Each element in a measurement system contributes errors, which should be separately specified if they significantly contribute to the degradation of total system accuracy. In an analog-to-digital converter, accuracy is tied to resolution; an 8-bit A/D, for example, can resolve to one part in 256 ($2^8 - 256$), so best accuracy as a percentage of full-scale range is theoretically 1/256, or about 0.4%.

acoustic coupler: A device that receives digital data, converts them to tones suitable for transmission over telephone lines, then couples the tones into a telephone line; it performs the reverse operation as well. Usually in the form of a cradle, into which a telephone handset is placed.

actuator: A controlled hardware device used to implement a change in a process.

A/D: *See* analog-to-digital converter.

adaptive tuning: In a control system, a way to change control parameters according to current process conditions; usually supplied as a subroutine.

[1]Courtesy of Sheldon Edelman, Consulting Editor, Instrument and Control Systems, Chilton Co., Radnor, Pa., May 1979, pp. 43–47.

ADC: *See* analog-to-digital converter.

address: An identifier, usually in binary or hexadecimal notation, which specifies a particular location in memory, in a register, or in any information source.

ALGOL: A high-level, procedure-oriented algebraic and logical programming language (from ALGOrithmic Language).

algorithm: A step-by-step procedure for solving a problem or accomplishing an end, sometimes used in reference to a software program.

aliasing: Generation of false (spurious) outputs in a sampling data system when the sampling rate is too low in comparison with the highest-frequency component of the sampled signal.

alphanumeric: Information consisting of alphabetic and special characters, and numerals.

ALU: *See* arithmetic logic unit.

analog: A reference to the representation of data by continuously variable quantities.

analog control system: Classically, a system that consists of electronic or pneumatic single-loop analog controllers, in which each loop is controlled by a single, manually adjusted device.

analog-to-digital converter (A/D, ADC): A device, or subsystem, that changes real-world analog data (as from transducers, for example) to a form compatible with binary (digital) processing.

APT: One of many special languages intended specifically for the programming of numerically controlled tools (from Automatically Programmed Tools).

arithmetic logic unit: The section of a central processing unit (CPU) that makes arithmetic and logical comparisons and performs arithmetic functions.

ASCII: A widely used code (American Standard Code for Information Interchange) in which alphanumerics, punctuation marks, and certain special machine characters are represented by unique, 7-bit, binary numbers; 128 different binary combinations are possible ($2^7 = 128$), thus 128 characters may be represented. Also a protocol.

assembler: A software program that translates assembly language into machine language.

assembly: The translation of mnemonic code into machine language by an assembler.

assembly language: A mnemonic programming language that approximates machine language.

BASIC: A high-level programming language designed at Dartmouth College as a learning tool called Beginner's All-purpose Symbolic Instruction Code.

baud: A unit of data transmission speed, sometimes variable, but most usually simply meaning bits per second; 300 baud = 300 bits per second.

benchmark: A test point for comparison purposes; in microprocessor-based equipment, a benchmark program is one used to compare aspects of performance among competing systems.

binary coded decimal (BCD): A way to express single-digit decimal numbers (0–9) in 4-bit binary notation.

binary notation: The expression of a number in a base-2 number system using only the digits 0 and 1; each digit's position represents a power of 2, rather than 10 as in the decimal system.

bit: One binary digit: the smallest element of digital data. (An abbreviation of "binary digit.")

bit-slice microprocessor: A microprocessor built-up in a modular way by its user, with the design based on the use of (usually) 4-bit-wide "slices" of CPU circuitry, leading to 4-bit, 8-bit, 12-bit, etc., CPU's.

bit switch: A switch that controls the logical state of only one bit, used for data entry usually only in the simplest of systems.

buffer: A temporary storage site that compensates for differences in data flow rates during transfers of data.

bug: An error, defect, problem, etc.

bus: A group of wires or conductors, considered as a single entity, which interconnects parts of a system; in a computer, signal paths such as the address bus, the data bus, etc.

bus protocol: *See* protocol.

byte: A group of adjacent binary digits, sometimes shorter than a computer's full-length word, and processed as a unit; usually, an 8-bit serial grouping. (*See* nibble.)

Cascade Control: An industry-standard process control program, which involves master/slave loops and several levels of control action.

central processing unit (CPU): The brain of a computing machine, usually defined by the arithmetic and logic units (ALU) plus a control section; often called a "processor," sometimes a "mainframe."

character: A symbol that represents information; the representation itself of the character in a computer-acceptable form (a symbol in ASCII, for example).

clear: To restore a device, or a circuit stage, to a prescribed state, often the zero state.

clock: A master timing unit in a computer system, used to synchronize and time the operations of the system.

COBOL: A high-level programming language for business applications (from Common Business-Oriented Language).

code: To write a program; or, the program itself; often, used interchangeably with "language." A representation of characters as in ASCII coding.

compiler: A program that converts a high-level, source (user-programming) language into machine language. (*See* interpreter.)

control section: That part of the central processing unit which fetches (retrieves), decodes, and executes (performs) the instructions dictated by the program.

conversational: A reference to a mode of system operation in which the system alternately accepts input from, and responds with output to, its operator.

CPU: *See* central processing unit.

crash: A malfunction in hardware or software that requires the computer to be reset or restarted.

current loop: A two-wire transmit/receive interface in which the presence of a 20-mA current level indicates data (a binary 1, or mark) and its absence indicates no data (a zero, or space). Normally used with Teletypes®, and the only communication method that uses a current signal.

D/A: *See* digital-to-analog converter.

DAC: *See* digital-to-analog converter.

DDC: *See* direct digital control.

deadbanding: A subroutine that allows for a zone of inactivity around the setpoint of a measurement (as in a process loop); when a measurement is within the "dead" zone, no control action is taken, and over-control is prevented.

development system: A piece of equipment for the design and test of microprocessor software and the integration of system software and hardware; the development system may either be based upon the microprocessor under investigation or merely emulate it.

digital: A reference to the representation of data by discrete pulses, as in the presence or absence of a signal level to indicate the 1's and 0's of binary data. Also, a type of readout in which the data is displayed as discrete, fully-formed alphanumeric characters.

digital status contact: A logical (on/off) input used mainly to sense the status of remote equipment in process control systems.

digital-to-analog converter (D/A, DAC): A device, or subsystem that converts binary (digital) data into continuous analog data, as, for example, to drive actuators of various types, motor-speed controllers, etc.

direct digital control (DDC): A method of control in which all control outputs are generated by the computer directly, with no other intelligence between the central computer and the process being controlled.

direct memory access: A mechanism for the high-speed direct transfer of information between memory and peripherals without CPU intervention.

diskette: *See* floppy disk.

disk memory: *See* magnetic disk.

distributed control: A system of dividing plant or process control into several areas of responsibility, each managed by its own controller (processor), with the whole interconnected to form a single entity usually by communications buses of various kinds.

distributed intelligence: *See* distributed processing.

distributed processing: The disbursement of computing power to several processors, all working within the same system, and operating either at the same level of control responsibility or as part of a hierarchical scheme of increasing power and broadening control.

DMA: *See* direct memory access.

documentation: The literature necessary to understand, operate, and maintain a processing system—operating instructions, service information, flowcharts, programs, applications information, etc.

dot-matrix: A method for the display of information, in which characters are formed within a grid (matrix) by the activation of appropriate junctions as formed by the rows and columns that make up the grid. An electronic dot-matrix display, for example, may consist of an array (often, 5×7) of ultra-miniature LED's which, when appropriately excited, illuminate to form the desired character pattern. Many types of hard-copy printers also use a dot-matrix approach to character formation.

dynamic range: The difference between maximum useable fullscale signal and minimum resolvable signal. Again, in analog-to-digital converters, for example, dynamic range is tied to the number of bits in an encoded datum; an 8-bit A/D has a dynamic range of about 48 dB. Because the dynamic range increases at a rate of about 12 dB per 2-bit increase in the coding, at least a 12- to 14-bit A/D is required to match the 80–90 dB dynamic range of a good transducer.

EBCDIC code: A standard code in which each character is represented by a unique, 8-bit binary code (from Extended Binary Coded Decimal Interchange Code).

emulate: To imitate a different computer system by a combination of hardware and software that permits programs written for one computer to be run on another.

EPROM: An erasable and reprogrammable read-only memory.

erase: To remove or clear information, usually used in reference to a computer memory or to equipment such as cathode-ray-tube (CRT) terminals, video screens, etc.

execute: To perform a specified operation or operational sequence in a program.

execution time: The time required to carry out an instruction, procedure, etc.

Feedforward: An industry-standard process control program, in which mathematically predicted errors are corrected before they occur; used mainly for process loops with long lags or response times.

fetch: To retrieve data or other information under program or other directed control.

firmware: Programs stored in ROM.

flag: A signal device in a microprocessor system that alerts the operator, or the system itself, to the occurrence of some desired or undesired event (often an interrupt).

floppy disk: A small flexible disk carrying a magnetic medium in which digital data is stored for later retrieval and use. (*See* magnetic disk.)

FORTRAN: A high-level language originally (and still) used for scientific number crunching, but which has also moved into the process control field; from FORmula TRANslation.

front end: In a process control system, the input end at which raw signals are converted to digital information for further processing.

function selection: An operator input technique that makes use of switches, usually hardwired, each labeled to activate a particular function; as a data-entry method, not much better than the use of bit switches.

global function command: A command that has only one meaning, regardless of the situation in which it is used.

handshake: A reference to or type of interface procedure that is reasonably easy to implement; its operation is based on a Data-Ready/Data-Received signal scheme that assures orderly data transfer.

hard copy: Permanent, printed version of information that is otherwise available only on a temporary basis (CRT-displayed data, programs or data ordinarily stored in memory, etc.)

hardware: The electrical and electromechanical equipment and parts associated with a computing system, as opposed to its firmware and software.

hexadecimal notation: The expression of a number in a base-16 number system using a combination of digits (0–9) and letters (A–F); the digit/letter combination is a shorthand notation, a compact way of representing a long binary number in 4-bit chunks.

high-level language: A "one-to-many" language, in that a single line of user (source) language produces many lines of machine language; BASIC and Pascal are examples of high-level languages.

illegal operation: An impossible-to-execute instruction.

instruction: A code set that defines some computer operation.

intelligent terminal: An input/output device with built-in intelligence in the form of a microprocessor, and able to perform functions that would otherwise require the central computer's processing power; sometimes called a stand-alone terminal.

interface: The place at which two systems, or a major system and a minor system (such as a computer and a peripheral), meet and interact with each other; the means by which the interaction is effected (as an interface card); also, to connect by means of an interface.

interpreter: A program that executes the instructions from the source (user) language, as each is encountered, and without converting the source language into machine language. An interpreted program is slow—as much as 20 times slower than an assembled program—but speeds up program development because the effect of source changes can be seen immediately.

interrupt: A signal that halts the operation of a running program to perform a special, higher-priority routine.

I/O: Abbreviation for input/output.

K: A shorthand notation meaning 1024 bits, bytes, or words of digital data. A "64-Kbit" memory contains 65,536 bits.

LED: *See* light-emitting diode.

light-emitting diode (LED): A semiconductor diode, the junction of which emits light when passing a current in the forward (junction ON) direction.

line printer: A hard-copy device that prints one line of information at a time.

loader: A software program that transfers data and other information from off-line memory to on-line memory.

location: In reference to memory, a storage position or register uniquely specified by an address.

loop: A sequence of instructions that is written only once but executes many times (iterates) until some predefined condition is met.

machine language: The internal, binary code by which a computer operates; the binary language executed by a computer. Also called object code and object language.

macro: A kind of programming shorthand by which a series of assembly language instructions, which the programmer intends to use more than once, is given a name. The named sequence is called a macro. When the program is assembled by a macro assembler, every named macro will be translated into the appropriate lines of assembly coding.

magnetic disk: A form of memory in which data is stored in a magnetic oxide that coats a plastic or metal disk. The data is recorded (written) and played back (read) by magnetic heads, which traverse the rotating disk under program control.

mass storage: Any very large capacity memory device.

memory: That portion of a computer system which stores information in a form understandable by the system hardware; also called storage.

micro: A microprocessor or microcomputer.

microcomputer: A complete computer in which the CPU is a microprocessor.

microprocessor: A usually monolithic, large-scale-integrated (LSI) central processing unit (CPU) on a single chip of semiconductor material; memory, input/output circuits, power supply, etc. are needed to turn a microprocessor into a microcomputer. Sometimes abbreviated as MPU, μP, etc.

microprogram: A set of elemental operations combined to form a higher-level operation or instruction. Once established, only the higher-level instructions need to be used to build new programs.

mnemonic code: Another shorthand, this time a set of symbols that aids the writing of assembly language programs; each assembly language instruction can be represented by a mnemonic.

Model Reference: An industry-standard process control routine used with loops that have a long lag time, but in which it is desired to use feedback control at intervals shorter than the lag time.

modem: A device that converts signals in one form to another form compatible with another kind of equipment. In particular, the device that changes digital data into a form suitable to be transmitted over telephone lines and vice versa. (From MOdulator + DEModulator.)

multiplexer: A device or circuit that samples many data lines in a time-ordered sequence, one at a time, and puts all sampled data onto a single bus; the electronic equivalent of a funnel. (A demultiplexer does the reverse job.) Important in, say, expanding the number of input channels available to, say, an A/D converter.

multiprocessor system: *See* distributed processing.

nibble: One-half byte (four bits).

object code: *See* machine language.

object language: *See* machine language.

octal notation: The expression of a number in a base-8 number system.

off-line: Not being in continuous, direct communication with the computer; done independent of the computer (as in off-line storage).

on-line: Directly controlled by, or in continuous communication with, the computer (on-line storage); done in real time.

page: A subdivision of computer memory of some given size; often, a block of information that fills such a page, and which can be transferred as a unit where needed.

parallel: When used with reference to digital data, the presentation of all data bits simultaneously, rather than as a serial sequence of bits over a time period. (Compare to serial.) One wire conductor per data bit is needed to carry the signal; 8 bits = 8 wires.

parity: "Even parity" is defined to mean that the number of binary 1's in a digital word is an even number; "odd parity" means that the number of 1's is an odd number. The concept of parity is a check on the accuracy of data. Memory boards, for example, have circuitry that maintains either one parity or the other during a write; if that same parity is not present when the data is read, an error is signaled.

parity bit: A bit added to a digital word before processing, to control parity, and used later as one check of the accuracy of the processed data.

Pascal: A high-level programming language derived from ALGOL and developed intensively by a group at the University of California at San Diego (thus, UCSD Pascal). In many ways simpler to learn and use than BASIC, Pascal can be made machine-independent, or "transportable." (Named for the French physicist, Blaise Pascal.)

peripheral: A supplementary piece of equipment that puts data into, or accepts data from, the computer (digitizers, plotters, printers, etc.)

PGA: *See* programmable gain amplifier.

polling: A method by which all equipment sharing a communications line can be periodically interrogated or allowed to transmit without contending for the line; often, a reference to a centrally controlled method of accessing a number of equipments.

port: A signal input (access) or output (egress) point.

process control language: A class of high-level programming languages oriented to users in the process industries, and requiring only a minimum of programming skill.

processor: *See* central processing unit.

program: A list of instructions that a computer follows to perform a task.

programmable gain amplifier (PGA): An instrumentation amplifier that changes its amplification (gain) under command from a digital code supplied through a programmed instruction in software. Very important in allowing a lower-resolution, lower-cost A/D to accept a wide-dynamic-range signal.

PROM: A user-programmable read-only memory.

propagation delay: The time required for a signal to pass from the input to the output of a device, typically nanoseconds (10^{-9} seconds), usually specified only for digital integrated circuits.

Proportional Plus Integral (P + 1): An industry-standard process control subroutine that is the software (digital) equivalent of a standard analog proportional controller.

protocol: A formal definition that describes how data is to be formatted, what the control signals mean and what they do, the pin numbers for specific functions, how error checking is done, the order and priority of various messages, etc. There are a number of protocols in use, among them being IEEE 488 (or Hewlett-Packard Interface) and CAMAC (or IEEE 583, 595, and 596), which use parallel interfaces to interconnect instrumentation (CAMAC also covers serial interfaces); ASCII, which covers a bit-serial protocol for computer/computer, computer/peripheral communications and low-speed instrumentation; BISYNC, DDCMP, and BASIC bit-serial byte-oriented protocols for computer/computer communications; and SDLC, HDLC, BDLC, X25, and ADCCP bit-serial bit-oriented protocols for computer/computer communications and all types of instrumentation. IEEE 488, CAMAC, ASCII, and SDLC are the protocols most frequently used.

prototyping system: *See* development system.

pulse input: In process control systems, a type of input used to measure pulse- or tachometer-type signals (speed, rpm, frequency, etc.).

RAM: *See* random-access memory.

random-access memory: A type of memory that allows the writing or reading of data in any order desired, and so the time required to obtain the data is independent of the location of the data.

read: The accessing of information from a storage device such as a semiconductor memory, tape, etc.; also, the transfer of information between devices, such as between a computer and a peripheral, particularly from external (secondary) storage to internal (primary) storage.

read-only memory (ROM): A memory that stores a special-purpose program, which cannot be changed by the computer.

read/write memory: Usually, a random-access memory.

real time: The actual time during which something takes place; the actual time of occurrence of an event.

refresh: To restore information that would otherwise be lost, so as to maintain its presence where desired. (As in certain types of storage after a read, on video screens with each scan, etc.) *See* transparent refresh.

register: Classically, the hardware for storing one machine word. When in the main memory, it's called a storage register, memory register, or location. A read/write memory in a CPU.

relocatable program: A software program so written that it can be moved to and executed from many different areas of memory.

reset: *See* clear.

resolution: The smallest detectable increment of measurement. Again, in A/D's, resolution is usually principally limited by the number of bits used to quantize the input signal; a 12-bit A/D, for example, can resolve to one part in 4096 ($2^{12} = 4096$).

restore: To return a word to its initial value.

ROM: *See* read-only memory.

routine: *See* program.

RS-232-C: A *de facto* standard, originally introduced by the Bell System, for the transmission of data over a twisted-wire pair less than 50 feet in length; it defines pin assignments, signal levels, etc. for receiving and transmitting devices. Other RS-standards cover the transmission of data over distances in excess of 50 feet.

run: To execute a program.

scrolling: The vertical movement of information on a CRT screen, caused by the dropping of one line of displayed information for each new line added; the movement appears as an upwards rolling if the new line is added at the bottom of the screen, and vice versa.

sensor: A device that produces a voltage or current output representative of some physical property being measured (speed, temperature, flow, etc.). Generally, the output of a sensor requires further processing before it can be used elsewhere.

serial: In reference to digital data, the presentation of data as a time-sequential bit stream, one bit after another. A great advantage of serial data is that it lends itself to transmission over simple twisted-pairs, such as telephone lines.

seven-segment display: Usually, a reference to a type of information display in which a character is formed by illuminating the appropriate members of a figure-8 arrangement of light-emitting devices.

simulate: To represent, by imitation, the functioning of one system or process by means of the functioning of another. (Compare to *emulate.*)

software: Generally used to mean computer programs; in its broadest sense, the term refers to the entire set of programs, procedures, and all related documentation associated with any system.

source code: *See* source language.

source language: In general, any language which is to be translated into another (target) language; usually, however, it refers to the language used by a programmer to program a system.

storage: *See* memory.

subroutine: A piece of software that may be used from several locations in a program. A program smaller than the main program, and called up from the main program to perform some specific task.

supervisory control: An analog system of control in which controller setpoints can be adjusted remotely, usually by a supervisory computer; also known as a digitally directed analog (DDA) control system.

supervisory interface: In digitally controlled systems, a class of interfaces in which the computer controls the setpoint of a local controller.

target language: In computers, any language (such as object code) into which another language (such as source code) is translated.

Teletype®: The trademarked name of a type of input/output terminal originally and still designed and manufactured by the Teletype Corp.; now often incorrectly used in a generic sense to indicate any similar piece of equipment.

Time Sampling: An industry-standard process control subroutine normally associated with DDC interfaces, and involving control actions at specified intervals; useful for loops with short time lags.

transducer: A device which converts something measurable into another form (often, a physical property such as pressure, temperature, flow, etc. into an electrical signal usually at a low level); frequently used interchangeably with 'sensor.'

transmitter: A device which translates the low-level output of a sensor or transducer to a higher-level signal suitable for transmission to a site where it can be further processed.

transmitter, process: A transmitter mounted together with a sensor or transducer in a single package designed to be used at or near the point of measurement; it generates a 4–20 mA signal; also called a field-mounted transmitter. It is usually also a two-wire transmitter.

transmitter, two-wire: A transmitter with only two external connections, one for 24- or 48-Vdc power, the other for a 4–20 mA output signal.

transmitter, three-wire: A transmitter with three external connections: two are used for plus and minus 24- or 48-Vdc power, the third for the 4 – 20 mA output.

transmitter, four-wire: A transmitter with four external connections: two are used for power (usually 120/240 Vac line power), the other two (positive and negative) for the output signal, which can be any standard industrial signal.

transparent refresh: A means by which a display can accept new data without disturbing current data; for a CRT display, transparent refresh implies a capability for animation.

volatility: With respect to memory, an inability to retain stored data in the absence of external power.

watchdog: In control systems, a combination of hardware and software which acts as an interlock scheme, disconnecting the system's output from the process in event of system malfunction.

word: A quantum of information, and the basic unit accessed by a computer. Common word lengths today are eight and 16 bits.

write: To put information into a storage device.

APPENDIX TABLES

Computer Input/Output Devices

Input	Output
Card reader	Card punch
Tape reader	Tape punch
Magnetic ink reader	Line printer
Scanner	Typewriter
Typewriter	CRT displays
CRT displays	Terminals to the CRT
Terminals to the CRT	Keyboard with terminals to the CRT
Keyboard with terminals to the CRT	Magnetic disks and tapes
Question displayed on CRT and choice answers; person who answers question touches answer button by means of sensor pickup signal	Magnetic cartridge
Time-sharing terminal	Mark-sensing device (IBM)
Magnetic disks and tapes	Time-sharing terminal
Optical reader	Optical character

Hospital Computer Applications

1. Patient monitoring systems
2. EEG analysis
3. Nuclear medical system
4. Cardiac laboratory
5. Automatic testing procedures
6. Cardiac computer laboratory
7. Preventive maintenance
8. Thermal dilution and impedance cardiograph to obtain cardiac output

Multiphasic Health Testing Measurements

Classification of Measurement	Measurement and Examination
Patient history	Diagnosis and prognosis Personal and family medical and psychological problems Social problems Nutrition problems
Physiological and clinical measurement	Physiological Blood pressure ECG Height and weight General appearance Chest x-ray if required Visual acuity and color-blindness tests Hearing test Respiratory spirometer test Pulse rate Cervical cytology Specialized physical tests Breast palpation Sigmoidoscopy/proctoscopy test Ultrasonic work-up Phonocardiogram Echocardiogram Fetal test Whole-body scan Barium enema and GI series Computer axial tomography Dental examination Dental x-ray and workup
Clinical laboratory measurements	Complete blood work-up and stool guiaic test

Microshock Effects Caused by 1-Second Contact with 60-Hz Leakage or Ground Loop Current

Path	Current Level[a]	Effect on Human System
Current path directly through heart via needle electrodes, catheters, cauterizers, etc.	0.000010–0.000020 A (equal to 10–20 μA)	Ventricular fibrillation can occur with currents, directly through the heart, in excess of 10–20 μA. No substantive evidence has been obtained on healthy human beings of the exact level of current required to trigger fibrillation. On dogs, however, 10 μA was sufficient.

[a]0–20 μA, safe for a normal heart; 20–800 μA, ventricular fibrillation threshold.

Macroshock Effects Caused by 1-Second Contact with 60-Hz Line Current

Path	Current Level	Effect on Human System
Body trunk Chest to chest Dorsal to sternum Limb to limb	0.001 A or 1 mA	*Perception threshold:* An individual just begins to sense the presence of the current
	0.016 A or 16 mA	*Let-go current:* The muscles clamp the victim to the source of electrical current and the person is barely able to release this hold
	0.050 A or 50 mA	*Pain threshold:* Sensation of pain; muscle reaction of pulling away can cause mechanical injury; fainting or physical exhaustion
	0.100 A or 100 mA to 3.0 A	*Ventricular fibrillation:* Over the range of 100 mA to 3 A the victim will go into ventricular fibrillation, which must be reversed for recovery
	3.0–6.0 A	*Muscular contraction:* Body muscle system contracts violently; all muscle action stops; diaphragm stops action, resulting in arrested breathing; heart stops beating; victim may revive spontaneously if removed from source of electrical energy

INDEX

Absolute refractory (EMG), 140
Accuracy, 18
Action potential (EMG), 142-43
Alarm standards, 259-62
Alarm system, in patient monitoring, 259-62
Alpha wave (EEG), 121, 126
Alveoli, 154
Amplifiers: 41-57
  Carrier amplifiers, 53
  DC amplifiers, 44-46
  Differential amplifiers, 47-50
  Electrometer amplifiers, 53
  Feedback, 50-52
  Input isolation, 42-44
  Instrumentation power supplies, 55-57
  Operational amplifiers, 52-53
  Phase splitters, 47
  Power amplifiers, 46-47
Analog-to-digital converter (ADC), 153, 163
Aorta, 63
Argon laser, 39
Arrhythmia, 73
Artifacts (electrodes), 21
A-scan, 199, 200
ATPS (ambient temperature pressure standard), 158
Atria (ECG), 68, 69, 71
Atrial rate (ECG), 80
Augmented limb leads (aVR, aVL, aVF), 76, 77, 80
Ausculatory gap, 100
Auto-analyzer, 178-89
Auto-listing, 20
Automated blood pressure systems, 100-103
Automated clinical blood devices, 176, 177
Automated hexiometer, 177-78
Automated pulmonary function measurements, 161-68

Bandwidth, 7
Barium sulfate, 226
Baseline artifacts, 136-38
Beta (current gain, β), 43, 46
Beta activity (during sleep for EEG), 133
Beta wave (EEG), 126
Bioplar (EMG), 138
Biotelemetry, 243-48
Bipolar junction transistor, 45
Bipolar limb leads, 75
Blood flow measurements:
  Cardiac catheterization, 112-18
  Cardiac output, 105-12
  Hemodynamics, 103-5
Blood gas analyzers, 53
Blood glucose levels, 176
Blood mechanism, 174-75
Blood oxygen content, 167
Blood plasma, 175
Blood pressure (arterial, venous systolic, diastolic), 82, 94
Blood pressure methods:
  Invasive, direct and indirect, 95
  Noninvasive, indirect, 95
Blood pressure systems: 94-103
  Semiautomated and automated systems, 100-103

Blood pressure systems *(cont.)*
  Sphygmomanometry, 95-100
Body temperature pressure standard (BTPS), 155-56, 166-67
Brain disorders (ultrasonics), 208-10
Brain waves (see Electroencephalogram)
Bridge circuit (see Wheatstone bridge)
Bronchi, 154
B-scan, 199, 203, 212, 213, 220, 221
Bundle of His (atrioventricular), 68, 69
Bytes, 19

Cadmium sulfide photocell, 38
Calibration wave, 138
Calomel (mercurous oxide electrode), 34
Cardiac angioplasty, 115
Cardiac catheterization, 112-18
Cardiac monitoring (see Coronary care unit)
Cardiac monitor standards, 259-62
Cardiac output (measurements), 105-12
Cardiac output (ultrasonics), 214-25
Cardiac pacemakers, 244-45
Cardiogram microcomputer, 113
Cardiogram, vector (VCG), 71
Cardiology (ultrasonics), 197
Cardiorespirograph, 168-69
Cardiovascular diseases (ultrasonics), 210-14
Carrier amplifier, 53, 54
Carrier-frequency-modulated patient monitors, 239
CAT (see Tomography, Computer Axial Tomography)
Cathode ray tube (CRT), 16-19, 64-67
Central (vertex) transients, 133
Chemical blood tests, 176-94
Chip muscle monitors (EMG), 133
Chronobiology (Human EEG), 134
Cinefluorograph, 226
Clinical EEG examination, 121-30
Clinical laboratory measurement, 173-95
CMOS (Complementary metal oxide semiconductor), 53
Collodion electrode technique, 126-38
Colorimeter, 179, 184-88
Computer axial tomography, 20-21, 228-32
Computer input/output devices (Appendix), 285
Computer-mode rejection (CMR), 48
Computer (respiration), 163-64
Computers (EMG), 151
Computer tomography, 232
Coronary care monitors, 238-43
Coronary care unit (CCU), 82-88
CPU (central processor unit), 21
CRT, 17-19
CRT (EMG), 136
C-scan, 199
Cuff (blood pressure systems), 97, 102-3
Cycle duration (ECG), 78

Damping in transducers (undampened, overdampened, critical dampened), 7
Darlington pairs, 43
dB (decibels), 14-15
DC amplifiers, 44-46
Delta waves (EEG), 121, 127
Depolarization (EMG), 141, 142
Dialyzer, 188
Differential amplifiers, 47-50
Differential white cell counter, 174
Diffusion, 161, 163
Display devices (oscilloscopes):
  dedicated, 64-65
  storage, 65-67
Displays, 16-21
  Auto-listing, 20
  CAT (computer axial tomography), 20-21
  CRT (cathode ray tube), 17-18
  Data storage parameters, 19
  Design parameters, 16-17
  Digital panel meters, 18
  Display parameters, 17
Distortion (amplifiers), 15, 16
Distribution, 161, 163
Doppler principle, 196
Dysrhythmia (see Arrhythmia)
Dysrhythmia monitors, 246-50

ECG (see Electrocardiogram, Electrocardiography)
ECG amplifiers, 239-42
ECG lead and waveform configurations, 75-82
  Continuous monitoring of the ECG in the Coronary care unit, 87-88
ECG lead system, 75
Echocardiography (heart), 109, 210-15
EEG (see Electroencephalogram)
Einthoven, William, 75
EKG (see Electrocardiogram, Electrocardiography)
Electrocardiogram (ECG), 69, 71, 88
Electrocardiography, 68-88
  Cardiac muscle physiology, 68-73
Electrodes, 5, 31-34
Electrodes (EMG), 144
Electrodes (platinum), 5
Electrodes (10-20 system for EEG electrode placement), 124, 125
Electroencephalogram (EEG), 33, 47-49
Electroencephalogramic measurements: 120-44
  Clinical EEG examination, 121-30
  Electronic problems and hazards associated with the EEG, 130-32

Sleep recordings and patterns on EEG, 132-34
Electrometers (amplifiers), 11, 13, 53
Electromyogram (EMG), 47-49
Electromyographic bioelectronic devices, 143-46
Electromyographic measurements, 136-52
Electroneurography, 136
Electrotherapy, 149, 150
EMG (see Electromyogram)
Examination (EMG), 146-50
Expiratory reserve volume (ERV), 157, 166

Feedback, 50-52
Feedback (negative), 50
FEV (see Forced expiratory volume)
Fiber-optic transducers, 39, 40
Fick principle, 105-7
Flame photometers, 53, 176
Flowmeter electromagnetic sine wave type, 112, 114
Fluoroscopy, 226
Forced expiratory volume (FEV_t), 157, 158
Forced expiratory volume (timed) (FEV_t), 158
Forced vital capacity, 159
Frequency response (transducers), 7-8
Functional expiratory volume (FEV), 157, 161
Functional residual volume (FRC), 166

Gain selection (EKG, EEG, EMG), 49
Galvanometric recorders, 59-60
Gamma camera (nuclear cardiology), 232
Graphic recorder, 58-59

Harvey, William, 103
Heart dimensions, 68
Heart rate, 169
Heart rate meter standards, 259-62
Heart sounds, 88-90
Heart sounds (PCG), 90-92, 111
Hematocrit, 175
Hematocrit centrifuge, 175
Hemodynamics, 103-4
Hemoglobin, 167
Hemotin, 174
Holter recordings, 250-51
Hospital computer applications (Appendix), 285

Impedance cardiograph, 112, 113
Impedance plethysmograph, 109-12
Infrared detectors, 38-39
Input impedance (amplifiers), 11-12
Input isolation (amplifiers), 42-44
Inspiratory capacity (IRC), 157
Instrumentation:
Displays, 16-21
Introduction, 1
Objectives, 1
Process, 1-22
Signal processors, 10-16
Transducer effects, 4-10
Intensive care (defined), 235-36
Intensive care monitoring, 238-43
Intensive care unit, 236-52
Interference, 60 Hz (ECG), 87, 88
Interference, 60 Hz (EEG), 130
Intra-balloon pumping, 115
Invasive sampling, 4

JFET (junction field-effect transistor), 53

KCl (see Potassium chloride)
K Complex (EEG), 133
Korotkoff sounds, 98-103

Laennec, Rene T. H., 88
Langmuir, Irving, 226
Lasers, 39
Leukocytes (see White blood cells)
Light beam galvanometer, 59
Linearity (transducers), 8-9
Lung volumes and flow rates, 155-59
Lymphocytes, 174

Macroshock effects (appendix), 287
Magnetic tape recorders, 63
Mastermind (Photovolt Corp.), 177
Maximum voluntary ventilation (MVV), 158
Measurement electrode, 34
Mechanics of breathing, 155
Memory capacity, 19
Memory (data storage), 19
Microelectrodes, 33
Microhematocrit, 175
Microprocessor, 177
Microprocessor-based control system terms (glossary), 270-82
Microshock effects (Appendix), 286
Minute ventilation (MV), 157
Monitoring (automatic), 82
Monocytes, 174
MOSFET (metal oxide semiconductor field effect transistor), 44-45
Motion artifact (blood pressure measurements), 100
M-scan, 199, 211
Multielement transducer system, 205
Multiphasic health testing measurements (Appendix), 286
Murmur:
heart, 90
Muscle cells, 141
Muscle phyncology:
depolarization, 71, 75

Muscle phyncology *(cont.)*
repolarization, 71, 75
Muscle potentials (see Electromyogram)
Myocardial infarction, 237
Myography, 136
Myoneural function, 141

National Aeronautics and Space Agency (NASA), 100-103
Needle electrodes, 33
Neonatal care units, 236
Neonatal respiration rate, 169-70
Nerve cells, 141
Neurology (EEG), 121
Neurology (ultrasonics), 197
Neurosurgery (EEG), 121
Nitrogen washout test, 166
Noise (electrodes), 33
Noise (transducer, physiological, white or thermal, ambient), 9-10
Noninvasive techniques, 4
NPN transistor, 45
Nuclear cardiology, 232-34
Null balance recorder, 60, 61

Obstetrics (Ultrasonics), 197, 216-23
Operational amplifier, 52-53
Opthalmology (ultrasonics), 215-16
Optical transducers, 37, 38
Oscillographic recorders, 58-59
Oscilloscope, 64-65, 136-38
Oscilloscope storage, 65-67
Oscilloscope (ultrasonography), 199
Output impedance (amplifiers), 12-13
Output impedance (transducers), 6
Oximeter, 167
Oxygen analyzers (see PO_2)

Panel meters (digital), 18
Parameters (data storage), 19
Parameters (design), 16-17
Parameters (display), 17
Peak expiratory flow rate, 159
Pediatric intensive care, 235
Pediatrics (EEG), 121
Peltier effect, 35
Percutaneous intra-balloon pumping, 115
Performance testing and safety, 253-69
Performance testing (blood pressure transducers), 255-59
Phase splitters, 47
pH meters, 35, 53
Phonocardiogram (PCG), 90-92, 111
Photic stimulation, 129
Photoconvulsive response, 129
Photodiode:
gas, 38
vacuum, 37
Photomultiplier (optical transducer), 38
Photoroentogenography, 226
Phototransistor, 38
Photovoltaic cell, 38
Physiological response to modern current (EMG), 139-43
Piezoelectric effect (ultrasonics), 195-96
Piezoelectric transducers, 36-37
Pneumotach-pressure transducer, 160
PO_2 (respiration), 167
Polarographic methods, 39
Polysomnography, 134
Potassium chloride (KCl), 22, 34, 35
Potentiometric recorder (null balance recorders), 60-61
Potentiometric transducer, 50-51
Power amplifiers, 46-47
Power distribution, 14-15
Power levels (ultrasonics), 223
Power supplies (instrumentation), 55-57
Precordial chest leads (V_1 to V_6), 76, 79
Pressor effect (blood pressure measurements), 100
Pressure conversion chart, 24
Pressure transducers, 25-31
P-R interval (ECG), 76, 77
Processors (signal), 10-16
Proebster, R., 136
Progressive coronary care unit, 244-45
Propagated impulse (EMG), 140
Pulso Echo Technique (Ultrasonics), 196
Purkinje fibers, 69-71
P wave (ECG), 76-77

QRS complex (ECG), 76, 77
Q-T interval (ECG), 78
Quartz crystal technology, 38
Quartz sensor, 36-37

Radiation therapy, 228
Radiography, 225-26
Radioisotope angiography, 232
Radiological and nuclear measurements, 225-52
Range for nominal amplitudes (ECG), 81
Real-time two-dimensional image, 201, 205
Recorder (auto-analyzer), 179, 180, 187
Recorders:
Galvanometric recorders, 59-60
Magnetic tape recorder, 63
Oscillographic recorders, 58-59
Potentiometric recorders, 60-61
X-Y recorder, 61-62
Recovery (EMG), 140
Red blood cells (erythrocytes), 174-75
Regulation (power supplies), 57
Relative refractory period (EMG), 140
Repolarization (EMG), 141, 142

Residual volume (RV), 157
Respiration (definition), 153
Respiration rate, 169-70
Respiratory measurements, 153-71
Resting potential (EMG), 139
Roentgen, 225

Safety hazards (cardiac monitors and biotelemetry), 264-68
Sample loading, 5
Sampling:
  dynamic, 3-4
  static, 4
Scanning (suggested cross-sections for abdominal, gynecological, and obstetrics), 218-19
Sector-scan system (ultrasonics), 206-8, 220-21
Semiautomated blood pressure systems, 100-103
Semipermeable electrodes, 53
Semipermeable membrane (electrodes), 34-35
Septum:
  atrial, 68-69
  ventricular, 68-69
Signal conditioner, hybrid (respiration rate), 169-70
Signal processors: 10-16
  bandwidth, 15
  distortion, 15-16
  electrometers, 12-13
  gain, 13-15
  gain-bandwidth product, 14
  input impedance, 11-12
  power distribution, 14-15
Silver-silver chloride (electrode), 34
Simulation techniques, 253-56
Sine wave electromagnetic flowmeter, 112, 114
Sleep recordings and patterns (EEG), 132-34
Sleep spindles (sigma), 133
SMA-12 test, 176-77
SMA II (Technicon), 189-93
Spectrophotometers, 53
Sphymomanometry, 95-100
Spirogram, 155
Spirometers, 161-65
Spirometric measuring devices, 159-61
Standard bipolar limb leads, 76, 77
Stethoscope, 88-90
  double-headed stethoscope, 90
Stroke volume, 105, 112
S-T segment (ECG), 78
Subthreshold (EMG), 140
Surgical case unit, 236
Synctium ventricular, 70
Systolic time interval, 90-92

Telemetry system (EMG), 150
Thermistors, 35, 36
Thermocouples, 35, 36
Thermodilution principle, 107-9
Thermography, 39
Theta wave (EEG), 121, 128
Threshold (EMG), 140
Thrombocytes (blood platelets), 174
Tidal volume (TV), 157
Time motion display, 205
T-M scan, 199-211
Tomography, 226, 227
Total lung capacity (TLC), 157
T-P segment (ECG), 78
Transducer (effects): 4-10
  damping, 7
  electrodes, 5
  frequency response, 7-8
  impedance (output), 6
  linearity, 8-9
  noise, 9-10
  sample loading, 5
Transducers, 23-40
  electrodes, 31-34
  infrared detectors, 38-39
  lasers, 39
  optical transducers, 37-38
  piezoelectric, 36-37
  pressure, 25-31
  semipermeable membrane electrodes, 34-35
  thermistor, 35-36
  thermocouple, 35-36
Transducers, fiber-optic, 39-40
T wave (ECG), 78

Ultrasonic measuring systems, 195-224
Ultrasonic performance testing (American Institute of Ultrasound in Medicine), 263-64
Ultrasonography, 198-205
Unipolar chest leads, 76, 77
Urology (ultrasonics), 197
U wave (ECG), 78

Ventilation, 161
Ventricles, 68, 69
Ventricular rate, 79, 80
Vital capacity (VC), 156, 157
V waves (EEG), 133

Washout, nitrogen (see Nitrogen washout test)
Wheatstone bridge, 28-30
Wheatstone bridge (carrier amplifier), 54
White blood cells, 174

X-ray, 225-28